RAIL HUM

T0227949

# Rail Human Factors
## Supporting the Integrated Railway

*Edited by*

JOHN R. WILSON, BEVERLEY NORRIS, THERESA CLARKE
AND ANN MILLS

Routledge
Taylor & Francis Group

LONDON AND NEW YORK

First published 2005 by Ashgate Publishing

Published 2016 by Routledge
2 Park Square, Milton Park, Abingdon, Oxfordshire OX14 4RN
711 Third Avenue, New York, NY 10017, USA

First issued in paperback 2016

*Routledge is an imprint of the Taylor & Francis Group, an informa business*

**British Library Cataloguing in Publication Data**
Rail human factors : supporting the integrated railway
  1.Railroads - Great Britain - Safety measures 2.Locomotive
  engineers - Training of - Great Britain 3.Locomotive
  engineers - Great Britain - Psychology 4.Human engineering
  5.Railroads - Great Britain - Signaling
  I.Wilson, John R., 1951-
  363.1'226'0941

**Library of Congress Cataloging-in-Publication Data**
Rail human factors : supporting the integrated railway / edited by John Wilson ...[et al.].
    p. cm.
  Includes bibliographical references.
  ISBN 0-7546-4382-4
    1. Railroad engineering. I.Wilson, John R., 1951-

  TF145.R215 2005
  625.1--dc22                                          2005045712

ISBN 13: 978-1-138-25249-3 (pbk)
ISBN 13: 978-0-7546-4382-1 (hbk)

# Contents

## PART 3: DRIVING – VISION AND VISUAL STRATEGIES

## PART 4: DRIVING – DEVICES IN THE CAB

## PART 5: DRIVING – SIGNS, SIGNALS AND SPADS

## PART 6: DRIVING – FATIGUE

# List of Figures

# List of Tables

# List of Contributors

| | |
|---|---|
| Lucy Adams | Serco Group plc |
| Arne W. Andersson | Uppsala University |
| Ian Andrew | CCD Design and Ergonomics Ltd |
| Rebecca Ashton | AEA Technology Rail |
| John Baker | Risk Solutions. |
| Martin Barnes | EMS |
| Jeremy Bevan | Health and Safety Executive |
| Louis C. Boer | TNO |
| Andy Bourne | London Underground Ltd |
| N. Brook-Carter | Transport Research Laboratory |
| P.C. Cacciabue | European Commission Joint Research Centre |
| Theresa Clarke | Network Rail |
| Nick Coleman | Human Engineering Ltd |
| Rob Cotterill | DNV Consulting |
| Matthew Cox | CCD Design and Ergonomics Ltd |
| Gary Davis | Davis Associates |
| Thomas de Boer | University of Groningen |
| D.W. de Bruijn | Intergo (The Netherlands) |
| Claire Dickinson | Health and Safety Executive |
| Trudi Farrington-Darby | Institute for Occupational Ergonomics, University of Nottingham |
| Abigail Fowler | AEA Technology Rail |
| Sarah D. Garner | Lloyds Register MHA Ltd |
| W.H. Gibson | University of Birmingham |
| W. Ian Hamilton | Human Engineering Ltd |
| Martin Hazell | CCD Design and Ergonomics Ltd |
| Peter Hellström | Uppsala University |
| I.A. Herrera | SINTEF Industrial Management, Norway |
| Paul Hollywell | Mott MacDonald Ltd |
| Caroline Horbury | London Underground Ltd |
| Anders Jansson | Uppsala University |
| James Jenkinson | Quintec Associates Ltd |
| S.O. Johnsen | SINTEF Industrial Management, Norway |
| Helen Jones | DNV Consulting |
| René Jorna | University of Groningen |
| Arvid Kauppi | Uppsala University |
| Lena Kecklund | MTO Psychology, Sweden |
| Derk Jan Kiewiet | University of Groningen |

| | |
|---|---|
| M.P.H.J. Koedijk | Intergo (The Netherlands) |
| Guangyan Li | Human Engineering Ltd |
| Barbara Long | Quintec Associates Ltd |
| Emma Lowe | Human Engineering Ltd |
| William Lukau | Lloyd's Register MHA Ltd |
| Edward Marshall | Synergy Consultants Ltd |
| Ronald W. McLeod | Nickelby HFE Ltd |
| Steve McLeod | Network Rail LNE |
| Andrew McNaughton | Chief Engineer, Network Rail |
| E.D. Megaw | University of Birmingham |
| Ann Mills | Rail Standards and Safety Board |
| Lucy Mitchell | CCD |
| Neville Moray | Nickelby HFE Ltd |
| Ged Morrisroe | Network Rail |
| Helen Muir | Cranfield University |
| Sarah Nichols | Institute for Occupational Ergonomics, University of Nottingham |
| Beverley J. Norris | Institute for Occupational Ergonomics, University of Nottingham |
| Eva Olsson | Uppsala University |
| A.M. Parkes | Transport Research Laboratory |
| John Pethick | Quintec Associates Ltd |
| Laura Pickup | Institute for Occupational Ergonomics, University of Nottingham |
| Simon Pledger | London Underground Ltd |
| Martin Reid | EMS |
| Daniel J.A. Rhind | Cranfield University |
| John Robinson | Network Rail |
| Katie J. Robinson | Rail Standards and Safety Board |
| D.P. Rookmaaker | Intergo (The Netherlands) |
| R. Rosness | SINTEF Industrial Management, Norway |
| Susannah Russell | Quintec Associates Ltd |
| Bengt Sandblad | Uppsala University |
| Andrew Shepherd | Synergy Consultants Ltd |
| Anthony Slamen | Network Rail |
| Stuart Smith | Institute for Occupational Ergonomics, University of Nottingham |
| Nicola Stapley | Human Engineering Ltd |
| Mike Stearn | CCD Design and Ergonomics Ltd |
| Lauren J. Thomas | Cranfield University |
| Shelley Thomas | CCD |
| Paul Traub | Lloyd's Register MHA Ltd |
| Claire Turner | Network Rail |
| Tjerk W. van der Schaaf | Eindhoven University of Technology |
| Darren van Laar | University of Portsmouth |
| Wout van Wezel | University of Groningen |
| J. Vatn | SINTEF Industrial Management, Norway |

| | |
|---|---|
| Guy H. Walker | Nickelby HFE Ltd |
| David Watts | CCD |
| C.E. Weeda | Intergo (The Netherlands) |
| Adam Whitlock | Quintec Associates Ltd |
| Johan Wikström | Uppsala University |
| John R. Wilson | Institute for Occupational Ergonomics, University of Nottingham |
| Maurice Wilsdon | Rail Standards and Safety Board |
| Tidi Wisawayodhin | Human Engineering Ltd |
| John Wood | CCD Design and Ergonomics Ltd |
| Linda B. Wright | ProRail, The Netherlands |
| Mark Young | Rail Standards and Safety Board |
| J.L. Zwartenkot | Intergo (The Netherlands) |

# Foreword
# The Human at the Centre of an Engineering System

Andrew McNaughton
*Chief Engineer, Network Rail*

An operational railway is a complex engineering system comprising a range of technologies, of varying maturity from Victorian through to the telemetries and knowledge age. The way this engineering system is configured and is integrated dictates whether it provides the levels of safety and reliability our customers desire, at a cost they are prepared to pay. The technologies are diverse and the individual elements may be at any stage of their life cycle, from conception and design through commissioning, servicing repair and maintenance to replacement. However, what they have in common is that they are either dependent upon or provide a service to competent humans – contractors, maintainers, operators. The human is at the centre of this complex system.

Until recently the system was engineered with limited and largely non-specialist consideration of the human factors and of the many interfaces between the people and the machines and technical systems they work with. The emergence of ergonomics, or human factors as a discipline in its own right within rail is a major step forward in the quest for a high performing and effective railway system.

Network Rail has placed this new (for it) discipline on the same footing as the established subjects such as track, signal, electrical and vehicle engineering and the like, and it is a professional headship alongside these in the company safety case.

The challenges facing the UK railway are microcosms of those facing railways across the world. Network Rail is seeking to improve its asset inspection processes exploiting technologies from defence and process industries to automate much condition monitoring and repetitive measurement, and then present to competent people the relevant information on which to make vital asset engineering decisions. The railway ergonomist is an essential player in optimising the entirely new human-system interfaces such advances require.

Railway maintenance requires technicians to assimilate information, make decisions and follow both rules and experience-based approaches to work; and to do this in a complete array of conditions, environments and times of day or night. The contribution of the railway ergonomist with a deep understanding of human factors, to help devise work methods and practices, is huge. Increasingly the maintenance technicians wield multimillion pound machines as well as shovels, use computer-based diagnostics as well as workshop tools – giving yet more new interfaces between human and equipment to consider!

Railway operators work with a great variety of command and control systems. At one end of the spectrum is the mechanical level signal box; at the other end is the VDU-based semi-automatic wide area control system; every technology era between the two is also well represented. The humans are much the same, but the way these systems interface with them is very different. Think of the way the human operator creates a picture of the real time railway from which to make decisions, the means of communication (receiving and transmitting) information. The design of the way the human interacts with the hard engineering system must rise to these challenges too.

So, with these few words I hope I have established why the discipline of ergonomics or human factors is vital for the railways as a whole, and is now an integral part of Network Rail and the engineering team, essential to our mission to improve all aspects of our railway engineering system.

# Introduction

The speed with which rail human factors has lost its status as the ugly duckling of the transport ergonomics world has taken many of us by surprise. It is not so long ago that if we were to talk of transport ergonomics then it would be assumed that we meant road transport, aviation and air traffic control and even perhaps the maritime industry. Rail human factors research work barely made an appearance in the journals and the conferences of our profession for much of the 1980s and 1990s. Now that has changed. For instance in the UK, both Network Rail (the operator of the infrastructure) and Rail Safety and Standards Board (the body overseeing safety and related issues on the rail network) have substantial human factors teams. The University of Nottingham is a member of the Engineering and Physical Sciences Research Council's Rail Research UK, a network of several universities where human factors is taken seriously and to be integrated along with engineering, operations and management concerns.

In addition in the past 18 months the European Network of Excellence for the railways, EURNEX, has started operation and one of its ten poles, out of which the new research agenda will be defined, is explicitly on human factors (there are in addition poles on safety and security, the environment and training and education with strong overlaps with human factors). Furthermore, all across Europe a growing number of leading human factors consultancies and universities are establishing or re-establishing programmes of work dealing with the railways. This is in cooperation with rail companies themselves who also have both strong traditions in ergonomics and also current innovative programmes – for instance in The Netherlands, Germany, France, Italy and across Scandinavia.

All this activity is all very well, but we need to be sure of the quality of what we do as well as the quantity. It was for this reason that Network Rail, RSSB and the University of Nottingham organised the First European Conference on Rail Human Factors in York, in October 2003. This was in association with RRUK and the Ergonomics Society. From the papers presented at that conference and from other contributions elicited since, we have brought together this edited book, the first to bring so many contributions on rail human factors and ergonomics between one set of covers. We have reflected as well as we can the issues which were seen as important by the people at the conference at the time, as well as those seen as important by the industry as a whole, and have managed to provide contributions relevant to most of the functions and stakeholders in the rail network. Whilst the majority of contributions are concerned with driving and signalling and control, we also cover functions such as maintenance, station operation and planning, as well as issues such as evacuation, violence to railway staff and safety culture.

What is perhaps most gratifying, along with the needs expressed by industry for the products of ergonomics and human factors work, is that this domain has started to now produce its own approaches, methods and tools rather than rely on the adaptation of those developed in different domains, and it is this that we will see increase into the future. Therefore we intend this book, and the conference series itself, to continue, and the second conference will be held in November 2005 with a second book to follow in late 2006.

The editors are grateful for the help of their colleagues at Network Rail, RSSB and University of Nottingham in both managing their programmes of rail human factors research and also in helping put on the conference and put this book together. They are also grateful for the assistance of Anne Floyde, Sarah Spencer and Lynne Mills to liaise with all the conference participants, with the authors of the chapters within the book and not least ourselves! At an institutional level we are also grateful for the support of our employers in putting the book together. The staff at University of Nottingham also acknowledge the contribution made by RRUK and EURNEX in funding some of the time spent in editing and producing the book. Finally thanks are due also to our publishers and editors at Ashgate, and particularly Alison Kirk.

We hope and expect that the book will be of interest not just to those working in the rail sector from a human factors point of view but also to the rail industry generally, and we also expect that it will be of interest and value to our colleagues in human factors working in other industries.

*John R. Wilson*
*Beverley J. Norris*
*Theresa Clarke*
*Ann Mills*
*February 2005*

# PART 1
# PERSPECTIVES ON RAIL
# HUMAN FACTORS

# Chapter 1

# Rail Human Factors: Past, Present and Future

John R. Wilson and Beverley J. Norris

## Introduction

Rail human factors research is growing rapidly in both quantity and quality of output. There was an early base of work carried out over many years, but the continual influences of safety concerns, new technical system opportunities, reorganisation of the business, needs to increase effective, reliable and safe use of capacity, and increased society, media and government interest have accelerated research programmes in several countries. One consequence is that ergonomics models and methods are being reassessed, and new ones developed, to meet the requirements for better human factors on the railways. This chapter picks up on these themes in the light of previous, current and future research.

Rail human factors research has, to an extent, been the poor relation of transport ergonomics, at least in comparison to aerospace (cockpit and air traffic control) and road driving. Good research has been carried out over the years – for instance in Sweden, the Netherlands, Japan and the UK – but this has been scattered around many different themes and sometimes appear to be more concerned to focus on an underpinning human factors issue than with the railways *per se*.

In the UK there was a good deal of rail human factors research carried out in the 1960s and 1970s largely through the British Rail Research Centre. Unfortunately, at least some of this research is no longer available to us, although the human factors research catalogue produced annually by Rail Safety and Standards Board (for example, RSSB, 2004) has managed to retrieve a considerable part of it. Although research did continue through the 1980s and 1990s, the small amount of this compared to that for other transport industries paralleled general lack of government interest in investing in the railways in any shape or form, again in the UK at least. For the human factors community as well, it is possible that the railways did not seem to provide an 'exciting' domain to work in as much as did aviation and air traffic control, and perhaps did not seem as high a priority to research funding bodies in terms of accident rates or passenger miles as road transport.

This neglect – by the human factors community, by researchers in other domains, and by society as a whole – has changed markedly since the mid or late 1990s. For many years, rail was a business that evolved slowly and where, despite occasional disruptions and major accidents, it appeared to run relatively smoothly. Recently, however, the Chief Engineer of Network Rail highlighted the influence of radically

changing public and government perceptions and relatively fast changing technical
systems in an industry where nothing much changed for 250 years (McNaughton,
2003). He highlighted also the very different demographics and workforce nowadays:
no-one any longer expecting to be in a job for 40 years, more like 2–3 years; much
greater population mobility; less willingness to work within a structure; new entrants
with less time and supervised experience to pick up skills; and a bimodal population
of 16–25 year olds – either with degrees or with few qualifications or competencies.
He saw ergonomics as helping bridge some of the gaps between what the industry
has and what the industry and wider society actually want. He also believed that
ergonomics can contribute greatly to ensure that existing and new systems meet the
needs and capabilities of passengers and staff.

Interest in the human factors of railway operations has never been greater amongst
governments, the media, the public, as well as operating companies and academics and
practitioners. Of course, fatal accidents (and subsequent inquiries) have encouraged
focus on safety, and on the contributions of human error, poor communications,
maintenance procedures and other central issues in ergonomics (see Farrington-
Darby *et al.*, 2005; Hale *et al.*, 2003, HSE, 2001a and HSE, 2001b). In parallel, the
need around the world to shift passenger miles from the roads to rail, the increased
potential performance of trains and the changing nature of railway ownership and
organisation have encouraged focus on a systems ergonomics view of total rail
network performance. This, and other contextual influences, are discussed next.


## Context for Human Factors in the Railway

Any useful ergonomics contribution must reflect its setting, and some of the relevant
context was alluded to above. The very environmental and internal factors which
generate the need for thorough human factors investigation in the railway network
also provide the very issues, difficulties and challenges for such research. The
context discussed below is certainly relevant to the UK, and probably in many other
countries as well.

There will be considerable changes in the technology used to identify where and
which trains are on the track, control their progress by keeping safe but efficient
separations, and communicate between train and signalling and control functions.
Design, implementation and operation of new systems will generate a host of new
human factors problems (and successes!) and will require fundamental and applied
understanding. Future control of the railway network will probably be much more
centralised, with many of the functions and operations currently carried out at
numerous small and large sites being brought into fewer centres. The integrated
functions may include route control, electrical control, signalling and possibly train
operating control. There will be opportunity for new display and communications
systems to reflect the increased information that the driver can have in-cab. As a
consequence of the integration of functions and the new technical systems, there will
be changes in the nature of the roles and organisation of work for staff in the control
centres. At the same time, the UK rail business works with a great variety of *legacy
systems*. Even the newer IECCs (integrated electrical control centres) can appear
dated alongside modern control rooms in other industries or rail network systems

in other countries. Alongside IECCs there are still NX (entry-exit) panel power control systems and old lever boxes on more rural routes. As with many industries another legacy is the incorporation of original techniques or tricks of the trade (such as memory aids and failsafe or interlocking systems) into modern systems through development of a computer equivalent, often without questioning whether this is an appropriate way to do things.

Any severe constraints on *investment* in transport and other infrastructure service industries will always make themselves felt in time, in terms of customer service, performance quality and reliability, as well as safety. No business has unlimited resources, and the careful identification of priorities for investment, and the total systems analysis of consequences of different investment levels, will help decision making. However, poor investment in the railway business in the past can lead to damage to the infrastructure and inadequate rolling stock. Little more needs to be said here other than to point out the impact that such lack of investment – and the public awareness of this – can have upon morale amongst employees. Related to this and to the change of ownership of the business, there is a danger that the culture within the railway business could change markedly, with less people aware of what it is to deliver rail transport as distinct from burgers or insurance.

Critically, and of great interest from a systems ergonomics viewpoint, the rail network and business create a system that must balance reliability of service, quality of service and safety of staff and passengers in a situation of limited capacity. Passenger numbers have increased in recent years, and companies wish to run more trains in the same envelope of time. However, faster long distance trains must share track with slower commuter and cross country services, which runs up against capacity limits and gives difficult timetabling; the slightest delay ends up with faster trains following slower ones, throwing the rest of the timetable into chaos. At the same time there are government requirements to increase freight capacity radically, and the only way to achieve this is to run more freight at night. However, accidents such as Hatfield and Potters Bar have thrown the spotlight onto maintenance – inspections, repair and renewal – which must take place at night or else take possession of the track and share the infrastructure with trains, which may decrease efficiency and increase safety risks.

With the change of organisation and ownership of the business, there is a need to preserve the best of the *culture* within the railways, generated over many years, whilst recognising and modifying culture which is not appropriate to modern systems of working. As many new people join, staff should all be aware of what it means to deliver rail transport as distinct from other products. Railway culture is highly evident in the *procedures* by which railways are operated ('the Rule Book'), and in the *tacit knowledge* held by experienced staff which includes them knowing which rules are vital, where there are 'grey areas' and even which rules are counter productive. Cullen (HSE, 2001) for instance found evidence of confusion amongst signallers over standing instructions (clause 12.2). During the 1990s, the break-up of the UK industry into (initially) more than 100 different businesses, and the initiation of major renewals such as West Coast Route Modernisation and European initiatives in train and track control (ERTMS), prompted considerable analysis of how the rail network ran and should run, what systems had to be in place to achieve this, and what the new railway culture should be. These debates continue.

*Safety* on the railways, especially in the UK but in many other countries also, has been looked at very differently in the last few years, by the general public, governments, the media, passengers and the industry itself. Like many other organisations or industries – the Health Service, for instance – there is a constant trade-off between safety, efficiency (embracing cost considerations), quality and reliability of the service, all in the context of a system with limited capacity. The reaction in the UK, for instance, to incidents such as Ladbroke Grove, and particularly following Hatfield and Potters Bar, is now increasingly recognised as something of an over-reaction which caused even more problems for the network. Although it is difficult for people to say this publicly, we cannot have 100 per cent safety, neither on the railways nor in general, and certainly not at a realistic price. This is why calculations are made for the value of a life or serious injury saved in order to make rational investment decisions. It is this debate that lies at the heart of difficult decision making over investment in train warning and protection systems (to reduce the likelihood of SPADs – signals passed at danger) and train control systems generally – with the more expensive systems perhaps meaning that the railways will be priced out of the market on worst case calculations. It is also at least questionable whether it is sensible to run the railways so carefully (for example, after Hatfield) that disgruntled passengers drive on the roads instead, which is arguably a less safe form of transport. Procedures which are brought in which are intended to be related to safety, if this is done reactively in a panic, may not actually improve safety and may impair effectiveness and performance, which may promote a culture of violations. We should also recognise that the behaviours of one group, say track workers, can impact on the safety of the network as a whole (for passengers, drivers, the public etc) as well as on the safety of themselves and their colleagues. To expand on this, in the specific case of safety rail ergonomics is concerned with: safety of rail company staff (for example, a trackside worker being hit by a train); reliability of rail staff which effects safety of passengers or other staff (for example, a driver committing a SPAD – signal passed at danger); organisation failure affecting passenger and staff safety (for example, planning a possession of the track, for maintenance); behaviour of the rail-using public (for example, passengers alighting a train); and behaviour of unauthorised network users (for example, children trespassing and sabotaging the track, and suicides).

In these circumstances, a key way to make improvements is to understand the performance of key railway staff and to provide improvements through better designed equipment, interfaces, jobs, communications, training and planning systems. Much of the work being carried out by human factors groups is aimed at supporting just such improvement, for the work of signallers, controllers, planners, drivers, maintainers etc. The human factors contribution must be multiple. First, we need to understand how the performance of stakeholders working in the rail network can effect (for good and ill) the performance of the network as a whole and the well-being of all other users or stakeholders. Secondly, we need to understand the potential effects of rail systems and jobs on the people working there, in terms of health and safety, attitudes and satisfaction, competency and skill development. Thirdly, we need to identify and communicate the characteristics of people which are relevant to design, implementation and operation of rail systems, equipment and jobs.

## Rail Human Factors Research

The rail system is a classic domain for human factors contribution. It includes work of all types, from vehicle control, to monitoring, to planning to physical work with tools. Its settings vary from vehicle cabs, to control rooms, to outdoors, to large buildings and spaces. The artefacts that stakeholders use vary from VDUs to handheld equipment, from signals to paper, from CCTV to hard wired controls. The stakeholders themselves include: signallers and controllers; drivers; station and on-train staff; planners, engineers and managers; track (maintenance) workers – mechanical and electrical, lookouts and safety controllers; passengers and the general public (the latter legitimately – for example, at level crossings, and illegitimately – suicides and trespassers).

Although the leading international scientific journals are yet to reflect this, there has been a real renaissance in rail human factors research in the past three or four years. Some critical human factors issues to do with the railways have been the subject of continual (or at least semi continuous) research down the years. Train driver vigilance, perception, recognition and acting upon signs and signals, investigations into signals passed at danger (SPADs) and design of signage and signalling systems are all subject to current research, but have been studied over many years (for example, Branton, 1979 and 1993a; Buck, 1963; Collis and Schmid, 2001; Embrey and Wright, 1999; McDonald and Hoffman, 1991; van der Flier and Schoonman, 1988; and Wilde and Stinson, 1983. From the point of view of signalling and control the interest was probably less in the 1960s and 1970s, but there have also been a number of contributions, for instance Carey, 1992; Collis and Schmid, 2001; Cordiner, Nichols and Wilson, 2000; Fay and Schnieder, 1998; Lenior, 1993; Luff and Heath, 2001; McDonald, 2001; Nichols, Bristol and Wilson, 2001; Olsson *et al.* 1996; Pickup *et al.*, 2005; Reid *et al.*, 2000; Wilson *et al.*, 2001; Vanderhaegen and Telle, 1998). (Note that those involved in controlling the movement of trains are called variously in different countries controllers, signallers and dispatchers.) A third reasonably long-standing and continuing theme of real human factors research has been in ride quality and passenger comfort – for example, Forstberg, 1997; Forstberg, 2000; Forstberg, Andersson and Ledin, 1998; van der Weide, 1999; Suzuki *et al.*, 1999; Branton, 1993b. A fourth area of research has been in safety and error, for instance safety culture (Clarke, 1998; Farrington-Darby *et al.*, 2005a) and errors including violations (Lawton, 1998).

Both Network Rail and Rail Safety and Standards Board (RSSB) in the UK have initiated large programmes of ergonomics/human factors research, admittedly more with consultancies than universities, but with fundamental contributions nevertheless (see elsewhere in this book). The topics for these research programmes reflect the fact that both organisations have multiple remits to deliver. For instance, the Ergonomics Team in Network Rail must create a foundation for ergonomics/human factors through small wins, providing key advice or research evidence to projects. Then they want to spread human factors via generation of new research deliverables and production of guidelines and standards. Once this process is more mature they want to embed human factors within the organisation and the industry network, through involvement in inquiries and working parties and through the collected evidence of the small wins and use by key engineering and operations staff of the research and guidelines (see also Kirwan, 2000 and Wilson, 1994). RSSB produce a useful annual catalogue of rail human factors studies, primarily but not wholly centred on safety concerns, and which contains details of studies over a number of years, some in the grey literature and harder to find (RSSB, 2004). They also carry out and commission a substantial programme of human factors research in partnership with many other companies in the rail industry.

Very recently, as part of a resurgence of interest in rail research generally, the EPSRC (Engineering and Physical Sciences Research Council) network of several UK universities has been launched as Rail Research UK (www.railresearchuk.org. uk). Although the network is dominated by engineering concerns, there is a distinct human factors component organised by the Universities of Nottingham and Leeds. RRUK, and therefore the human factors group within it, are also members of the new European Network of Excellence, EURNEX (www.eurnex.net). At a national level across Europe, publicly funded rail research programmes in France, Sweden and the UK explicitly identify human factors, those in the Netherlands, Germany, Portugal and Spain emphasise safety in many forms, and in Austria and Finland a human factors perspective is implicit (ERRAC, 2004).

The most recent overview of rail human factors research and practice has been provided by the First European Conference on Rail Human Factors, held in York, UK in October 2003 (Wilson *et al.*, 2005). Over 50 papers covered a wide range of topics, and open discussion at the meeting revealed much else going on. One difficulty is that much good research has until recently either been commercially confidential or else carried out by consultancies with less vested interest in (scientific journal) publications. To rectify this, papers given at the conference are being published in this book and selections of papers from the conference will appear in extended form in *Applied Ergonomics* and *Cognition, Technology and Work*.

## Transfer of Human Factors Knowledge

An interesting consequence of the increasingly large human factors rail research agenda and programme may be advances in ergonomics/human factors theory and methodology. Real problems in particular domains have long driven (partially if not wholly) advances in the human factors discipline: for example, user trials in consumer ergonomics (long predating user tests in human computer interaction (hci)); vigilance in naval studies; fatigue, attention and signage in road driving; human reliability in the nuclear industry; mental workload and situation awareness and cognitive ergonomics generally in aviation; and cognitive models in several of the above as well as hci. All these domains had real user needs and reasonable funding programmes, as well as interesting problems for human factors specialists to address.

The railways throw up just about every type of human factors concern and therefore research approach. As this chapter was being written one of the authors attended, within the space of half a day, meetings about signalmen's injury potential in pulling levers, use of analytical models in understanding driver behaviour at signs and signals, virtual teamwork in signalling and control, and the impact of organisational interfaces on reliable and efficient planning of maintenance. All branches of ergonomics – physical/biomechanical, cognitive and social/organisational – are therefore of great relevance.

One problem faced by ergonomics/human factors is the transferability of its approaches and knowledge. This is often alluded to in the transfer of design guidance. A good example is in human computer interface design where it is all too easy to end up with guidelines that are too general or too specific as we try to transfer them between system designs. A more global issue is transfer across domains, cultures and settings. This has been faced by those seeking to understand and improve systems in, for instance, manufacturing and non-aerospace transport where the most insightful models and methods may come from cockpit ergonomics or nuclear human factors. Examples of such transfer in rail human factors, given our desire to minimise any reinvention of wheels, are in areas of safety culture, working to procedures or rules, notions of expertise, understanding situation awareness and measuring workload, but there are many more.

A corollary of this is, however, transfer in the reverse direction. We believe that rail human factors has much to tell us about the transfer of ergonomics models and methods, from the domains of aerospace and continuous process control (especially nuclear power plants) into the more grounded home of heavier industry and transport, manufacturing and construction. What much of the current rail human factors research has in common is that models or methods which have been in the ergonomics/human factors community for some years are being re-examined and made more relevant to use on the railway as well as hopefully improved for general use. Also, methods from other disciplines, anthropology/ethnography, sociology and engineering, are being adapted for rail human factors use especially where there is emphasis on field study.

All the recent investment in rail ergonomics/human factors research must have practical benefits of course. The bodies financing it want to see advice, recommendations, guidelines and standards emerge, that can have a demonstrable effect on the efficiency, effectiveness, reliability, quality and safety of the railways. Such outcomes can be used in different ways by the different rail partner companies.

At the same time, this concentrated research effort by many different groups in one application domain should provide advances in our fundamental knowledge. Already we are seeing adaptation and development of: human error identification and human error probability systems and tools; train driver cognitive and performance models; a suite of workload tools with underpinning justification; control room ergonomics audit tools; qualitative methods to investigate safety culture and deeper influences on performance; understanding of the development and role of experience; a set of electronic ergonomics analysis tools; and models to understand joint cognitive systems and distributed complex socio-technical systems. Many of these advances are reported elsewhere in this book.

## Future Rail Human Factors Research

The future rail human factors research programme – in the UK from the direct knowledge of the authors but also in the Netherlands, Italy, Germany, Australia, Sweden and many other countries where the authors have colleagues – is very diverse. For instance, UK research (and in many cases subsequent application) is currently underway in:

- impact of new control and communications systems on train driving;
- train driver attention, fatigue, and use of reminder appliances;
- route knowledge and driver experience;
- signal and signage siting and sighting;
- signaller and driver mental workload;
- interface design in signalling centres and train cabs;
- experience and expertise in signalling and control;
- use of CCTV;
- passenger behaviour at stations and on the train;
- behaviour at level crossings;
- new tools for human error identification;
- team working in signalling and control centres;
- inspection by patrol and automated systems;
- organisation of possessions;
- reliability and effective performance in track maintenance;

and many more areas.

This programme, and its equivalents across other countries of Europe and further afield, is very exciting. It is also potentially valuable for the rail business, for society, and, by transfer of theory, methods and knowledge – for human factors generally.

## Acknowledgements

The authors wish to acknowledge the funding for their rail research provided by, amongst others, Network Rail, RSSB, Rail Research UK, Serco and EURNEX. They are also grateful for the collaboration over the years with their colleagues at Nottingham, especially Sarah Atkinson, Trudi Farrington-Darby, Sarah Jackson, Sarah Nichols, Laura Pickup, Brendan Ryan and Stuart Smith.

## References

Branton, R. (1979), 'Investigations into the Skills of Train Driving', *Ergonomics*, 22: 155–64.
Branton, R. (1993a), 'Train Drivers' Attentional States and the Design of Driving Cabins', in D.J. Oborne, R. Branton, F. Leal, P. Shirley and T. Stewart (eds), *Person-centred Ergonomics*, London: Taylor and Francis: 97–110.
Branton, R. (1993b), 'Ergonomics Research Contributions to Design of the Passenger Environment', in D.J. Oborne, R. Branton, F. Leal, P. Shirley and T. Stewart (eds), *Person-centred Ergonomics*, London: Taylor and Francis: 111–22.
Buck, L. (1963), 'Errors in the Perception of Railway Signals', *Ergonomics*, 6: 181–92.
Carey, M.S. (1992), 'A Position Paper on Human Factors Approaches for the Design of VDU Interfaces to Computer-based Railway Signalling Systems', HSE Contract Research Report 46.

Clarke, S. (1998), 'Safety Culture in the UK Railway Network', *Work and Stress*, 12: 285–92.

Collis, L. and Schmid, F. (2001), 'Human-centred Design for Railway Applications', in J. Noyes and M. Bransby (eds), *People in Control: Human Factors in Control Room Design*, London: The Institution of Electrical Engineers: 273–91.

Cordiner, L.A., Nichols, S. and Wilson, J.R. (2001), 'Development of a Railway Ergonomics Control Assessment Package (RECAP)', in J. Noyes and M. Bransby (eds), *People in Control: Human Factors in Control Room Design*, London: The Institution of Electrical Engineers: ch. 11.

De Waard, D. (1996), 'The Measurement of Drivers' Mental Workload', PhD thesis, University of Groningen, Haren (Traffic Research Centre, The Netherlands).

Embrey, D. and Wright, K. (1999), 'Railway Group Standard for Controlling the Speed of Tilting Trains through Curves – A Human Reliability Analysis of the Detection of Multiple Speed Signs and the Adoption of Permitted Speeds', Report No. R99/05, Human Reliability Associates, October.

European Rail Research Advisory Council (2004), 'Rail Research in the EU', Report of the ERRAC, Brussels, May.

Farrington-Darby, T., Pickup, L. and Wilson, J.R. (2005), 'Safety Culture in Railway Maintenance', *Safety Science*, 43: 39–60.

Fay, A. and Schnieder, E. (1998), 'Information and Knowledge – Valuable Assets used for Train Operation and Control', *Proceedings* of Computers in the Railway, Madrid.

Förstberg, J. (1997), 'Motion-related Comfort Levels in Trains: A Study on Human Responses to Different Tilt Control Strategies for a High Speed Train', Swedish National Road and Transport Research Institute, Report No. VTI 274–1997.

Förstberg, J. (2000), 'Ride Comfort and Motion Sickness in Tilting Trains: Human Responses to Motion Environments in Train Experiment and Simulator Experiments', Railway Technology, Department of Vehicle Engineering, Royal Institute of Technology, Stockholm, Report No. TRITA–FKT 2000:28.

Förstberg, J., Andersson, E. and Ledin, T. (1998), 'Influence of Different Conditions for Tilt Compensation on Symptoms of Motion Sickness in Tilting Trains', *Brain Research Bulletin*, 47 (5), 525–35.

Hale, A.R., Heijer, T. and Koornneef, F. (2003), 'Management of Safety Roles: The Case of Railways', in *Proceedings of 3rd International Symposium on Safety and Hygiene*, Porto, Portugal, March 2003.

Health and Safety Executive (2000), *The Management of Safety in Railtrack*, Sudbury: HSE Books.

Health and Safety Executive (2001a), *The Joint Inquiry into Train Protection Systems* ('The Uff–Cullen Report'), Sudbury: HSE Books.

Health and Safety Executive (2001b), *The Ladbroke Grove Inquiry: Part 1* ('The Cullen Report'), Sudbury: HSE Books.

Kirwan, B. (2000), 'Soft Systems, Hard Lessons', *Applied Ergonomics*, 31, 663–78.

Lawton, R. (1998), 'Not Working to Rule: Understanding Procedural Violations at Work', *Safety Science*, 28, 77–95.

Lenior, T.M. (1993), 'Analysis of Cognitive Processes in Train Traffic Control', *Ergonomics*, 36, 1361–8.

Luff, P. and Heath, C. (2001), 'Naturalistic Analysis of Control Room Activities', in J. Noyes and M. Bransby (eds), *People in Control: Human Factors in Control Room Design*. London: The Institution of Electrical Engineers, 151–67.

McDonald, W.A. and Hoffman, E.R. (1991), 'Drivers' Awareness of Traffic Sign Information', *Ergonomics*, 34: 585–612.

McDonald, W. (2001), 'Train Controllers Interface Design and Mental Workload', in J. Noyes and M. Bransby (eds), *People in Control: Human Factors in Control Room Design*, London: The Institution of Electrical Engineers: 239–58.

McNaughton, A. (2003), Opening address by Chief Engineer of Network Rail to the First European Conference on Rail Human Factors, York, 13–15 October.

Nichols, S., Bristol, N. and Wilson, J.R. (2001), 'Workload Assessment in Railway Control', in D. Harris (ed.), *Engineering Psychology and Cognitive Ergonomics* (Vol. 5, Aerospace and Transportation Systems), Aldershot: Ashgate: 463–70.

Norris, B.J. and Wilson, J.R. (2003), *Proceedings of the First European Conference on Rail Human Factors*, Centre for Rail Human Factors, Institute for Occupational Ergonomics, University of Nottingham, York, 13–15 October.

Olsson, K., Aldrin, C., Karlsson, S. and Danielsson, M. (1996), 'Technology Driven Changes of Train Dispatch Centres', in *Human Factors in Organisational Design and Management – V*, Elsevier Science: Amsterdam: 49–54.

Pickup, L., Nichols, S.C., Clarke, T and Wilson, J.R. (2005), 'Fundamental Examinations of Mental Workload in the rail industry', *Theoretical Issues in Ergonomics Science* (in press).

Rail Safety and Standards Board (2004), *Human Factors Research Catalogue*, London: RSSB.

Reid, M., Ryan, M., Clark, M., Brierley, N. and Bales, P. (2000), 'Case Study – Predicting Signaller Workload', in *Proceedings* of the IRSE Younger Members Conference 2000.

Suzuki, H., Tanaka, A., Omino, K. and Shiroto, H. (1999), 'Discomfort Due to the Floor Tilt of Railway Vehicles', *Japanese Journal of Ergonomics*, 35: 323–32.

Traverso, C. (2000), 'Railways Agree ERTMS Specifications', *International Rail Journal*, May, 29–32.

UIC (1998) 'ERTMS (ETCS/EIRENE), MMI: The Man Machine Interface of the European Training Control System and the European Radio System for Railways', report of Union Internationale des Chemins de Fer, Paris.

Van der Flier and Schoonman (1988), 'Railway Signals Passed at Danger: Situational and Personal Factors', *Applied Ergonomics*, 19: 235–241.

Vanderhaegen, F. and Telle, B. (1998), 'Consequence Analysis of Human Unreliability during Railway Traffic Control', in *Proceedings* of Computers in the Railway, Madrid.

Van der Weide (1999), 'Assessment of Comfort Requirements Related to Curves and Turnouts', Arbo Management Group/Ergonomics, Utrecht, Report No. 2056e, March.

Wilde, G.J. and Stinson, J.F. (1983), 'The Monitoring of Vigilance in Locomotive Engineers', *International Journal of Accident Analysis and Prevention*, 15: 87–93.

Wilson, J.R. (1994), 'Devolving Ergonomics: The Key to Ergonomics Management Programmes', *Ergonomics*, 37: 579–94.

Wilson, J.R., Cordiner, L., Nichols, S., Norton, L., Bristol, N., Clarke, T. and Roberts, S. (2001), 'On the Right Track: Systematic Implementation of Ergonomics in Railway Network Control', *Cognition, Technology and Work*, 3: 238–52.

Wilson, J.R., Norris, B.J., Clarke, T. and Mills, A. (eds) (2005), *Rail Human Factors: Supporting the Integrated Railway*, Aldershot: Ashgate.

# Chapter 2

# The RSSB Human Factors Programme

Ann Mills

## The UK Railway Industry Structure

The Rail Safety and Standards Board's (RSSB) role is to provide leadership in the development of the long-term safety strategy and policy for the UK railway industry. The company was established on 1 April 2003 (it was formerly called Railway Safety), implementing one of the core sets of recommendations from the second part of Lord Cullen's public inquiry into the Ladbroke Grove train accident. The company manages railway group standards, measures and reports on safety performance and provides safety intelligence, data and risk information to inform safety decisions on the railways. It also manages the industry's research and development programme, and promotes safety within the rail industry and to the travelling public. The Rail Safety and Standards Board is a not-for-profit company owned by the railway industry. The company is limited by guarantee and has a members' council, a board and an advisory committee. It is independent of any single railway company and of their commercial interests.

## Human Factors in the UK Rail Industry

Human factors, despite popular belief, is not a newly emerging discipline in the railway industry. British Rail had a sizeable team in the 1980s and commissioned a significant amount of research work. Following privatisation, however, there was a lull in the numbers of human factors professionals working directly within the industry and it was not until the Ladbroke Grove accident in October 1999 that there was a renewed interest in the discipline. Many of the recommendations made by Lord Cullen in his report following the inquiry relate specifically to the need for human factors.

The RSSB's Human Factors team has grown from one individual recruited in 2000 to a team of six professionals, all with first degrees in psychology and (as a minimum) a Masters degree in either Human Factors or Ergonomics. Network Rail and the HSE/HMRI have sizeable human factors teams and many of the other railway companies regularly contract buy-in human factors expertise from an increasing pool of consultancies (many of whom were traditionally working in other industries).

### RSSBs Human Factors Strategy/Work Streams

The Rail Safety and Standards Board's structure is based on two directorates – policy and standards and safety management systems. The technical services department, within the safety management systems directorate, comprises a pool of technical experts who provide high quality advice to other departments and direct to industry on specialist engineering and operational matters. As well as the traditional disciplines (vehicles and plant engineering, electrification, track and structures, signalling and telecommunications, operations), the human factors team is seen as another technical railway discipline.

The role of technical services and the human factors team is to give vital input to support decisions on safety strategy, to write the content of railway group standards and other risk controls, to represent the UK in European standards development, to provide technical input to vehicle acceptance body management, to attend formal inquiries, to provide the technical lead for the research programme and to provide technical advice (and education) to railway group members. Human factors is fully integrated into all of the company's activities.

*Standards*

RSSB manages the process whereby the industry sets railway group standards and facilitates and coordinates the industry input to technical standards for interoperability (TSIs). A review of the full railway group standards catalogue has been completed from a human factors perspective to identify where gaps exist in relation to human factors control measures. This review has allowed the team to prioritise areas in urgent need of human factors input as well as identifying where new human factors standards are required. Much of the human factors input for standards will undoubtedly be in informing existing standards, but there may also be a need for new standards or guidance documents purely addressing human factors needs. Such needs have been identified in the current railway group standards suite of documents and new standards/guidance notes or good practice guides will be proposed where necessary. An example of this is a guidance note currently being drafted on cab design.

In addition to the railway group standards the main source of information about operational procedures is the Rule Book. Human factors projects that involve task analysis or task design are restricted to examining what should be done according to the rules. Rather than assuming that current procedures are sufficient though, it is the intention to model the tasks of drivers, signallers, and trackside workers from a bottom-up, user-centred viewpoint when supporting the operations team's ongoing work on the Rule Book. Recently, for example, this approach has identified gaps in the person in charge of possessions (PICOPs) task within engineering possessions.

The European Union has rail at the top of its transport agenda with specific initiatives covering open access to infrastructure, building a single European supply market and creating a Trans-European Network (TEN). All these initiatives have the potential for significant impact on the management of safety on the UK rail network through directives and harmonisation of standards. RSSB's human factors team's involvement in developing these standards is significant and is set to grow in the future.

*Safety Intelligence Data*

RSSB is the central administrator of the safety management information system (SMIS). It is an Oracle-based database of all safety incidents and accidents occurring on Network Rail's controlled infrastructure or within Network Rail's railway safety case. All Railway Group members (RGMs) have access to the system to input their events and to review safety performance. These data are used to measure performance against the objectives laid out in the Railway Group Safety Plan and to develop the Safety Risk Model (SRM), which provides a structured representation of the causes and consequences of potential accidents arising from railway operations and maintenance. The SRM is used to inform a variety of safety related decisions such as standard justification and cost benefit analysis.

RGMs in the past year have been able to provide additional information in SMIS using a taxonomy of human factors cause categories. The team is currently reviewing the data collected in the past year and is spearheading approaches to improve the quality and quantity of these data. It is hoped that this work will help industry improve its understanding of the contribution of human factors to risk, improve the quality and refine the focus of human factors research and assist in better integration of human factors issues into the risk model.

*New Systems*

Since February 2004, the technical services department has included a small new systems team, which is leading the development of operational and engineering concepts, principles, and rules, for the national European Traffic Management System (ERTMS) and GSM–R projects. This work, at the leading edge of railway signalling, communications and operations practice, is geared to specific implementation programmes on behalf of the Strategic Rail Authority (SRA) and Network Rail (Cambrian and Strathclyde early deployment schemes). The human factors team provides support and resources to the new system team to ensure that the development of both implementations schemes understand and incorporate human factors into the design needs, that is, the characteristics, abilities and limitations of the human end users.

*Formal Inquiries*

Having completed a review of human factors in the formal inquiries process in 2003, the strategy for 2004 focussed on known gaps of human factors application and tool development to support the integration of human factors knowledge. The tools are intended to be used by non-experts such as independent chairmen and panel members. Furthermore the team has attended and sent observers to assist on over 15 formal inquiries and will continue to assist with formal inquiries on request. Having completed the development stage for new human factors support tools the goal for the future is to assist those involved with the inquiry process to apply human factors knowledge and tools in practice. The team have presented an overview of human factors tools and resources and issued a summary information sheet to independent chairmen. We intend to reinforce this message and encourage take up through an interactive training day in 2004. In the future our aim is to ensure a successful transfer of human factors knowledge to the new Rail Accident Investigation Board (RAIB).

*RSSB Rail Safety Research Programme*

In 1996 the Davies report identified that several parts of the privatised railways found it hard to invest in research. It recommended that the government should provide research funding in the short term. The Rail Safety and Standards Board Research Programme (RSSBRP) was set up in 2001 for an initial period of five years, with funding from HM Government to fulfil the Davies report recommendations – at least as far as safety is concerned.

RSSB is managing the RSSBRP on behalf of the industry and its wider stakeholders; in carrying out this duty, our vision is 'to deliver research that identifies achievable ways of improving safety, as a contribution to meeting the expectations of society for a safe railway'. There is universal agreement on the need for research. The challenge has been and continues to be, to build a programme that delivers achievable ways of improving safety in the most important areas.

The RSSBRP is in its third year of an initial five year programme, with research being conducted across three broad topic areas (engineering, management and operations) divided into 13 themes, ranging from the engineering of the wheel-rail interface through human factors and operations research to policy issues such as the tolerability of risk. It should be acknowledged that whilst there is a separate human factors theme there are many research projects across the broad spectrum of the research themes that have a significant human factors element that is being investigated.

*Human Factors Research*

At the time of writing approximately 350 research projects have been completed or are in contract. Of those, approximately 100 have had, or are due to have human factors input. The research which the team has provided technical input into has to date covered the full breadth of the human factors discipline and railway operations. The list below provides examples of ongoing work:

- an investigation to provide an understanding of the risks of current railway shift patterns and develop strategies for risk reduction and control;
- development of a set of railway specific tools to measure workload;
- an investigation of the factors affecting rule compliance in the industry and the development of a set of tools that organisations can use to improve rule compliance;
- adaptation of HEART human error quantification methodology to the railway industry;
- an investigation of train driver visual strategies;
- communication and management of passengers post-incident;
- development of a model of route knowledge acquisition and identification of the most appropriate methods to support knowledge acquisition;

- human factors centred approach to junction signalling;
- development of signal sighting tools;
- identification of the main opportunities for improving safety through facilitating effective teamworking within the rail industry;
- evaluation of a novel warning device for user worked crossings;
- an investigation of the role and impact on driver behaviour of current and future train protective devices aimed at preventing or mitigating the consequences of driver human error.

For a complete list of completed and on-going research please refer to the Rail Safety and Standards board, website www.rssb.co.uk/r_and_d.asp. Many of the deliverables resulting from the research programme are also available to download from this site.

As a result of extensive industry consultation the main topics of future research in the human factors theme will be maintenance and inspection, alarms and alerts, automation, CCTV, and a human factors good practice guide for the railway. The team is also involved in specifying projects in the operations, track and rolling stock, signalling and telecoms, safety culture, level crossing and public behaviour themes.

*Leadership and Technical Advice*

RSSB has a key role in developing a wider industry understanding and application of human factors principles. Each year the team publishes a human factors research catalogue on CD Rom that aims to help the industry to learn about the human factors tools, techniques and principles available, and help to ensure that research effort is not duplicated. The catalogues also provide a means to disseminate the research reports produced from the RSSB research programme, as well as continuing to raise awareness of human factors work and the associated benefits within the railway industry.

We also produce a range of publications aimed at increasing awareness and understanding of human factors issues, providing sound advice about how to mitigate human factors risks. Recently, in conjunction with Emma Lowe at Network Rail, the team published a leaflet on fatigue that was supported by materials that could form the basis of safety briefings for safety critical workers.

As a result of many companies approaching the team to support their efforts in developing human factors awareness training, the team is currently developing a series of seminars aimed at providing the basis of human factors theory coupled with exposure to practical tools and best practice.

In addition to providing general education the team provides input and advice to Railway Group companies and industry-wide groups such as the National SPAD focus group and the Rail Industry Advisory Committee. Recent examples of work include a review of axle inspection techniques, cab evaluations, HAZOPs relating to new safety devices, TPWS (train protection warning system) reset and continue, in-cab CCTV systems and GSM–R cab fitment.

**Conclusion**

Following privatisation there was a down turn in the number of human factors specialists active within the rail industry. The accident at Ladbroke Grove in October 1999 however has proven to be a catalyst leading to a marked increase both in interest and the application of human factors as a discipline within the UK rail industry. Indeed RSSB, Network Rail, HSE (HMRI) and London Underground all now have specialist human factors teams (quite a sea change considering that at the time of the Southall rail disaster (1997) there were no human factors professionals employed in the rail sector). The first European conference on rail human factors has further raised the profile of human factors as a discipline that should be integrated into all rail activities.

Having championed the cause for human factors in the rail industry, we as human factors professionals now need to ensure the discipline delivers against its bottom line promises to reduce risk and improve safety. This will only be achieved if we are able to follow our own principles and ensure that the services and tools we deliver are fit for purpose and user centred. Our users, the railway industry, now believe that human factors have a key role to play in safety management. The challenge for human factors practitioners is to demonstrate the benefits of our discipline in practice.

Chapter 3

# The Ergonomics Programme at Network Rail

Theresa Clarke

Over recent years, we have increasingly understood the need to account for human as well as technical and economic factors and the interaction between diverse system elements. Recent inquiry reports, including those of Lord Cullen, have highlighted the importance of understanding and managing the human role in the operation of complex rail systems.

The importance of human factors to safety at every level and function of the railway is increasingly accepted. They are also important for reliable performance and efficiency. The contribution of ergonomics to information interfaces, trackside signals and signage, job and team design, communications and processes, will improve performance of all staff involved in the operation of the network, which in turn will improve system reliability as well as safety. The actuality of service to customers is improved through the contribution of ergonomics to improved station and rolling stock design, better signs and information and more comfortable ride quality, for example. Our understanding of ergonomics in the railway context will also help with the business decision making, and will contribute to maximising the network's efficient use through more effective planning, maintenance and operations.

The Ergonomics Group (now Ergonomics National Specialist Team) was established in July 2001 with an objective to embed the principles of ergonomics into the organisation so that consideration of human factors becomes an integral part of the way we conduct our business. This is achieved through the development of policy and standards, improving awareness and skills, and support in the resolution of tactical problems.

A number of themes are central to our thinking:

- human centred engineering – the design of systems, processes, working practices and methods around the capabilities of the individuals expected to use them;
- the systematic integration of ergonomics principles and practices into the development and operational lifecycles of the railways.

## Ergonomics Policy, Strategy and Standards

The first release of the ergonomics policy, part of the asset engineering suite of policies alongside track, structures, electrification and plant, signalling and telecommunications, was in September 2002. It covers a number of areas including:

- control room, workspace and human machine interfaces;
- work organisation, job and task design;
- external interfaces such as driver – infrastructure interaction and stations and related facilities;
- communication and system connectivity.

Underpinning this policy document there are two further classes of document:

- the delivery strategy and plan for the policy comprising:
  - aims;
  - goals both strategic and operational;
  - rolling year on year activity plan;
- company standards and specifications. Currently these comprise 'Incorporating Ergonomics within Engineering Design Projects' (and associated Guidance Notes) and 'Control Room Design, Specification: Process and Guidance'. These are shortly to be joined by 'Human System Interaction Design Guidelines', and a number of guidance notes on tools and methods for such as workload measurement are in production.

Work is progressing in a number of core areas. The availability of standards and guidance, for instance the inclusion of 'Incorporating Ergonomics into Engineering Design Projects Standard' into the GRIP process, has provided a mechanism for ensuring that ergonomics activity is appropriately scoped and incorporated into project remits. Working closely with other engineering asset heads and project teams the Ergonomics Group is involved in a range of projects: high profile programmes deploying novel technologies and solutions such as GSM–R (Global System for Mobile Communications [Radio]); complex renewal schemes such as the West Coast Route Modernisation; and those such as concentrator replacement and modifications to a variety of generations of control facility and technology.

The inclusion of ergonomists at programme level, as well as within delivery units within major projects, is increasingly accepted in ensuring effective and efficient consideration of system operability, performance and safety.

To support this involvement and to promote consistency a number of assessment tools and techniques have been developed. This includes workload assessment tools, surveys such as REQUEST and the Baseline Ergonomics Assessment tool set (see other chapters in this book).

Understanding the driver infrastructure interfaces is essential to the efficient running of trains. What information is provided and the form of that information are prerequisites to enabling the driver to drive to the limits and constraints of the infrastructure. The performance modelling of driver cognitive tasks using a range of tools and framework models (developing in sophistication and usefulness as our understanding increases both as a discipline and across the industry), is enabling the early modelling of human (driver) performance and the consideration of potentially conflicting task demands in a systematic and structured way as the earliest stages in a design. Simulation, as a way of illustrating the concepts and then refining the design, and the availability of increasingly richly rendered animated images, can be used in signal sighting, testing the drivability in particularly complex layouts, and then for training and briefing of drivers.

The Network Rail Ergonomics Group have viewed ergonomics in maintenance in terms of very broad functions or stages. Each of these seems to present significantly different requirements for human factors study and to offer different routes for ergonomics improvements:

- organisation system;
- planning;
- inspection;
- set-up;
- work at the site;
- feedback and monitoring.

To date a scoping and feasibility investigation has identified those critical issues to do with reliable, efficient and safe maintenance which are people-related, and where ergonomics can make a contribution. This initial feasibility work has been wide ranging, covering everything from organisational boundaries and planning to behaviour and equipment at site.

More recently the Ergonomics Group have been supporting the WCRM Safety Group in their investigations into incidents which have occurred during possessions. This work has involved focusing in particular on the Engineering Supervisor role and evaluating how they can best be supported in their challenging and changing work environment. A detailed report, including recommendations will be produced shortly. Over the coming year the Ergonomics Group aim to build on this work and provide support in other areas of maintenance such as on-track inspection and the move towards more automation in this area.

A number of pieces of work are being undertaken to understand human error. The work has focused on developing an understanding of the various operator roles and responsibilities, how their tasks can lead inadvertently to error and ways in which these errors can be mitigated. This has involved looking at specific patterns of irregular working (in conjunction with operational safety) and the development of better-designed tools, equipment and systems. In particular a number of assessment tools have been developed which aim to identify human error and its causes; it is hoped that these tools will be used by both operators and their managers and not just by qualified ergonomists in order to identify and mitigate human error.

Additionally a cross-industry programme focused on communication has been heavily supported by human factors expertise and culminated in a number of cross-industry workshops aimed at improving understanding and communication.

Whilst in many respects it is early days for the discipline within the Network Rail organisation, the Ergonomics Group is now well positioned and is positively contributing solutions to the many engineering and operational challenges.

# PART 2
# DRIVING – TRAIN DRIVER BEHAVIOUR

# Chapter 4

# Driver Performance Modelling and its Practical Application to Railway Safety

W. Ian Hamilton and Theresa Clarke

## Introduction

This chapter reports on the development and main features of a model of driver information processing capabilities. The work was conducted on behalf of Network Rail in order to meet a requirement to understand and manage the driver's interaction with the infrastructure through lineside reminder appliances. The model utilises cognitive theory and modelling techniques to describe driver performance in relation to infrastructure design features and operational conditions. The model is capable of predicting the performance time, workload and error consequences of different operational conditions. The utility of the model is demonstrated through reports of its application to the following studies:

- research on the effect of line speed on driver interaction with signals and signs, and the calculation of minimum reading times for signals (time);
- the development of a human factors SPAD hazard checklist to identify and manage SPAD error traps, and a method to resolve conflicts between signal sighting solutions (error);
- research to understand the demands imposed on drivers by ETCS (European Train Control System) driving in a UK context (workload).

The chapter also reports on an ongoing project to specifically validate the model's utility as a tool for assessing cab and infrastructure drivability.

In the past three years Human Engineering Limited has undertaken a range of studies for Network Rail concerned with driver visualisation of lineside objects (Li, 2001, 2002, 2003). The primary motivation for this work has been to develop data and a better understanding of driver performance that can inform the creation and management of an ergonomic interface between the train driver and the rail infrastructure (Hamilton and Colford, 2002). The principal benefits from these studies will be in supporting SPAD risk management strategies and in providing essential guidance on cab and infrastructure design standards.

The research work has utilised a combination of the development of a cognitive model of driver performance for wider application and computer-based simulations of signals and signs, presented to participants as if they were being approached at line speeds. Approximately 250 drivers have so far participated in the experimental trials, amassing a substantial amount of data on how performance is affected by the design features and positioning of lineside signals and signs. Some of this information has already been distilled into a draft Guidance Note on factors affecting signal and sign conspicuity (Network Rail Company Guidance Note, 2003).

The approach of conducting experimental studies to evaluate design options is, however, time consuming, costly, and can be logistically impracticable. In response to these limitations, the work has also focused on developing a more theoretically coherent approach to understanding driver performance that can be applied earlier and more widely in design development.

## Modelling Driver Performance

This objective is to be accomplished through the development and validation of a quantitative method capable of producing site-specific track layout, signal and sign positioning recommendations using a cognitive task analysis (CTA) and a model of driver performance capabilities. The model can be applied to significant points along a route. As the model is an information processing model the definition of a significant point will be wherever the task information needs to change or wherever availability of information from the infrastructure changes. A simplified and abstract example is illustrated in Figure 4.1.

Figure 4.1 shows how the CTA approach can be used to model the availability of information as the train proceeds along its route. In frame i, the driver can see signal S1 and gradient sign G1, whereas the remaining signals and sign can only be anticipated according to the driver's knowledge. Once the driver can no longer see signal S1 (frame ii), the information contained in the signal is only available as a memory; at the same time, signal S2 comes into view and so the information is available as a perception, rather than by anticipation (shown in frames ii and iii). A similar transition occurs as the gradient sign G1 is passed (frame iv).

This approach has utility as an evaluation method. For example, since memory, perception and anticipation have different levels of capacity and reliability, modelling the driver's processing of information in this way provides a method of highlighting where each aspect of the driver's capacity is being over-used, well-used, or under-used.

The modelling will use as input the timing of key events when driving a proposed line layout (derived from tools such as VISION, Thameslink 2000). These events will then be associated with the driver's decisions and actions required to negotiate the route. In addition input parameters would include features of the stock to be used, permissible train speeds, track and visibility conditions, etc. The outputs will be data on driver performance times, workload and error proneness that can be used to evaluate the feasibility of the driving task. In other words the intention has been to develop a general tool for drivability assessment.

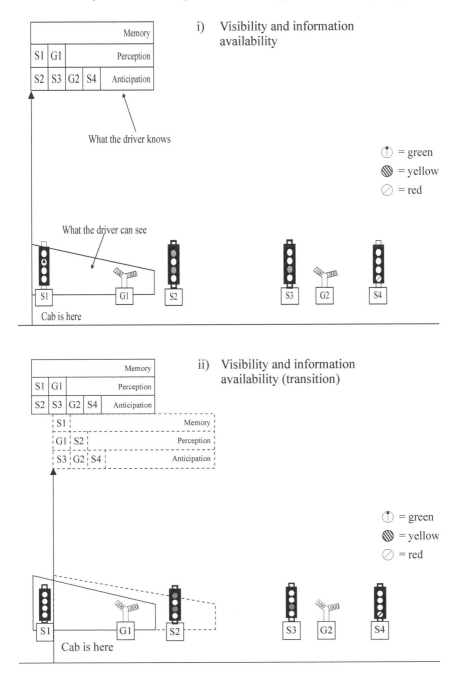

**Figure 4.1    How information availability can be modelled**

iii)  Visibility and information
     availability

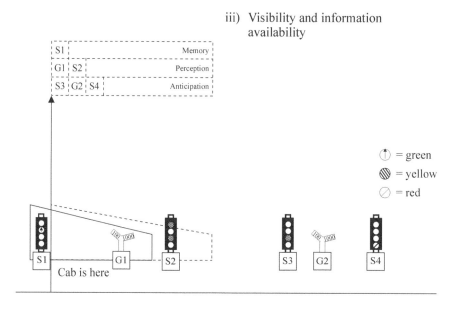

iv)  Visibility and information
    availability (transition)

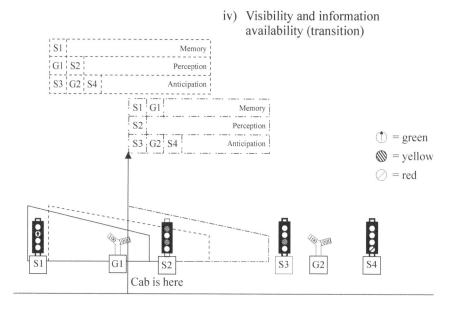

**Figure 4.1 cont'd**

**A General Model of Skilled Performance**

Fundamental to the approach is the idea that route features, signs and signals can be considered as parts of a single information stream to the driver as the train proceeds along its route. At the same time the driver is interacting with the controls and displays provided within the cab. The driver's task is to integrate the various sources of information in order to accomplish a strategy to achieve the goals of:

- moving the train within the limits of the movement authority;
- achieving an efficient safe speed profile; and
- making scheduled stops.

The driver's ability to do this will be governed by the interaction between the usability of the lineside and cab interfaces and the driver's fundamental information processing capabilities. This interaction is illustrated in Figure 4.2.

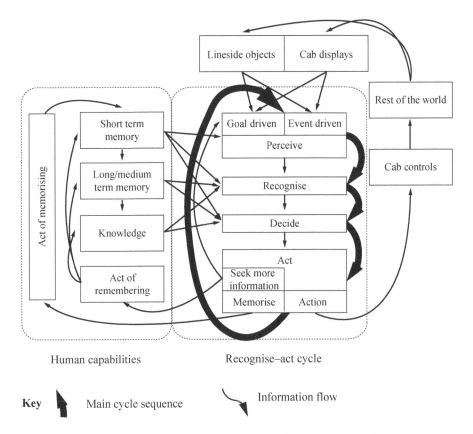

**Figure 4.2    Human capabilities and the recognise–act cycle in the CTA model**

As illustrated in Figure 4.2, the model is only a general model and clearly must be elaborated with appropriate theories of skilled performance and cognition before it can be useful in making predictions about task performance. This has been accomplished using established human factors techniques and theory. How this has been done is described next and followed by a series of application examples that illustrate the use of the method.

## Building Performance Predictions

*Cognitive Task Analysis of Strategy*

Any attempt to predict human work behaviour must have at its heart a rational model of task strategy including rules for how this is shaped by events – this is a cognitive task analysis. Since we are dealing with skilled task behaviour it will be goal directed and follow certain methodological, or procedural rules. Ergonomics provides numerous techniques for task analysis that are suitable for this purpose. For instance both hierarchical task analysis (Kirwan and Ainsworth, 1992) and the GOMS model (Card, Moran and Newell, 1983), represent the goals, strategies and decision-making processes of a skilled operator.

The GOMS method has been chosen for use here because it also has associated with it a model of human information processing that is useful in deriving quantitative estimates of behaviour execution times (see below). This approach is not taken because it is believed to be an absolutely accurate and precise description of how drivers work; rather, it provides a useful approximate model. The approach acknowledges that driver knowledge can vary but assumes a minimum essential competency for all drivers. In this way the model can be used to specify the knowledge that the drivers must possess to be able to drive the route.

Applying the cognitive task analysis technique involves specifying the goals to be achieved in the task, then decomposing the methods to be followed for their accomplishment. In practice this is done by creating a hierarchy of goals and sub-goals – such as that shown in Figure 4.3.

The sub-goals are selected by a plan (rules) determined by operating conditions. For example, on passing a caution signal the driver should decelerate the train. But when and by how much this is done will depend on the aspects shown, railhead conditions, train performance characteristics, etc.

Sub-goals have associated action scripts (methods) and these comprise behaviours. The behavioural level of description is based on a fixed taxonomy of behavioural codes (Table 4.1). These are defined to be as general purpose as possible and also have associated workload parameters and performance time algorithms. The utility of these is explained below.

*Task Performance Time Prediction (PerfCalc)*

Once the strategic aspects of the task have been captured in the cognitive task analysis, it is possible to predict the sequence of behaviour that will occur in performing a task. If this is to be used to examine and assess the demands imposed on a driver

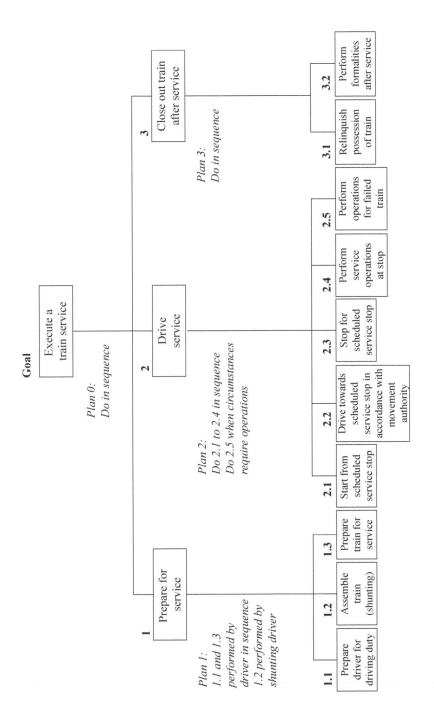

**Figure 4.3    Goal structure of a cognitive task analysis**

**Table 4.1   VACP behavioural codes**

| | **VACP behavioural codes** | | |
|---|---|---|---|
| **Visual**<br>1-Register/detect<br>*(Detect)* | **Auditory**<br>1-detect/register sound<br>*(Detect)* | **Cognitive**<br>1-Automatic<br>*(Simple reaction)* | **Psychomotor**<br>1-Speech<br>*(Speak)* |
| 2-Read<br>*(Text symbology)* | 2-Verify auditory feedback<br>*(Verify feedback)* | 2-Sign/signal recognition<br>*(Recognise)* | 2-Discrete actuation<br>*(Reach or switch)* |
| 3-Scan/search/monitor<br>*(Search)* | 3-Orient to sound (general)<br>*(Locate)* | 3-alternate selection<br>*(Choice reaction)* | 3-Manipulate<br>*(Manipulate)* |
| 4-Inspect/check<br>*(Inspect check)* | 4-Interpret semantic content<br>*(Interpret speech)* | 4-Estimation, calculation,<br>conversion<br>*(Calculate)* | 4-Discrete adjustive<br>*(Adjust or move)* |
| 5-Discriminate<br>*(Compare identity)* | 5-Orient to sound (selective)<br>*(Verify and locate)* | 5-Evaluation/judgement<br>(single aspect)<br>*(Decide)* | 5-Continuous adjustive<br>*(Control)* |
| 6-Track/follow<br>*(Trace)* | 6-Discriminate sound<br>characteristics<br>*(Compare)* | 6-Encoding/decoding, recall<br>*(recall or prepare)* | 6-Serial discrete manipulation<br>*(Keying)* |
| 7-Visually locate/align<br>*(Align track)* | 7-Interpret sound patterns<br>*(Analyse)* | 7-Evaluation/judgement<br>(several aspects)<br>*(Judge)* | 7-Symbolic production<br>*(Write or draw)* |
| | | | 8-Walk<br>*(Walk)* |
| | | | 9-Run<br>*(Run)* |

the sequence of tasks needs to be represented as a timeline. Often, however, data for task performance times are unavailable and so estimates or predicted task timings are needed.

The engineering psychology literature (Card, Moran and Newell, 1983; Boff and Lincoln, 1988; Williams, 2000; Kieras and Meyer, 2000) provides a set of empirical data and standard equations, such as Fitt's Law and Hicks Law, that can be used to derive performance times for elementary behaviours. The data for these timings are provided as a range from 'fast', to 'median', and through to 'slow' and are detailed in Table 4.2.

**Table 4.2    Timings for elementary behaviours**

| Element | Symbol | Timings (s) | | |
|---|---|---|---|---|
| | | **Fast** | **Median** | **Slow** |
| Perceptual processing | $T_P$ | 0.05 | 0.1 | 0.2 |
| Cognitive processing | $T_C$ | 0.025 | 0.07 | 0.17 |
| Motor processing | $T_M$ | 0.03 | 0.07 | 0.1 |
| Cognitive iteration time | $I_C$ | 0.092 | 0.15 | 0.157 |
| Eye movement time | $E_M$ | 0.07 | 0.23 | 0.7 |
| Movement iteration time | $I_M$ | 0.07 | 0.1 | 0.12 |
| Typing, keystroke time | $T_{PTi}$ | 0.158 | 0.3332 | 1.154 |
| Keying, keystroke time | $T_{PKi}$ | 0.3 | 0.577 | 1.091 |
| Writing, time per character | $T_{PWi}$ | 0.545 | 0.732 | 0.952 |
| Quadratic coefficient for speech | y | 0.0095 | 0.0145 | 0.0194 |

The following is an example of how these elemental parameters can be used to construct performance time algorithms for the behavioural codes shown in Table 4.1.

*Psychomotor Function No. 2 – Discrete Actuation*

The task of reaching and activating a switch includes the parameters of perceptual processing, cognitive processing, motor processing, and Fitt's Law (which governs the time to move the hand to the switch). Hence we obtain:

$$Algorithm: T_{Pa} = T_P + T_C + T_M + I_M \, log_2 \, (D/S + 0.5)$$
$$Where: T_{Pa} = time\ to\ perform\ action$$

As each behavioural code in the cognitive task analysis can be related to a performance time algorithm like the one above, predicted activity times can be generated for each task. By following the activity sequencing described by the cognitive task analysis, these can then be used to generate a wholly predicted timeline for a particular driving task. Once a baseline analysis has been performed, the tasks can be modified according to further design iterations, to enable any effects on performance to be determined. This way, many design iterations can be evaluated, with the most promising candidates identified.

## Workload Prediction

The method is also capable of calculating the workload for the scenario timelines. As there is no absolute behavioural scale of workload, nor a single universal predictive technique, the method relies on calculating a range of workload statistics. They describe workload with respect to the conceptual three-dimensional space represented by Figure 4.4, and described in the following paragraphs.

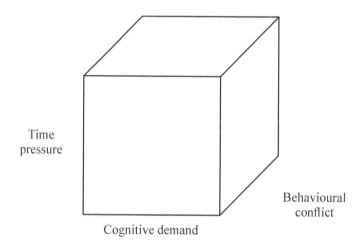

**Figure 4.4    The three dimensions of workload**

*Workload as Time Pressure*

Performing an action will take a certain period of time to complete. Where more than one action is required to be performed in the same time span, human behaviour dictates that, rather than perform two actions simultaneously, a person will time share between activities. This will become an issue when the tasks that have to be completed have to be done in a set period of time (for example where the driver has to cancel more than one alarm before the brakes are applied).

*Workload as Cognitive Demand*

Performing an action places a demand on one or more of the resources within the cognitive system. The more complex an action, the more demanding it will be for attention. This greater demand can be expressed as either greater effort required on the same resources or as more resources being required. The sum of demands on resources, imposed by an action, accounts for workload even in the absence of any other action. With more than one action, action demands also sum to give a total workload demand. This summation is useful for prediction purposes but it is not strictly valid. Demand is usually assessed subjectively and we cannot know if the scale is linear. Cognitive theory offers no guidance on the measurement of demand but merely suggests that adding new demands should mean more workload.

*Workload as Behavioural Conflict*

When there is more than one action their separate demands on resources may also give rise to competition for the same resource. For example two actions may both involve some visual scanning and manual activity. This competition is expressed as a conflict rating. Theory suggests that some resources hardly compete at all (for example, visual and manual) while others conflict totally (for example, visual and visual).

Workload expressed as conflict can help to identify behavioural competition which renders effective time sharing impossible. This is useful information for system design and is a strong argument for treating conflict as a separate workload statistic.

From the scientific literature it is known that the time/demand statistics have correlations with subjective workload measures while time/conflict statistics correlate with objective performance measures. The range of techniques used includes the commonly accepted set of predictive workload statistics listed here. It is not possible here to explain these in detail and so the reader is directed to the source reference in each case:

- *T1*     count of tasks current at each time interval (Parks and Boucek, 1989);
- *T2*     ratio value of total task time against total time available for the task (Parks and Boucek, 1988);
- *Dsme*  sum of rated demands for concurrent tasks at each time interval (Aldrich, Szabo and Bierbaum, 1989);
- *Dvacp* sum of rated demands by modality at each time interval;
- *VACP*  rule based model to examine conflicts between behaviours in each of the visual, auditory, cognitive and psychomotor categories (McCracken and Aldrich, 1984, as modified by Schuck, 1996);
- *Qh*     a set of rules and parameters to analyse performance limitations due to working memory and divided attention capabilities (Hamilton and Boughton, 2003).

*Error Prediction*

The method also allows error prediction. Once again standard techniques are used for error identification and are applied to the task analysis data. The SHERPA method (Kirwan, 1994) is used for the identification of generic error sources. The GEMS technique (Reason, 1990) is used to extend the investigation of error sources to include those that arise from limitations in task knowledge and information processing capacity.

**Examples of Application**

*Signal Reading Time Prediction*

A cognitive task analysis of the signal reading task was developed to determine component behaviours – see Newman (2003). Performance time equations were then used to calculate how long these components take. From this an estimate was made of how long the signal reading task would take under both normal conditions and in the presence of certain sighting hazards. This work supported the validation and extension of the minimum reading time (MRT) form (WS Atkins Rail, 2002).

*Human Factors Error Analysis For SPAD Hazards Checklist*

In a bid to provide greater coverage and objectivity of human factors in SPAD occurrence a checklist of human factors hazards for SPADs was compiled from evidence of SPAD incidents (Lowe and Turner, 2003) and this book. In the development of the checklist the driver performance model was used to define the error mechanisms by which various hazards can act to influence SPAD occurrence. Understanding the mechanism of effect has helped to develop mitigation solutions for many of the identified hazards.

*ERTMS Studies of Driver Workload*

The European Rail Traffic Management System (ERTMS) is the arriving standard for railway interoperability in Europe. Part of this standard describes the European Train Control System (ETCS) future standard for in-cab signalling for trains in Europe. The driver's interface was produced several years ago following an ergonomics study of the task of train driving. Each national railway organisation has the responsibility of fitting this European Standard to the needs, practices and traditions of the national railway.

The work reported by Hamilton and Colford (2003) is the early human factors assessment of the compatibility between the ETCS standard and UK national railway practice. Separate detailed task analyses were produced of train driving in the UK and train driving using ETCS. From these task analyses, workload models were prepared and compared. Then a combined task analysis for hypothetical future UK–ERTMS train driving was produced. This formed the basis for predictive numerical workload assessments of train-driving scenarios. These have been used to identify performance and operational issues that must be addressed during any implementation of ERTMS in the UK.

*Ongoing Validation*

The driver performance modelling approach described here is currently undergoing an extensive validation study. Cognitive task models have been devised for driving scenarios and performance and workload predictions calculated. These scenarios will be recreated in a train driving simulator and run repeatedly with a sample of drivers. Their performance times on task behaviours and their measured and reported

ERTMS: brake demand – Qh

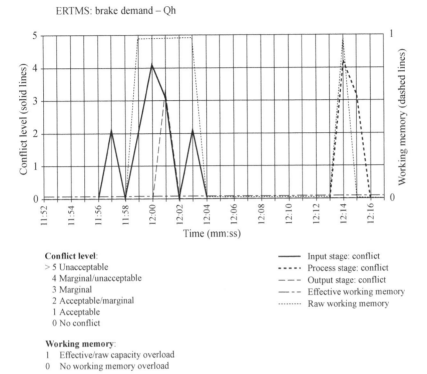

Conflict level:
> 5 Unacceptable
  4 Marginal/unacceptable
  3 Marginal
  2 Acceptable/marginal
  1 Acceptable
  0 No conflict

——— Input stage: conflict
- - - - Process stage: conflict
— — — Output stage: conflict
— - - Effective working memory
········· Raw working memory

Working memory:
1  Effective/raw capacity overload
0  No working memory overload

**Figure 4.5    Sample *Qh* workload graph for ERTMS**

workload will then be compared with the predicted data. The results of the subsequent comparison will be reported early in 2004.

**Summary and Conclusion**

This chapter has reported on the development and main features of a driver performance model that is intended for use in assessing infrastructure and cab drivability. The method utilises cognitive task analysis data for driving scenarios to be used with a set of information processing models to predict operator performance time and workload. Error identification techniques can also be applied to investigate human reliability issues with the data. The utility of the approach has been illustrated by reference to three separate studies each of which highlights a particular aspect of the method's predictive power. A separate scientific validation study is also underway.

This cognitive task analysis and human performance model approach is recommended as a suitable assessment tool for the following reasons:

•    it is centred on the requirements of the driver and therefore on the ergonomic usability of the infrastructure for the driving task;

- it supports a common approach to the design and sighting of both signs and signals;
- the location of information (for example in-cab versus outside cab) is a parameter of the model rather than part of the structure of the model. it can therefore continue to be used as an increasing portion of the information presentation shifts to in-cab displays;
- train speed is a parameter of the model, linked to signal and sign visibility, so the model can be used to evaluate increasing line speeds;
- the model includes the actions of the driver as well as the driver's perception and cognition, so it covers the whole of the cycle from appearance of a sign or signal to the effect on the train of the driver's reaction;
- it provides a language for the description of root causes of error in SPADs, their isolation and the prescription of a mitigating measure based on the diagnosis. the same language can be used to track the effectiveness or not of the prescribed countermeasure;
- the output from workload and human error analyses can be used to de-risk design solutions, to ensure that human performance limitations and the potential for error are taken into consideration at the earliest opportunity within the system design process.

## References

Aldrich, T., Szabo, S. and Bierbaum, G.R. (1989), 'The Development and Application of Models to Predict Operator Workload during System Design, in Human Performance Models, NATO AGARD (D-Model Demand)', in G.R. McMillan, D. Beevis, E. Salas, M.H. Strub, R. Sutton and L. Van Breda (eds), *Applications Of Human Performance Models To System Design*, New York: Plenum Press.

Boff, K.R. and Lincoln, J.E. (eds) (1988), *Engineering Data Compendium: Human Perception and Performance*, Ohio: Harry G. Armstrong Aerospace Medical Research Laboratory, Wright-Patterson Air Force Base.

Card, S.K., Moran, T.P. and Newell, A. (1983), *The Psychology of Human Computer Interaction*, New Jersey: Lawrence Erlbaum Associates.

Hamilton, W.I. and Boughton, J.S. (2003), 'Development of a Theoretically Coherent Workload Prediction Technique', Amsterdam People in Control Conference.

Hamilton, W.I. and Colford, N. (2002), 'An Approach to Addressing Lineside Ergonomics for the Train Driver', Technical Report prepared by Human Engineering Limited for Railtrack plc. Ref. No. HEL/RT/02690/RT1.

Hamilton, W.I. and Colford, N. (2003), 'Developing a Baseline Operational Specification of Driver Requirements for UK ERTMs', IEA Conference, Korea.

Kirwan, B. (1994), *A Guide to Practical Human Reliability Assessment*, London: Taylor and Francis.

Kirwan, B. and Ainsworth, L.K. (eds) (1992), *A Guide to Task Analysis*, London: Taylor and Francis.

Kieras, D.E. and Meyer, D.E. (2000), 'The Role of Cognitive Task Analysis in the Application of Predictive Models of Human Performance', in J.M. Schraagen, S.F. Chipman and V.L. Shalin (eds), *Cognitive Task Analysis*, Mahwah, NJ: Lawrence Erlbaum Associates, Inc.

Li, G. (2001), 'The Influence of Backplate Design and Offset on Railway Signal Conspicuity – Human factors support to the railway signal design for Great Western Zone, Railtrack plc', Technical Report prepared by Human Engineering Limited for Railtrack plc, Ref. No. HEL/RGWZ/01543/RT2.

Li, G. (2002), 'Evaluation of Signal Design Schemes for the Paddington Throat in a Virtual Reality Environment', Technical Report of Test Phase 2 of Human Factors Support to the Fibre Optic Signal Assessment for Great Western Zone, prepared by Human Engineering Limited for Railtrack plc, Ref. No. HEL/RGWZ/01633/RT3.

Li, G. (2003), 'Driver recognition and detection of lineside signs and signals at different approach speeds – Findings of signal/sign trials in a virtual reality environment', Technical Report prepared by Human Engineering Limited for Network Rail, Ref No.HEL/RT/02785.

Lowe, T. and Turner, C. (2003), 'A Human Factors SPAD Checklist', *Proceedings* of the First European Conference on Rail Human Factors, York.

McCracken, J. and Aldrich, T.B. (1984), 'Analysis of Selected LHX Mission Functions', Technical Note ASI 479–024–84(b), Anacapa Sciences.

Network Rail Company Guidance Note (2003), 'Factors Affecting Signal and Sign Conspicuity', Draft Issue prepared by Human Engineering Limited for Network Rail.

Newman, M. (2003), 'Validation of Minimum Reading Time Requirements for Signal Sighting', HF Rail Conference, York.

Parks, D.L. and Boucek, G.P. (1989), 'Workload Prediction, Diagnosis and Continuing Challenges', in G.R. McMillan, D. Beevis, E. Salas, M.H. Strub, R. Sutton and L. Van Breda (eds), *Applications Of Human Performance Models To System Design*, New York: Plenum Press.

Reason, J. (1990), *Human Error*, Cambridge: Cambridge University Press.

Schuck, M.M. (1996), 'Development of Equal-interval Task Rating Scales and Task Conflict Matrices as Predictors of Attentional Demand', *Ergonomics*, 39 (3): 345–57.

Thameslink 2000 (2002), 'Report Core Section Drivability: Driving Simulator Trials – Phase I Pilot Study', MPD/TL2/HFOG/REP/014, 27 May.

Williams, K.E. (2000), 'An Automated Aid for Modelling Human–computer Interaction, Cognitive Task Analysis', in J.M. Schraagen, S.F. Chipman and V.L. Shalin (eds), *Cognitive Task Analysis*, Mahwah, NJ: Lawrence Erlbaum Associates, Inc.

# Chapter 5

# Acting or Reacting?
# A Cognitive Work Analysis Approach to
# the Train Driver Task

Anders Jansson, Eva Olsson and Lena Kecklund

## Introduction

The main objective of this study was to bridge the gap between analysis and design within the context of the train driver system. A cognitive work analysis approach (CWA) was used in order to apprehend the constraints imposed on the train driver. First, a thorough understanding of the train driver task was gained through a mixture of methods. Analyses showed that the driver has to assimilate information from a number of sources, which sometimes create attention conflicts. Despite this, it was concluded that the main weakness of the information environment is not too much information, but that the drivers find it difficult to obtain relevant information. In fact, the drivers sometimes found themselves driving in an informational vacuum. The results further suggest that there are two different driving styles in terms of how the drivers interact with the Automatic Train Control (ATC) system (equivalent to the UK ATP system). One group of drivers adopts a feedback style, waiting until they receive an indication or an alarm from the system before acting. Another group adopts a more feed-forward driving style, and try to plan their driving in order to keep one step ahead of the ATC system. Obviously, today's ATC system supports the feedback style since this gives optimal driving, taking into consideration the limits of the system. However, when there is some sort of disruption to the ATC system a more active driving style is required, something which could be difficult to switch over to for drivers that are used to the feedback style. According to experiences gained from accident investigations carried out in different areas, as well as recent research into safety, errors and accidents can be traced back to underlying faults at different system levels. Providing drivers with more information, integrated and presented in a way that support the driver's natural way of driving and thinking ahead, is seen as the most important measure in order to achieve a design of the information environment that meets the demands of unexpected and unusual events. In short, we believe that the driver's situation awareness is an important parameter to consider in future driver desk solutions. Second, based on the constraints found in the CWA-analyses, and by using the methods for user-centred system design (UCSD) and rapid prototyping, a graphic representation of the track was developed. It was designed with the ambition to enhance the strategy for feed-forward planning found within one group of drivers, and as a complement to the future European MMI-standard (ETCS–ERTMS). During the design process, a group of five drivers were involved, and within a short period of time four design iterations were completed. Early tests show that this part of the interface supports the feed-forward decision strategy used by the drivers who prefer an active driving style, and as a result, the Swedish Railway Administration (Banverket) has decided to support a new project (LINS).

## A System Safety Approach

A system safety approach is adopted in order to comprehend the train driver task, that is, we believe that the driver's information environment, work environment and organisational support functions, and the demands of different societal functions, combine to create the conditions which sets the level of safety as established by current research in the field. This means that different functions and systems interact to influence the safety in the train driver system. By the term train driver system we mean the function, technology, driver and organisation involved in the operation of an individual train. Results from this system safety approach can be found elsewhere (Kecklund, 2001; Kecklund *et al.*, 2003). In this chapter, however, we will focus on the individual driver's use of information.

## Scope and Purpose

The aim of the study is to carry out a survey and an analysis of the train drivers' work situation, in particular the use of information, and how this affects driver behaviour. New approaches to understanding work and safety have to emphasise the need for a cross-disciplinary approach and an in-depth understanding of the work domain (for example Vicente, 1999). Such an approach requires an identification of the behaviour-shaping features and cognitive constraints imposed by the environment and an analysis of actual work practices (for example Rasmussen, Pejtersen, and Goodstein, 1994). A first purpose of the study is to carry out an elaborate analysis of the driver's use of information in order to reach a thorough understanding of what kind of cognitive task it is to drive a train. Moreover, we expect the driver's behaviour to be somehow adapted to today's ATC system, and a second purpose of the study is therefore to describe how such system dependent behaviours manifest themselves.

According to experiences gained from accident investigations carried out in different areas, as well as recent research into safety, errors and accidents can be traced back to underlying faults at different system levels (for example Reason, 1990, 1995, 1997; Turner and Pidgeon, 1997). A third purpose of the study is therefore to measure the level of knowledge of the ATC system within the drivers. Results from this part of the study have also been published elsewhere (Olsson, Kecklund, and Ingre, 2001).

A final purpose of the study is to develop and test the part of the ETCS-MMI where the driver can receive information that is relevant for planning and preparation. We wanted the drivers to be able to prepare themselves with any kind of information they thought might be necessary in order to have a more complete view of the traffic situation further on. This relates to the discussion of situation awareness (Endsley and Garland, 2000), and also to the discussion of automation surprises (Bainbridge,

1987). It also has its counterpart on the traffic control side, where Kauppi *et al.* (2003) propose future control strategies in terms of re-planning. In order to keep the driver in the loop, and to avoid automation surprises, it seemed necessary to try to improve the quality of the drivers' situation awareness within a system safety perspective.

## Method

### Data Collection

First, observational studies were conducted. The aim of these studies was to prepare for and make possible the video recordings later on. A number of different train cabs were inspected and three different driver-desk solutions were chosen in order to achieve a representative set of cab designs for the study. Video recordings were then made on four different real-schedule routes, one long-distance route in a fairly new train cab, one middle-distance route in an older train cab, and two routes in commuter train cabs. The video recordings were then analysed, allowing in-depth analyses of the information environment. During the video recordings, three cameras were used. One camera was focusing on the signals and signs along the tracks, as well as on the nearest surrounding environment. A second camera was focusing on the hands and the face of the driver. The third camera was focusing on the ATC-panel on the driver desk, in particular the two small displays with information from the ATC system indicating speed limits (Figure 5.1). The ATC-panel was the same in all four cab-designs that were part of the study. All together, six professional train-drivers participated in this part of the data collection.

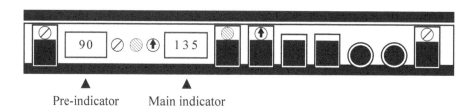

**Figure 5.1    The ATC-panel on the driver-desk with speed-limit indicators at left**

  The video recordings were then shown to another group of drivers who had not been involved in the data collection process before. All together, seven professional train drivers participated in this part of the data collection. They were asked to comment and think-aloud while they were watching their (anonymous) colleagues driving. These think-aloud processes were tape-recorded. After the think-aloud process, these seven drivers were interviewed on a number of subjects, and finally, they completed a recognition test, investigating their domain-specific knowledge.

A vast amount of data was also collected to identify the problems that can impair the train driver's work performance and give rise to risks in the train driver system. These data were collected by use of two questionnaires including self-reported errors. In total, 200 professional train drivers answered both questionnaires. In addition to this, a thorough analysis of records from accident investigations was completed.

In the final phase of the study, five drivers from three different operators (companies) were involved in a user-centred design process, evaluating and changing, through a rapid prototyping process, a design solution exploring one part (D) of the ETCS-MMI (Figure 5.2). Through this part of the interface they could evaluate and plan their driving in the near-future. These five drivers had not been involved in the previous analyses.

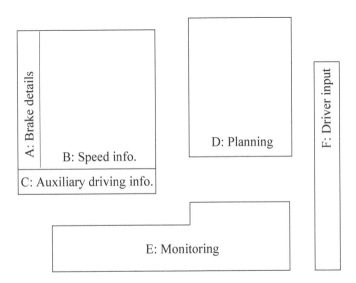

**Figure 5.2    The different areas in the ETCS-MMI**

**Results**

*Drivers' Use Of Information*

The analyses show that the driver has to use and integrate information from several sources. There are obvious information channels as the trackside signals and boards, the ATC system, the route book and timetables, the rulebook and different kinds of real-time safety messages. However, analyses of the think-aloud protocols and interviews show that the driver also uses information from other sources, such as the surroundings near the track.

The requirement of information also seemed to be quite different from one part to another along the route. Three different phases could be identified. Between two stations, the driver is focusing on the speed-limit and the adjustment of the speed of the train. When approaching the next station, the driver's attention shifts towards the surrounding environment (people on the platform) and the braking condition of the train at this particular station. Finally, when leaving the station, the driver is focusing on the possibilities to leave the station as soon as possible (Jansson *et al.*, 2000).

Most drivers who participated in the study claimed that domain-specific knowledge (route-knowledge) is essential if one wants to reach the goals of driving smoothly and to keep on time with the timetable. In conclusion, the task of driving a train is a blend of situation awareness, automated cognitive processes, recognition, working memory limitations and dynamic decision making (Jansson *et al.*, 2000).

*Drivers' Interaction with the ATC System*

The main problems with the interaction with the ATC system can be summarised as a lack of complete information to support the driving task, inadequacies concerning how the information is presented and updated, and its insufficient integration. This creates conflicts of attention. The drivers reported several errors made while gathering data from the ATC system and when inputting the data into it. Many drivers wanted to see the distance to the target point presented in the cab. Many drivers reported problems with the understanding and design of some ATC functions, in particular the supervision of release speed.

*Drivers' System Knowledge*

The results of the questionnaire show that many drivers do not have a complete understanding of how the ATC system, signals and rulebook interact and that some drivers´ held a number of dangerous misconceptions. Also, the system promotes a driving style that is feedback driven. Many drivers adopt this style. This means that the driver is less involved in an evolving situation further down the track. The results indicate that the drivers' knowledge and their mental models of the entire system must be improved (Jansson *et al.*, 2000; Olsson *et al.*, 2001).

Based on the recommendations presented below as well as the in-depth knowledge of the train drivers' task acquired in the previous parts and areas of the project, ideas for a new train driver interface were developed in an interactive Java-prototype. The requirements for the new European Train Control System (ETCS) were considered in the development of the prototype. The new interface was designed to support an extended view of the traffic situation ahead and a group of train drivers spent eight days with the researchers in developing this illustration of the ideas for a new train driver interface. This work has also been reported elsewhere (Stjernström, 2001).

*Recommendations for the Use of Information and the Interaction with the ATC System*

The following general principles should serve as guidelines for the design of a new information environment for the train driver.

1   *Provide more information*
    The first principle is that the driver should have access to far more information when driving the train than is currently the case. The train is often driven in an informational vacuum, a condition that many drivers have grown accustomed to but which prevents train driving from becoming a more active job. More information is not the same as more information channels, however, since the pick-up of information is a fast and efficient cognitive process, but the handling of different instruments and menus is not. Consequently, the addition of further channels to those already available is not recommended. Rather, any new information should be integrated and given a more uniform presentation. In such an environment it should be possible to provide the drivers with more information. Ultimately, the question of information is also a question of how active the driver's role is to be and in what way the different functions are to be automated.

2   *Support the drivers' natural way of working*
    The second principle relates to the way in which the manner of presentation and the content of the information should support the driver's natural way of working (that is, his or her natural understanding of the driving situation and natural ways of handling information in the cab). The driver's understanding of the work in terms of mental models is an important point of departure in this context. It can be hard to identify and define such a mental model, but the strategies that drivers use in their work probably give a good indication of where they need support. If information could be presented in a form that the drivers can easily integrate in their existing models, this would be the best solution. However, new information, possibly via new information channels, always requires a certain process of adjustment on the part of the person handling the information, which can take different amounts of time for different people. It is therefore important for the drivers to be involved at an early stage in the design process (see next section).

3   *Involve the drivers at an early stage*
    The third principle concerns involving the driver at an early stage in the development process. Design proposals in the form of prototypes that drivers' representatives can evaluate and propose changes to are probably the best and most accessible route to an information environment that also works in practice. Such a process of evaluation and modification should be carried out in stages ('iterations') to make sure that important points of view are successively incorporated into the solution finally proposed.

4   *Integrate information – and information sources*
    A fourth principle is that the information in different kinds of documents, often carried around by the driver, should be presented in the cab. The information should be gathered in a single presentation to save time and to make it easier for the driver to assimilate, integrate and mentally process. It is best to have as much information as possible collected in the same medium. A uniform presentation also reduces the need to search for information in several places which demands less time and attention from the driver and makes the driver's job easier. One of the things requested by the drivers is a new way to present information on how well they are adhering to the timetable. This should also be supported as it is part of a driver's natural driving strategy.

5   *Support learning with dynamic information*
    The fifth principle is that information should be presented in a different format
    from today's. Continual updating of information greatly facilitates learning and
    an understanding of what is happening. How a driver would make use of such
    dynamic information is hard to specify, but there is reason to believe that it would
    lead to a more active way of driving and a greater situational awareness amongst
    the drivers. It might also help to improve the human safety barrier at a late phase
    in a risk situation where the only barrier available is the driver him or herself.
6   *Provide new ways to present information*
    The sixth principle is that a certain amount of the information presented could
    be shown in such a way that the driver has it in front of his or her eyes when
    observing the immediate environment. This could be a presentation that combines
    an eye-catching format with the stimulation of senses other than the visual. New
    technology is available to do this, and further developed it should be possible
    to adapt to what is required, and what would be functional, in the train driver's
    environment.
7   *Distinguish between planning and execution*
    The final principle recommended is that as clear a distinction as possible be made
    between the planning and execution of action. Technology is often superior to
    humans in terms of pure routine computations, while it is important that the train
    driver retains active control over the action to be applied, that has been applied
    or possibly should be applied. It should, for example, be possible to automate
    the setting of ATC values to make sure that the correct values are always given
    with the help of information about the train. This should be compared with
    today's routines, where this is done manually. However, some thought should be
    given to whether such an automated procedure can be made reliable; it requires
    information about the train being input in a way that does not involve the driver's
    memory. Letting the technology carry out computations of a recurring nature
    lessens the mental load on the driver, in this case by removing the need to perform
    repetitive tasks, something that people are generally poor at doing.

*A Prototype of the Planning Area in the ETCS-MMI*

Based on the constraints found in the analyses, a graphic representation of the track
was developed (Figure 5.3). In the interviews and from the analyses of the think-
aloud protocols, but also from the recognition test, it was clear that most drivers felt
their route-knowledge, that is their domain-specific knowledge, was a good starting
point when it came to the task of finding a useful design metaphor.

  The drivers claimed that their route-planning, as well as their ability to brake
smoothly when approaching a station, were dependent on the quality of their
route-knowledge, for instance being able to remember a particular station. It
was presumed that a graphic representation of the route should help the driver to
remember and recognise, not only the most obvious items in any station context,
but also useful pieces of information such as up-coming speed-limits and line
inclinations. Therefore, it was designed with the ambition to enhance the strategy
for feed-forward planning found within one group of drivers and by using the
methods of user-centred system design (UCSD) and rapid prototyping. The graphic

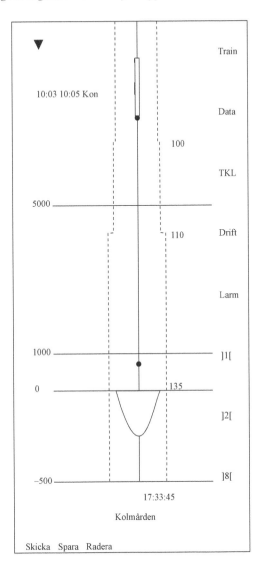

**Figure 5.3    The planning area developed and evaluated in the study**

representation was presented as a complement to the future European MMI-standard (ETCS–ERTMS).

During the design process, a group of five drivers were involved, and within a short period of time four design iterations were completed. The drivers were asked to comment and change every aspect in the prototype.

Through the design iterations it became clear that the graphic representation supports the feed-forward decision strategy used by the drivers who prefer an active driving style, and consequently, we believe that it also enhances the quality of the drivers' situation awareness at any moment.

## Conclusion

The train driver system employs a number of barriers, one of the most important of which is the ATC system. This system constitutes an effective barrier against individual errors and mistakes, such as the driver missing a speed or signal restriction due to fatigue or lack of attention. After all, one can expect that people in this type of job will make such errors and mistakes. Incidents do occur, and driver performance is not always perfect. From a human factors system safety perspective, however, it is the aggregate effect of various kinds of problems and failures that constitutes the major risk in the train driving system. Examples of such combinations are drivers showing signs of fatigue because of trying working conditions, lack of job motivation, lack of system knowledge, and a reactive driving style in situations with degraded information systems or other technical failures. The last of these aspects has been the focus in this chapter.

Among the drivers, the ATC system is seen as a necessary support system. This can be illustrated by a quote from a train driver 'I would probably not be able to manage the high driving tempo today without the support of ATC system' (Olsson *et al.*, 2001). However, the ATC system investigated in this study was originally not intended to be a support system for the driver, but a surveillance system with technical capabilities to help reduce potential for accidents. Thus, when the drivers want to communicate with the system, the interaction is kept at a rather low level. The result of this lack of interaction has been that most drivers have developed a reactive driving style. In most situations such a strategy is sufficient, but when the ATC system is out of order, the same strategy can be dangerous.

The inadequate knowledge about the ATC system and its relation to the rest of the signalling system, the lack of driver information regarding the reason for ATC intervention, and fatigue all combine to create a situation in which there is a distinct risk that serious mistakes will be made. This can occur during a temporary loss of ATC supervision resulting from malfunction or maintenance activities or at places where ATC is inactive or not installed. These conditions are particularly dangerous for those drivers who have adopted a feedback-related style of driving, which means that the latter condition involves a greater risk of the drivers acting too late, if at all, as they wait in vain for signals from the inoperative ATC system. The consequences that different driving styles can have for safety and train driver behaviour in actual driving situations should be investigated further.

This study has generated a large knowledge base concerning the train driver system, behaviour shaping features and cognitive demands and constraints imposed by the information environment. One of the most important cognitive constraints was lack of adequate information for planning ahead. It must be assumed that the driver must always be able to handle complex and demanding driving situations that deviate from the normal driving task.

Broadly speaking, the role of the train driver in future train driver systems can take one of two directions. On the one hand, the driver can be regarded an error-prone component, whose presence is necessary only because it is not possible to automate all functions in the train. This is what Bainbridge (1987) called 'the irony of automation'. On the other hand, the driver can be seen as an independent agent, prepared to take necessary actions and decisions if presented with appropriate information in a dynamic mode. In this study we have developed and evaluated a part of a future driver interface, and the test results so far show that the drivers' decision strategies can be improved.


## References

Bainbridge, L. (1987), 'The Ironies of Automation', in J. Rasmussen, K. Duncan, and J. Leplat (eds), *New Technology and Human Error*, London: Wiley.

Endsley, M.R. and Garland, D.J. (eds) (2000), *Situation Awareness Analysis and Measurement*, Mahwah, NJ: Lawrence Erlbaum Associates.

Jansson, A., Olsson, E. and Kecklund, L. (2000), 'Att köra tåg: Lokförarens arbete ur ett systemperspektiv' (The Driving of a Train: Analysis of the Train Driver's Work by Use of a System Perspective), Technical Report 2000–031, Uppsala: Institutionen för informationsteknologi, Uppsala University.

Kauppi, A., Wikström, J., Hellström, P., Sandblad, B. and Andersson, A.W. (2003), 'Future Train Traffic Control, Control by Re-planning', paper presented at the First European Conference on Rail Human Factors, York, England, 15–17 October.

Kecklund, L. (2001), 'Risks and Proposals for Safety Enhancing Measures in the Train Driver System', *Final Report on the TRAIN-project*, Borlänge, Sweden: Banverket.

Olsson, E., Kecklund, L. and Ingre, M. (2001), 'ATC och informationsergonomi – enkätundersökning av lokförare i Stockholm hösten 1999' (The Train Driver's Information Enironment and ATC – A User Perspective), *Technical Report 2001–013*, Uppsala: Institutionen för informationsteknologi, Uppsala University.

Kecklund, L., Olsson, E., Jansson, A., Kecklund, G. and Ingre, M. (2003), 'The TRAIN-project: Effects of Organizational Factors, Automatic Train Control, Work Hours and Environment: Suggestions for Safety Enhancing Measures', paper presented at the Human Factors and Ergonomics Society meeting, Denver, Colorado, October.

Rasmussen, J., Pejtersen, A.M. and Goodstein, L.P. (1994), *Cognitive Systems Engineering*, New York: Wiley.

Reason, J. (1990), *Human Error*, Cambridge: Cambridge University Press.

Reason, J. (1995), 'A Systems Approach to Organizational Error', *Ergonomics*, 38: 1708–21.

Reason, J. (1997), *Managing the Risks of Organizational Accidents*, Aldershot: Ashgate.

Stjernström, R. (2001), 'User-centred Design of a Train Driver Display', *Technical Report 2001–016*, Uppsala: Institutionen för informationsteknologi, Uppsala University.

Turner, B.A. and Pidgeon, N. (1997), *Man-made Disasters*, Oxford: Butterworth-Heinemann.

UIC (1998) 'ERTMS (ETCS/EIRENE), MMI: The Man Machine Interface of the European Training Control System and the European Radio System for Railways', report of Union Internationale des Chemins de Fer, Paris.

Vicente, K.J. (1999), *Cognitive Work Analysis: Toward Safe, Productive, and Healthy Computer-based Work*, Mahwah, NJ: Lawrence Erlbaum.

# Chapter 6

# The Role of the Future Train Driver

Susannah Russell and Barbara Long

## Introduction

The future role of the train driver will have to evolve in conjunction with the likely increases in technology in trains and in the railway infrastructure. The tasks, equipment and operating systems that could be in place in ten to 30 years time may require a different set of train driving skills. The recruitment and training strategies will need to develop to reflect such changes.

Research undertaken by Quintec Associates Limited, sponsored by the Rail Safety and Standards Board (RSSB), is evaluating the human factors implications for the train driver resulting from likely changes to the rail industry. Looking at key attributes of drivers in passenger and freight trains, conceptual models are being developed. Through interviews, observations and focus groups with rail and technology specialists, the train drivers' role in the future is evaluated. This chapter provides an overview of human factors analysis of current train drivers and the implications for the future.

The role of the train driver has changed gradually over many years, although there are significant similarities between the current driving tasks and those of many years ago. Train cab design has also changed gradually over the years as technology has developed. This 'evolutionary' approach has at times failed to take full account of the interactions between the driver and new train systems, especially automated driver safety aids, such that there are potential human factors conflicts between the driver and the systems in the cab.

The advent of new signalling systems, new communications systems, and enhanced driver safety aids means that the driver's role and the tasks that are performed are likely to change significantly in the years to come.

This project aims to make explicit the role of the train driver in the medium to long term (10–30 years). The intention is that future technological and infrastructure developments can adopt a more robust and informed process of proactive human factors integration rather than the current 'reactive' or 'evolutionary' processes of change influenced by specific project needs and innovations.

Conceptual models of current train driver activities have been developed, so that in the future these can be compared with the predicted train driver activities and the future model ideal. This chapter will present a summary of the 'base' model of current drivers' tasks and the related human factors issues, and the drivers' opinions about future needs for technology, training and selection. Future work will develop a 'predictive' and an 'ideal' model to assess the relationships between current tasks and those of the future.

**Model Development**

A hierarchical task analysis of current train driver tasks was used as the starting point for the development of the base model (Human Engineering Ltd, 2003). By using information from the Rule Book (Safety Standards Directorate, 1999) the task analysis was augmented to represent different types of train driver, passenger train drivers and freight train drivers; to provide greater detail and understanding for the actions required for these tasks; and to add a set of non-standard operating scenarios.

Additionally, many human factors documents were reviewed which aided a more in-depth understanding of some of the pertinent design issues that had already been identified. The primary source of literature was the Railway Safety Human Factors Research Catalogue (2002) which proved a valuable source of human factors information. Some Railway Group Standards (Safety Standards Directorate, 2002; RSSB, 2002) were also reviewed and as well as some training videos (RSSB, RED 1 and 2; Spa Films, 1998; RSSB, RED 3), providing a good source of initial information about train driving activities and associated human requirements to help develop the Base Model. Further understanding and information was gained from various visits, meetings and attendance at conferences and workshops[1] to build up a full picture of the issues in the present and future rail industry.

In order to elicit further information, a structured interview template was developed. This was a live document, allowing modifications as a result of interviewee feedback. One-to-one interviews were used to capture information from 16 train drivers, varying by age, train operating companies (TOCs) and experience in the different types of driving. The driver interviews lasted between 45 minutes and one-and-a-half hours.

The sections in the interview template were as follows:

- after collecting demographic data, the first part of the interview asked driver opinions about the future prospect of driverless trains;
- next, a driver influences section asked the train drivers to talk about the information and other factors that are important for their job and the types of personality characteristics that they believe are required for train driving tasks. this looked not just at what is available, but asked them to think about what could be changed;
- then the interview report asked questions about drivers' training, if there is anything lacking that they would like to see in the future and how particularly they would like to be able to use technological advances, such as simulators;
- next, in the technology list section, drivers are asked about how they think possible future technologies could be applied in the rail industry. This was discussed in much detail;
- finally, the task analysis addressed drivers' experiences with these tasks and how the task breakdown as given may differ from their actual experience. They were also asked to think about the activities that might cause them problems, in terms of prone to error, high workload etc., and identify any technological or operational improvements that could be made to these tasks.

The issue of 'driverless trains' was also explored in interviews with 258 passengers. The data collected from both the passengers and train driver interviews will support the development of a future train model in future stages of the research.

**Base Model: Current Passenger and Freight Train Driving**

The background data and the structured interviews enabled the development of the base model using TRAP™ (training requirements analysis programme[2]). Human factors analyses were conducted on each sub-task in the task analysis:

• Interface analysis to establish the inputs and outputs required for each sub-task;
• DIF (difficulty, importance, frequency) analysis rates the difficulty, importance and frequency of each sub-task and provides a score between 1 and 5, where 1 is a high priority task and 5 is low. Specific supporting data were used (Air Affairs Ltd, 2000);
• KSA analysis establishes the knowledge, physical and mental skills, and attitudes required to perform each task. Specific supporting data were used (Air Affairs Ltd, 2000; Occupational Psychology Centre, 2001);
• workload scores are defined based on the amount of data to which the driver has to attend; how much of the data require interpretation; and the complexity of the response required. Workload is scored out of 100 per cent;
• human error analysis was conducted firstly looking at the likelihood of the error occurring, and then consequences of the error. Suggested error consequences were recorded in the TRAP model and human error criticality scores were then assigned to each sub-task based on these assessments. Specific supporting data were used (Porter, 1992).

Some interesting outcomes of the base model are discussed, highlighting some of the key differences between the passenger train drivers (PTDs) and freight train drivers (FTDs) models.

*Interfaces*

Freight train drivers' tasks are more affected by the variety of loads carried and therefore require more adaptable speed control skills. There is less modern technology in freight trains than in the newer passenger trains, although older passenger trains are quite basic. For example, all passenger trains have the driver reminder appliance although most freight trains do not. Freight trains always run driver only operations, and have fewer station stops to learn than passenger trains.

For both PTDs and FTDs, some TOCs now authorise their drivers to carry mobile phones, which significantly improves communications with their control centres and relevant signal boxes. However, the Rule Book states that the driver should first try to use the cab radio, then the signal post telephone, then if they still cannot get through to the signal box they may use 'any other phone'.

*KSA*

Route knowledge is critical because of the amount of time that it takes to build up, and it impacts on all parts of preparing for and driving the train. The depth of route knowledge is really tested in the non-standard operating scenarios. PTDs require the additional skill of having to know the station stopping requirements, although FTDs require an increased amount of time required for slowing down due to heavy and long loads and they tend to cover much greater areas than PTDs, therefore requiring more extensive route knowledge.

FTDs tend to have greater knowledge of their locomotives and vehicle equipment, often conducting more comprehensive checks of the engine with a walk through the engine compartment.

The physical skills for train driving are primarily hand/eyecoordination tasks, such as speed control and manipulation of other cab equipment, for preparing and driving the train. The speed of response may need to be greater in high speed and heavier freight trains, than slower passenger trains or empty freight trains

Surprisingly, both PTDs and FTDs reported the importance of reading, writing and verbal communication skills as they aid the driver in preparing for the journey and reporting incidents and defects. Some faults, for example brake dragging, are more common for FTDs, therefore they are likely to have more skill at diagnosing and reporting this fault due to increased familiarity.

Many mental skills are required for interpreting cab equipment, the external environment, signals and warning noises. Memory, comprehension, vigilance and concentration are also critical elements of train driving.

The key skills identified in good PTDs and FTDs are:

- vigilance, important in monitoring gauges as well the external environment;
- safety consciousness, for example important in waiting for right away signal;
- diligent checking, for example important in checking the running signal is clear after having received the OK indication from the train manager;
- time consciousness, for example important in maintaining route and times;
- other attributes which may not initially be thought to be important for a train driver include being a good verbal communicator, paperwork completer and reader to support the verbal, reading and writing skills required;
- another important attribute may be compassion, which is required for dealing with an incident or when responding to passenger enquiries.

*DIF (Difficulty, Importance, Frequency)*

Those activities with tasks of lower DIF scores tend to be part of the psychological preparation process for the more difficult activities. Many tasks score 2 on the DIF scale, and the range in these activities is large, for example discussing incidents or causes for concern on route and checking the repair book. These tend to be fairly important tasks, which may have serious consequences. However, errors in these preparatory tasks can often be noticed and prevented during some later activity. When errors occur here, they are at the beginning of a chain of events that may or may not lead to an incident.

Tasks scoring 1 are safety critical tasks. These include testing equipment, such as traction systems and driver safety device, as well as the primary driving tasks. Errors in these activities cannot always be prevented by some later activity, as they are at the end of a chain of events that could lead to an incident.

High DIF scores, that is scoring 1–2, should be judged with some care. In developing the 'Ideal' train model a combination of all elements, including high DIF and high human error potential and/or workload, will be addressed as a priority.

*Workload*

High workload tasks are considered to be those scoring above 60. Those are tasks involved in primary driving activities, such as maintaining speed, monitoring the environment, adhering to signals, maintaining the route and times and stopping the train. These tasks are all done in conjunction or quick succession; on their own they may not have such high workload, but the combination of the tasks affects the workload rating. These activities require the full range of knowledge, such as route knowledge, knowing the timetable and vehicle equipment, and mental skills to be able to remain vigilant for lengthy periods of time, whilst being able to interpret signals and many aspects of the environment. These tasks require good physical hand/eye coordination for speed control, and to be able to achieve these tasks effectively a professional driving attitude is required (vigilance and diligent checking for line-side information and signals) to ensure safe running of the train whilst not compromising route times.

PTDs reported greater stress caused by having to adhere to stricter timetables. Also, having to deal with passengers, and associated technologies such as power operated doors, creates more unpredictability, particularly in the event of having to manage a train evacuation. FTDs' most critical tasks tend to be as a result of having a longer, heavier load, pulling onto the line requires a lot of care and stopping the train is a complex activity relying heavily on route knowledge as they have to prepare early to slow before approaching the signal.

*Human Error*

Human error analysis helps identify tasks that have high human error. This includes tasks that might not have high workload scores, but with the possibility of latent errors can still have catastrophic effects. These may be tasks that are easy to perform and do not often go wrong, and therefore complacency can result.

Checking daily notices and late notices are the first line of information about a future event, but the time gap and likelihood of needing the information means that it is rarely attended to in any detail. PTDs and FTDs receive most of their information in large paper-based documents, which make it difficult to scan quickly for information relevant to your route.

Other tasks are more obviously critical to human error and have high DIF scores, such as the primary driving tasks. These require high vigilance and concentration and the tasks are complicated by the possibility of other factors occurring whilst still having to conduct the primary driving task to the same level, for example, cancelling the audible Automatic Warning System for a clear signal. This indicates simply that they can continue doing exactly what they were doing. This leads to habituation which can lead to human errors.

All the abnormal scenarios have higher error possibilities due to the infrequency of occurrence and lack of familiarisation with required actions. Tasks such as driving in adverse weather conditions add to the already high workload task of general driving, having to consider the rules for safe driving and physical skills needing to adapt to the change of circumstances. Also, verbal communication with the signaller, because of the interaction between two humans, has an increased likelihood of human error.

## Selection

There are four main psychometric tests that are currently (in February 2003) used to select train drivers in the UK in addition to conducting interviews.[3] These tests are:

1  concentration (under time pressure);
2  vigilance and reaction times (under time pressure);
3  trainability of rules and procedures;
4  mechanical comprehension.

Other tests under development for possible future use are faultfinding, personality assessment for safe drivers, planning skills and communications skills.

By assessing the output of the human factors analyses in the base model and making comparisons with current selection methods, it can be seen that the selection tests do not identify all the characteristics that are required to be a good train driver, even with current UK trains.

In the interview PTDs and FTDs were asked to rate the importance of:

1   working well under pressure;
2   enjoying working alone;
3   good vigilance;
4   good rules and procedure following;
5   good information interpretation;
6   good verbal communication;
7   good written communication;
8   good concentration;
9   comfortable with computer technology;
10  positive attitude to change.

The results showed 100 per cent agreement with vigilance and concentration, in that all drivers rated this to be a highly important trait that was required to be a good driver, and these are already tested. Most drivers rated working alone, following procedures and interpreting information as either very important or fairly important. These are supported by the current selection methodology in the trainability of rules and procedures. Most PTDs rated working under pressure very highly and those tests of tasks under time pressure cover this skill already. However, FTDs reported that they needed to be patient, relaxed and laid back as they often have to take second place on the railways to the strict timetables of passenger trains, and are sometimes put at the back of the queue in terms of priority for access to stations and across junctions.

Both PTDs and FTDs rated verbal and written communication skills highly and written skills are not currently tested for. Related to this is the need for computer skills. Most drivers rated the current need for computer skills as rarely important or not at all important. However, most noted the increasing requirements for understanding computers as part of a normal working life. Most felt that drivers should already be selected with some computer skills, as these would be a requirement in the very near future.

Another more surprising finding was that drivers rated a need to have a positive attitude to change. It was felt that the day-to-day working of a train driver required acceptance of many different scenarios and that things were often unpredictable; therefore, if drivers had a positive attitude to change, their performance during these unpredicted tasks would be better.

## Training

As part of the interview, drivers were asked about their experiences of driver training. Although this was not a topic that was studied in depth, drivers were generally fairly satisfied with the level of initial training that they had received, and also the training that new recruits were receiving.

The most frequently mentioned item was gaining enough route knowledge during training. The older drivers discussed the significant amount of time that they spent as a trainee/second driver, and how these experiences made them a better driver today. Younger drivers or driver trainers reported that many new drivers do not get to experience their full route in difficult driving conditions during their training.

FTDs noted a significant difference in their route knowledge requirements in that they tended to cover a wider area than PTDs, although it was often less detailed because they were not concerned with the same number of stations and stopping points.

Some suggestions were made regarding improved training, namely: single line working; driving in adverse weather conditions (particularly in low adhesion conditions and night time driving); emergency situations and fire training (particularly the use of fire extinguishers). Also reported was that a yearly performance measurement test would not be unwelcome, computer training should begin now to familiarise drivers with information technology prior to its introduction. Drivers feel that they should be able to prioritise where they need additional training.

As part of the interview, drivers were asked to give comments about how they felt simulators should be used in the future. Those drivers who were against the use of simulators altogether tended to be the older drivers. Of those that thought that simulators would be good, most favoured full motion simulators over route or procedural simulators. However, many saw the benefits that office- or Internet-based systems could give them in being able to have free training time to update their familiarity with routes or procedures.

Nearly all drivers felt that simulators should be used for training drivers for abnormal operating scenarios, as these were obviously difficult to train for in the real world. Many drivers stated that although they knew the theory, there were significant gaps in their experience regarding emergency procedures or driving in difficult conditions. These elements could be easily simulated in a full motion simulator, and would have a significant impact on drivers' confidence to cope with unpredictable operating scenarios.

## Future Technology

The technology list is a way of exploring the advantages and disadvantages of such technology in order to assist in building a representation of the 'ideal' best practice and more futuristic model. Towards this end, drivers were asked to provide their views on the pros and cons of such technology and provide feedback on how it might be used. The technology list was divided into different sections to support the understanding of how technologies could be used.

**Table 6.1    The technology list**

| Input technology<br>*Where do you get your information?* | Input technology<br>*What information do you get?* | Outputs<br>*How do you activate/control/communicate with the systems?* |
|---|---|---|
| • Single function computer<br>• Multifunction displays<br>• Computer with windows type display<br>• Intranet-linked computer<br>• Internet-linked PC<br>• Simple audio warnings<br>• Complex audio warnings<br>• Direct voice output<br>• Personal digital assistant<br>• Pager<br>• GPS<br>• Head-up displays<br>• Head mounted/visor Displays<br>• Cab phone<br>• Mobile phone<br>• CSR<br>• GSMR | • Track monitoring equipment on sleepers<br>• Track monitoring equipment on trains<br>• Train health monitoring equipment<br>• Weather/environmental information<br>• Infrared obstruction notification<br>• Trackside cameras for pedestrian/obstruction<br>• On train cameras to monitor passengers<br>• Other camera equipment<br>• Passenger counting equipment<br>• Electronic ticketing<br>• Platform doors | • Haptic controls<br>• Direct voice input<br>• Touch screen<br>• Roller ball<br>• Finger ball<br>• Mouse<br>• Keyboard<br>• Multifunction keypads<br>• Automatic route control<br>• Auto stop at stations/platforms<br>• Auto speed control<br>• Auto acceleration and deceleration<br>• Auto signal identification and response |

The results showed that preferences point toward having some kind of computer system, functionally specific to the train drivers' tasks, with automatic download for purposes such as maintenance defaults going straight to the relevant depot. Drivers thought that a multifunction display would be the best method of information presentation. Most were enthusiastic about having electronic formats for a back-up to information that is currently produced on paper, such as the Rule Book, operating notices and fault finding information. Many felt that a personal digital assistant (like a

Palm Pilot) was a great idea, as well as having some improved verbal communication through some advanced radio or telephone system.

Direct voice output was preferred over more bleeps and tones in the cab, as they already felt that this was becoming an audio overload. Head-up displays produced a variety of responses, though all felt that some kind of signal reminder system would be a benefit. Train, track and weather monitoring would provide useful information, but this was not suitable for the driver in the cab and this information should be sent back to the maintenance depot.

Infrared is probably not technologically advanced enough to be useful for obstruction notification. However, cameras at safety critical points on the track, such as level crossings would be useful. The touch screen was definitely the best option identified for control, and direct voice input was also thought to provide good input without needing more resources from the driver.

Automation discussions brought up many issues, with drivers concerned about:

- dealing with emergencies and unpredictable events;
- lack of reliability with automation and satisfaction with infrastructure;
- the role of the train driver, and losing his skills.

Drivers felt that some level of automation would be good where it provided an aid to their role, but full automation would make the driver a systems monitor, and that this was not considered to be a good role for a future driver.

## Conclusion

The base models for PTDs and FTDs have been developed through interactions with drivers, and opinions have been collected about the future of training, selection and technology. Route knowledge is critical, as are high vigilance and concentration. The primary driving activities create the highest workload on a daily basis, with the non-standard operations reaching peak workload levels due to the unpredictability of the occurrence and the unfamiliarity of the required responses.

It appears to be important that train driver selection tests should be expanded to include good verbal and written communication skills, and a positive attitude to change. FTDs specifically should not be tested for working under time pressure, but for being patient.

Many technological advances were viewed positively, such as multifunction displays, direct voice output with supporting information displayed electronically. Increasing automation may allow the driver to move away from their primary position although current train drivers cannot envisage why this would be appropriate.

Quintec Associates' continued research for the RSSB will provide information about how to integrate technology and people when thinking about the role of the future train driver, through the development of two additional models:

- a future 'predictive' model of the role of the train driver will be established based on key forecast 'changes' in technology, such as the introduction of ERTMS;

- a best-practice 'ideal' model of the role of the train driver will be established using human factors integration principles with future technologies, whilst considering the output from the base model to optimise the use of the human in the future of train driving.

The contrasts and overlaps between all three models ('base', 'predictive' and 'ideal') will be analysed, enabling the identification of skills gaps, training needs, selection criteria, high-level implications for workload and high-level impact on human error. It will also help determine whether the predictive model is forecast to achieve this ideal model.

## Notes

1    Romney, Hythe and Dymchurch Railway Familiarisation Course April 2003; Railway Safety Seminar, April 2003, IMECHE; New Trains Conference, June 2003, IMECHE.
2    TRAP™ is a Microsoft Windows-based tool developed by Quintec Associated Ltd for use in the analysis of tasks and training needs. It utilises standard windows features, with specific functions to capture human factors analysis data and provide a detailed audit trail.
3    Occupational Psychologist Centre at Watford, meeting held on 18 February 2003.

## References

Air Affairs (UK) Ltd (2000), 'Train Driver Training Needs Analysis – Summary Report', PR00013, 21 December.

Human Engineering Ltd (2003), 'Stage 1.2 Hierarchical Task Analysis of the UK Train Driving Task', HEL/RS/02768a/RT3, Issue 1, January.

Porter, D. (1992), *A Detailed Task Analysis of Four Types of Train Driving*, Issue 1, Cheshire: SRD.

RSSB (2002), *Human Factors Research Catalogue*, CD-Rom, June.

RSSB (2002), Approved Code of Practice – Train Driving, (F) GO/RC3551, Issue 3, October.

RSSB, A Series about SPAD Reduction: Red Programmes 1 and 2, Vision Consultancy, video, details available at http://www.rssb.co.uk/natini_video.asp.

RSSB, Red Special Edition 3: SAS SPAD, SOY SPAD Vision Consultancy, video, details available at http://www.rssb.co.uk/natini_video.asp.

Safety Standards Directorate (1999), *Master Rulebook for Personalised Rule Book*, Series, GO/RT3001–GO/RT3013, London: Railtrack.

Safety Standards Directorate (2002), *Train Driving*, (D) GO/RT3252, Issue 4, October, London: Railtrack.

Spa Films (1998), 'Train Protection and Warning System: Introducing TPWS', TPWS/VID/1, video.

# Chapter 7

# Investigating Train Driver Behaviour: The Use of Lineside Information When Regulating Speed

Trudi Farrington-Darby, John R. Wilson and Beverley J. Norris

## Introduction

A qualitative study gathering data on driver behaviour was performed with UK freight train and passenger train drivers of varying years driving experience. This chapter reports on the method and findings of one aspect of that, namely train drivers' use of lineside information (specifically permanent speed restriction boards) in the regulation of speed. The findings suggest that train drivers of all levels of experience use permanent speed restriction signage but the reasons why and the way they use them will differ depending on their task goals and the experience they have of the route they are driving.

Train driver behaviour is becoming of increasing interest as drivers are considered as main users when designing or changing work environments (both proximal environments such as train cabs and more distal environments like the infrastructure). Driver behaviour has long been an area of active research for road vehicle drivers. Published work investigating road vehicle drivers' use of signage has yielded interesting results, some of which may be transferable and add to the understanding of train driver behaviour.

Road sign registration by the driver is reported as linked with the perceived relevance of the signs' information to the driver's current situation and ability to reach his or her goals (Rumar, 1994). Subconscious registration of information (Fisher, 1992; Crundall and Underwood, 2001) suggests that drivers may discard irrelevant information before they realise they have registered it and before it requires further processing ( Avant *et al.*, 1986; Saitta and Zucker, 2001). These ideas together with findings that people recall what is of use to them (Fisher, 1992) may be sufficient evidence that studies of sign usage based simply on 'show and recall', and studies that lack the contextual richness required to drive discriminatory processing of visual information, may not be showing the true use of signs for driving tasks.

There is also an emphasis in much of the driver behaviour literature on the role of experience on driving road vehicles. This literature should be used with caution as the term experience can mean different things depending on how it is defined. Drivers' information requirements and how they use information will change with experience in driving or the route they are driving. As experienced drivers develop route familiarity they tend to refer to the signs less (Sprenger *et al.*, 1999). Experienced drivers' search for roadside information extends further afield (this is also the case with increased complexity of scene for both experienced and novice drivers (Pottier, 1999)) whereas less experienced drivers tend to focus nearer the vehicle.

A study by Krose and Julesz (1990) found that if an observer is informed about a target position before the target is presented performance in finding the target is better. This may have some relevance in the searching of information that is positioned where drivers have been told it will be (new information) or where they know it will be from recall of a well-known route.

## Study Background

This particular study came about as a result of the changes that are occurring on the West Coast route. New faster tilting trains are being introduced and this introduction of new rolling stock is being accompanied by infrastructure engineering changes. With the new trains travelling at higher speeds than are currently permitted on the rail network, traditional speed restrictions showing permissible speeds (many of the trains would still be using these) would not be valid for the faster trains. The newer trains may follow permissible speeds (PS) in some places (not all sections of track would allow for the higher speeds for these new trains) but will also follow a system of signs indicating enhanced permissible speeds (EPS).

The questions that were being asked at the start of the study were:

1   how many signs should there be on one sign post before there were too many (number);
2   how far from the required action should the signs be placed (position) bearing in mind that the infrastructure limits positioning and sighting of signs for drivers;
3   what information should be shown on the signs and how should it be arranged in terms of placement, colour, size, pictures or letter etc. (sign profile)?

Other studies were being performed looking at the design of the EPS signs and there has also been work looking at recognition and recall of signs, that is dealing with questions two and three. This left question 1.

Question 1, like 2 and 3, is very specific. Anecdotal information from train drivers in the past planted the idea that drivers may not need permanent speed restriction signs to regulate speed because they have route knowledge. If this was true were we worrying about nothing? If it was not, then some understanding of why and how train drivers use signs was required. The need for this understanding forms the basis of the study described in this chapter.

## Methodology

To develop an initial understanding of train drivers' use of lineside speed restriction signs to regulate speed, a qualitative approach was adopted using drivers as informants. The whole process included two cab rides, a literature review focusing on what was known about train drivers and what could be transferred from research on road drivers, and semi-structured interviews with train drivers.

*Cab Rides*

An initial cab ride was taken on a short suburban route to familiarise the researcher with some of the practical aspects of a cab environment and the infrastructure. A second cab ride in a high-speed train and on a longer journey up the West Coast route was made to place some of the concepts that were being discussed in interviews into a real world context.

*Interviews*

A total of 11 train drivers were interviewed. The sample included eight passenger train drivers and three freight train drivers. The experience measured in years ranged from one week to 32 years for the passenger train drivers. The experience was above 22 years for all the freight train drivers as they have usually come from passenger train driving in the past. The range of experience of the drivers interviewed is shown in Table 7.1.

**Table 7.1    The range of train drivers interviewed**

| Driver number | Years of driving experience | Train type currently driven |
|---|:---:|:---:|
| 1 | 4 years | Passenger |
| 2 | 32 years | Passenger |
| 3 | 2 years (8 years on a rural railway line) | Passenger |
| 4 | 29 years | Passenger |
| 5 | 32 years | Passenger |
| 6 | 32 years | Passenger |
| 7 | 2 years | Passenger |
| 8 | 1 week | Passenger |
| 9 | 24 years | Freight |
| 10 | 22 years | Freight |
| 11 | 28 years | Freight |

*Interview Questions*

The questioning during interviews involved various approaches to eliciting information from drivers. The questioning took the form of situation descriptions and follow up probe questions and direct questions. Another approach involved asking drivers to compare their current use of signs with why and how they used them when they were new to a route and new to driving.

*Analysis*

This chapter focuses on the results from coding of data that concentrated on clues to the how and the why of the use of signs in regulating speed. A theme of experience playing a crucial role in the use of any information for speed regulation emerged and so special attention was paid to coding information that expanded on that topic. In the case of train driving experience is not just simply the years spent driving a train but is related more to the number of times a route had been driven. In this study drivers with many years train driving experience also had the most experience of the routes they were driving. They were also able to recall what it was like to be a newer driver when they moved onto new routes.

**Findings**

On a global level the findings of this study suggest that train drivers use lineside information as cues that inform speed regulation strategy choice and development. Strategies are considered in this context to be chosen from a catalogue of possible strategies for the regulation of speed. These are based on what is deemed most likely to achieve goals based on the conditions external to and within the driver.

Some of the main factors that were suggested from this study to influence the choice of cues and strategies for speed regulation for train drivers are:

- the driver's experience and knowledge of the route;
- the driver's overall experience of driving a train;
- the confidence a driver has in his/her knowledge;
- how the driver learns the route;
- what is considered important and relevant to the driver as an individual;
- the environmental conditions the driver is travelling in;
- the type of train being driven.

Train drivers all reported using permanent speed restriction signs in order to regulate speed but not all of them reported relying on them.

Figure 7.1 shows the findings from this study in terms of *why* and *how* drivers use permanent speed restriction boards for speed regulation.

All the *whys* and *hows* of using signage by train drivers are thought to require the driver to register the sign at some level. Reading sign content may not be as important for drivers who are using or reacting to the location of a sign rather than its content, and recognising the sign from its characteristics other than the speed value displayed on it.

There is some overlap between the cells in Figure 7.1. Newer drivers may use signs for location orientation and for speed information at the same time. In fact if a train driver were not anticipating a speed change he or she is unlikely to be anticipating what it is. If a train driver has no idea where he or she is along the route then they are unlikely to know where and what the next speed restriction is.

As drivers move from novice to experienced drivers on a route they are travelling on they move from the left to the right hand side of Figure 7.1. They will also rely

**Moving from novice to experienced train driver**

**Moving from novice to experienced train driver**

**HOW**

| WHY | *No anticipation – bottom up processing* | *Anticipation but not full recollection of details* | *Good recollection and anticipation but not full confidence* |
|---|---|---|---|
| **LOCATION** **Orientation** | 20 sign appears and orientates the driver that they are approaching the next station. | Driver passes a series of bridges but is not sure where he is as these could be any bridges. He knows there is stretch of track with 50 and then 60 between 3rd and 4th bridge. When he sees those he knows where he is. | The driver knows where he or she is but a 70 sign after a neutral section sign confirms that he or she is at position x. |
| **LOCATION** **Cue for action** | 50 sign appears to a driver that knows they will have to reduce speed at some point and reminds the driver to brake at 60 to brake at that point. | Can't recall exactly where to perform an action such as a drop in speed. Sign shows exact location of restriction. The driver takes a marker off that sign. | The driver knows where to brake between the pub and the station but the AW board tell confirms that braking position. |
| **SPEED** **What speed** | Sign appears. The driver did not anticipate a change in speed or what it might be. | The driver knows where the speed change is but cannot recall the exact speed. | The driver knows to change to 50 at position x but wants to confirm the change is to 50mph. |

**Figure 7.1**   **'Why' and 'how' train drivers use permanent speed restriction signs in the regulation of speed**

less on speed restriction boards for cuing position and actions and will adopt more of their own landmarks. In that respect they move more towards the lower right hand corner of the figure.

## Discussion

*Why Do Drivers Use Permanent Speed Restriction Signage in their Regulation of Train Speed?*

This depends on the goals of the task at the time. If the goal at the time is to ascertain or confirm the location on the route, the sign may be used to determine where a driver is which in turn helps them to access their route knowledge and places them in an area of the track of known speed restriction. In this respect the train driver is using the visual cue of a sign just as he or she would any other visual cue, for example a signal number, a bridge, a building or any proprioceptive cue such as a curve to the left after a slow junction crossing or an auditory cue such as the sound of three bridges close together. In this case the train driver could use the sign as a positional navigator as much as he or she could use any other kind. It is in a way personal to that driver. He or she may choose to use that cue because they have always used it, they were taught it by their driving instructor for that route, it is used as a main landmark by other drivers in conversations about that route; or perhaps they do not always use it, it just happens that other cues were missed due to a lapse in concentration.

The goal of the driver may be to initiate an action (for example, brake, accelerate or check speed) or to begin to look for where they might want to perform an action. If used in this way the train driver is using a sign as a location cue but also as a cue for action. If the goal is to check speed the drivers may use a sign to cue a check of the speed they changed to at some point prior to reaching the sign. This can be seen as using the sign as a reminder to check back on an action that was taken previously. Unless the information on the sign is being used this is purely a positional cue that suggests an action and could be any cue provided the same meaning was placed on it. For example, at the pink pub the driver checks the train speed is still at 70mph.

If the goal of the driver is to determine speed from the sign, then the lineside speed restriction sign is used for its content. The driver may be interested in the location but also has a specific goal of obtaining the information contained within the speed board, namely the speed and the position of the speed on that board (some boards have two speeds, the position of the speed on the board is used as indication of the type of train it pertains to, most typically freight and passenger trains are distinguished by this method).

The three reasons given by drivers for their use of permanent speed restrictions can occur at the same time – that is, the driver may be interested in knowing where he or she is, when to take action and what action to take or they may occur individually.

The importance of individual goals in the use of lineside signage by train drivers described in this study is seen also in the findings from studies on road drivers (Fisher, 1992; Rumar, 1994). The reasons *why* signs are used (that is, the goals) may be the same for drivers at different levels of route and driving experience but *how* drivers use them will differ with experience.

*How Do Drivers Use Permanent Speed Restriction Signage in their Regulation of Train Speed?*

Newer drivers on a route tend towards the left hand side of Figure 7.1. They use signs from a position of poor or no anticipation. In order to meet their goals for location or instruction information they rely more on the information presenting itself. This bottom up form of information processing is what occurs in other forms of driving where road drivers encounter a new route. It only really occurs with novice drivers when they are learning a route or if the route changes after they have learnt it. It is this type of processing of visual information that is tested in an unfamiliar environment or simulation with studies that evaluate the recall of presented information or the time to recognise signs. More experienced drivers who usually anticipate the appearance of visual cues or information are unlikely to reproduce natural real world results in such settings. This type of study also assumes that the goal of the driver's task is met through the use of the information on the signs being presented. In that respect the findings from this study suggest that goals and context are key when evaluating the processing of signs and potentially other forms of lineside information in such studies. Krose and Julesz's (1990) work with road drivers showed improved target detection performance with prior knowledge of where to expect to see the target, much as an experienced train driver might hold.

Like road drivers (Pottier, 1999), newer train drivers describe concentrating on the lineside information nearer them. The information they receive is all relatively new to them with little being anticipated and there is no need to look further afield for a more global picture. As the driver becomes more familiar with the route and encounters it in different conditions (the factors that affect speed regulation are encountered more and in different combinations) the train driver maps his or her own cues or markers for action, navigation and information. The information they rely on becomes more of the unofficial cues that lie further from the lineside. The map of a route with information relevant to individual drivers, were it to be drawn by the driver, would potentially look very different to the one they would draw when they first learnt the route. Their repertoire of strategies to deal with different factors affecting the choice of actions to regulate speed would be expanding. These more experienced drivers have more information in memory and have better recall. In the earlier stages of consolidating route knowledge they may have low confidence and occasionally lack accurate recall; the use of the signs is not a complete reliance but more a case of filling in occasional blanks in order to achieve the goals for regulating speed.

The really experienced drivers describe almost automatic, subconscious processing (Crundall and Underwood, 2001; Fisher 1992) of information and regulation of speed. This group of drivers is probably the more difficult group to elicit information from regarding their use of speed restriction signs (Berry and Dienes, 1993). This poses a methodological issue for further investigation of expert knowledge and behaviour in this group.

Experienced train drivers use signs in a more confirmatory capacity. It is important to mention here that this group of drivers, when asked if they would be confident that they could maintain their current level of safety and efficiency when driving should all the permanent speed restriction signs be removed from the railway network, all replied yes, but only for the routes they knew well. When signs were used in a confirmatory capacity to ascertain location or to cue actions many other lineside cues could be used in the same way. When the signs were used for content information (speed) they alone could be used in that way.

Experienced drivers use their top-down processing strategies more to drive the search for the information than to establish from where and what they anticipate it to be. This was also described to be the case when degradation of visual cues occurred.

## Conclusion

Qualitative research in the field, whilst deeply informative, is not without its difficulties. Train drivers are usually driving so many of them were unavailable for interview during their normal working hours. Accessing participants for significant lengths of time proved to be challenging. The sample size for this study is relatively small compared to more quantitative studies, but by the end of the interviews many themes were being repeated, suggesting that this number of participants was sufficient. A practical point at which to cease interviewing occurred when the same themes began to recur with different interviews. That was the reason for stopping the study at the number of drivers that it involved.

Newer drivers are usually being trained and were less available than other drivers. This means that the range of experience, whilst broad, is clustered more towards the more experienced end. This problem was overcome to some degree by the use of comparisons and observations of newer drivers made by some of the more experienced drivers.

Different geography and types of trains were not fully represented. Fortunately several of the drivers interviewed were able to comment on other jobs where they drove other locomotives and encountered other types of infrastructure.

The potential changes to signage proposed on the West Coast route has raised a number of human factors questions about train driver performance. In order to answer specific questions from projects and to better understand incidents involving train drivers a greater understanding of the how and why of driver behaviour is required. Additionally a framework or model of driver behaviour is needed, upon which to hang study findings and to drive future studies in terms of their hypotheses and methodologies.

The findings described in this chapter are only a small section of the information gathered regarding driver behaviour during a series of interviews with train drivers. The main purpose was to ascertain why and how they used lineside permanent speed restrictions to regulate their speed.

The findings suggest that train drivers use speed restriction signs depending on their goals and their level of experience of a route (the latter in turn is closely tied in with confidence in knowledge and recall) and that there are aspects of how drivers use signs that are transferable between train and road drivers.

It is proposed that novice train drivers use signs in a way that is mostly bottom-up processing for the purposes of determining their location on a stretch of a route, performing actions through the use of signs as location cues and determining speed changes. As drivers obtain driving experience on a specific route they require signs to ensure a full picture of where they are, where they want to act and what speed they need to travel at. They anticipate sign location and content but utilise the signs to complete the picture of information they have and ensure accuracy by filling in the gaps if and when they have poor recall. Very experienced drivers of a particular route reported using signs in a confirmatory capacity but for the same goals of determining location, where to perform actions and what speed to be travelling at.

This chapter reports on the information provided by 11 train drivers during semi-structured interviews. It is hoped that the findings will form the basis of future research, provide answers to some of the questions that are raised during railway projects, and assistance in understanding some of the methodological issues for study design when investigating driver behaviour and the use of visual cues.

## Acknowledgements

The time and effort of the drivers' standards managers and drivers given in this investigation was key to the study's success and is much appreciated.

The collaborative effort between the University of Nottingham's Institute for Occupational Ergonomics and Network Rail's West Coast Route Modernisation has meant that this research could be performed with operational staff as the main participants and will be fed back to those who influence the design of the railway infrastructure.

## References

Avant, L.L., Thieman, A.A., Brewer, K. and Woodmar, W. (1986), 'On the Earliest Perceptual Operations of Detecting and Recognising Traffic Signs', in A. Gale (ed.), *Vision in Vehicles*, Proceedings of the Second International Conference on Vision in Vehicles, Nottingham, September 1985, Elsevier: North-Holland: 77–86.

Berry, D. and Dienes, Z. (1993), 'Towards a Working Characterisation of Implicit Learning', in D. Berry and Z. Dienes (eds), *Implicit Learning: Theoretical and Empirical Issues*, Hove: Lawrence Erlbaum Associates: 1–18.

Crundall, D. and Underwood, G. (2001), 'The Priming Function of Road Signs', *Transportation Research*, Part F4: 187–200.

Fisher, J. (1992), 'Testing the Effect of Road Traffic Signs' Informational Value on Driver Behaviour', *Human Factors*, 34 (2): 231–7.

Krose, B.J.A. and Julesz, B. (1990), 'Automatic or Voluntary Allocation of Attention in Visual Search?', in D. Brogan (ed.), *Visual Search*, London: Taylor and Francis: 21–8.

Pottier, A. (1999), 'Evaluation of Functional Fields of View in the Real Driving Situation', in A. Gale (ed.), *Vision in Vehicles*, Proceedings of the Seventh International Conference on Vision in Vehicles, Marseilles, September 1997: 233–41.

Rumar, K. (1994), 'Human Factors in Road Traffic', in G. Jansson and S. Berstoem (eds), *Perceiving Events and Objects: Resources for Ecological Psychology*, Hillsdale, NJ: Lawrence Erlbaum Associates: 319–28.

Saitta, L. and Zucker, J.D. (2001), 'A Model of Abstraction in Visual Perception', *Applied Artificial Intelligence*, 15 (8): 761–76.

Sprenger, A., Schneider, W. and Derkum, H. (1997), 'Traffic Signs, Visibility and Recognition', in A. Gale (ed.), *Vision in Vehicles*, Proceedings of the Seventh International Conference on Vision in Vehicles, Marseilles, September 1997: 421–5.

# Chapter 8

# Analysing and Modelling Train Driver Performance

Ronald W. McLeod, Guy H. Walker, Neville Moray and Ann Mills

## Introduction

Arguments for the importance of contextual factors in understanding human performance have been made extremely persuasive in the context of the process control industries. This chapter puts these arguments into the context of the train driving task, drawing on an extensive analysis of driver performance with the Automatic Warning System (AWS). This chapter summarises a number of constructs from applied psychological research which are thought to be important in understanding train driver performance. A 'situational model' is offered as a framework for investigating driver performance. The model emphasises the importance of understanding the state of driver cognition at a specific time ('now') in a specific situation and a specific context.

The work reported in this chapter was carried out as part of a study for the Rail Safety and Standards Board (RSSB). The aim was to understand and assess the risks of driver unreliability associated with extended uses of the Automatic Warning System (AWS) on the UK rail network. The study is summarised in McLeod *et al*. (2003a and c).

Despite what, to all everyday experience, appears to be an extremely powerful attention-getting device, combined with a highly visible visual reminder, there is a significant and recurring risk of AWS failing to prevent experienced train drivers passing signals at danger. According to Vaughan (2000), 'No-one ever thought that the driver would ignore the warning. It was utterly taken for granted that a driver would always take notice of the signal' (p. 12). In connection with the crash at Watford Junction in 1996, Hall (1999) wrote that; 'It seems inconceivable that a driver can acknowledge receiving two warnings and yet take no action to apply the brake to stop the train at the red signal, yet it happens' (p. 97).

Although relatively straightforward conceptually, the ways in which AWS is used in the UK introduce a number of complexities for the driver. These reflect the complexity and variability of UK signalling, as well as the various ways AWS is used. These include:

- AWS does not adequately discriminate between the possible states giving rise to the alarm. The system therefore depends on the driver having an adequate appreciation of the existing situation to be able to correctly interpret the alarm;
- there are a variety of situations where the sunflower can refer not to the immediate past signal, but to a signal some time prior to that;

- the timeframe of AWS activity (that is, the time period over which any AWS signal is 'active' in terms of conveying information about the track ahead) can vary from a few seconds to possibly many minutes.

Train driver performance is dominated by cognitive and perceptual factors. Much of the existing human factors research in the area draws on relatively simple models of human decision making and information processing. Traditional, hierarchically-based forms of task analysis are at the core of much of this research. From the perspective of cognitive psychology and cognitive engineering, these approaches are held to be fundamentally flawed. Specifically, they are not able to capture the contextual and situated nature of human performance in general, and train driver performance in particular.

Vicente (1999) offers a compelling critique of traditional approaches to task analysis. Similarly, Hollnagel (1998) provides a detailed critique of human reliability estimation techniques from the perspective of cognitive engineering. The arguments made by both Vicente and Hollnagel are extremely well made, and may be considered definitive. There are doubts about whether the techniques they propose actually overcome these limitations (cognitive work analysis (CWA) in the case of Vicente and the cognitive reliability and error analysis method (CREAM) in the case of Hollnagel), or whether they will translate from a process control context to train driving. Nevertheless, we believe that many of the limitations and weaknesses of traditional forms of task analysis and human reliability estimation they identify are equally valid for understanding train driver performance.

Other than in the sense of identifying factors likely to influence or shape performance little if any of the existing research has tried to understand the mechanisms by which contextual and situational factors influence driver performance. The relative simplicity of the psychological basis of the existing work, and the persistence of SPAD incidents, suggests that more comprehensive insights are needed.

## Recent Thinking in the Psychology of Real-world Human Performance

There is now a large, established and well respected body of thinking and research that focuses on understanding human cognition and performance in the real world contexts in which it occurs. Recognising the psychological complexity of the AWS problem, a number of ideas from this wider knowledge-base were reviewed. These go beyond the immediate literature on human factors aspects of AWS. The aim was to introduce a number of areas of thinking about cognitive performance in real world tasks that seemed relevant to understanding the risks associated with train drivers use of AWS. The following sections very briefly introduce some of these areas. A more detailed review, with full reference list, is available in McLeod *et al.*, 2003b.

### Strategic Behaviour

The importance of strategy and strategic behaviour in human performance has been recognised, at least in the research literature, for many years. In particular, strategy is essential in helping the human maintain performance in situations of stress, very high or very low workload or when subject to other influences. The concept has however rarely been used explicitly to try to understand or explain human behaviour and performance in real-world situations. Recent research by Merat *et al.* (2002) into train driver eye movements provides contemporary insights into strategic behaviour in train drivers.

*Situation Awareness*

Situation awareness is possibly the most widely cited construct in understanding human performance in complex real-time systems. Not surprisingly, it is also considered a key psychological construct in safe train driving.

Loss of situation awareness leads to what Sarter and Woods (1997) have described as automation 'surprises'. In connection with the Purley rail crash in 1989, Vaughan (2000) describes 'Driver Morgan of the Littlehampton train ... completely at a loss to understand what had happened' (p. 91). Vaughan notes that driver Morgan must have acknowledged and cancelled the two previous AWS warnings leading up to the red signal, but failed to initiate an appropriate braking manoeuvre until it was too late.

A number of factors can potentially interact to determine the train driver's understanding or belief about the current situation and therefore how an AWS alarm is interpreted. These include: the nature of the alarm (bell or horn); visibility of signals and magnets on the track ahead; the driver's interpretation of the nature of the preceding alarm; expectations about the current location (from route knowledge, as well as WONS, PONS and Late Notices); the train speed to be achieved, and by what time or by what location it is to be achieved.

Changes in any of the factors can potentially lead to a change in the driver's interpretation of a particular AWS alarm. In combination, they have the potential to cause the driver to misinterpret what an AWS alarm refers to, and what action to take.

*Situated Behaviour*

The term 'situated behaviour' refers to a broad group of areas of psychological thinking. They all seek to take the understanding of human performance out of the laboratory context, and to ground it in the real-world environment in which it occurs. The common themes are that the context and particularly the situation at the time, have an extremely powerful influence on how the human performs in real world tasks.

The work of Rasmussen *et al.* (1995), Vicente (1999) and many others on the analysis of cognitive work[1] takes the view that it is not possible to understand human performance without locating it within organisational and technological constraints. They demonstrate why traditional normative analysis approaches are not well suited to understanding complex socio-technical systems because they underestimate (if they capture them at all) the context dependent aspects of human performance.

*Distributed Cognition*

Ecological-based research such as Ed Hutchins's ground-breaking *Cognition in the Wild* (1995) and much of the varied research into distributed cognition emphasises the critical role that the situation and environment play in cognitive performance. The distributed cognition concept recognises that cognitive performance draws on, and is supported by, many artefacts of the organisation and surrounding environment.

It is, for example, clear that the placement of a prominent AWS magnet in the middle of the track can help to cue the driver in advance of a warning sounding. The driver is also likely to take advantage of many other environmental cues. These might include TPWS antennas, hot axle box detectors, other incidental AWS sounds (such as a relay's clicking) and a myriad of other track/trackside artefacts.

### Cognitive Control Modes and Subjectively Available Time

Hollnagel (1998) has developed a model of cognitive performance (COCOM) that recognises and seeks to account for the critical role of context in cognition. His ideas of cognitive control and subjectively available time seem potentially important in understanding train driver behaviour. Route knowledge supports anticipation and future-orientated behaviour. It allows the driver to think ahead, and helps control the allocation of cognitive and perceptual resources based on expectations about the future. It also helps the driver in spotting and interpreting cues and other information. In the COCOM framework, the effect of route knowledge might be to increase subjectively available time, thereby allowing the driver greater cognitive control over performance. Among the more important aspects of tactical behaviour is the (often unconscious) scheduling of visual attention, which switches the eyes to and from the different sources of information.

### Ecological Optics

Following the lead of J.J. Gibson (for example, 1979), there is a very large body of knowledge about the ways in which humans, as well as many other species, make use of information directly available from movement of the eyes of an observer in a visual field. A sizeable body of this work has concentrated on how information directly (that is, without requiring cognition) available from what is known as the 'optic flow' is used to control the timing and coordination of movement through the world. The information is derived directly from pattern and motion perception, and guides action without the need for thought, calculation, or conscious decision-making. This model of the visual control of movement has been shown to apply to many areas, including walking, running, driving cars, and flight (both by humans and birds).

From the perspective of ecological perception, the train driver is in a relatively unique and paradoxical situation. Seated at the front of a fast moving vehicle, the driver is on the one hand enclosed in an extremely powerful and compelling optic flow. But, apart from very few situations (such as approaching buffers), the information that their senses would normally draw on to control movement is of no relevance to their driving task. The driver has no control over directional movement, and control of speed is entirely mediated by cognitive processes such as perception of signs, and route knowledge. Whether and how the driver balances the (cognitive) use of the speedometer and the (direct) use of the optic flow in estimating train speed is not known.

The implications – if any – of this paradoxical situation for driver performance and reliability are not known, and, so far as we are aware, have never been investigated explicitly. It is however possible that it could, for example, contribute to the 'driving without attention' phenomenon (discussed next).

*Attention*

There are at least two quite distinct meanings of the word 'attention' that are relevant to the study of train driver performance. Both have been intensively investigated for many years in laboratory and practical situations.

The first is often called 'vigilance', and is typical of 'watch-keeping' tasks, where an observer tries to detect signals that arrive at infrequent intervals during which nothing much happens. Often the vigilance decrement can be avoided by an occasional message to the observer that requires an answer. The experimental proactive AWS system (Dore, 1998) provided a means to achieve this in requiring, effectively, an 'answer' from the driver as to the signal aspect that was approaching. The vigilance decrement is also dependent on the extent of trust that the driver has in AWS. The occurrence of a string of frequent signals that do not require a response may also cause the vigilance decrement to increase.

The second meaning of attention is in 'dynamic selective attention'. In real situations, particularly in the case of vision (though also in hearing), attention tends to be directed to one source of information at a time. To some extent attention can be shared between two or more tasks, although performance will usually be degraded.

It is interesting that although evidence given to the Southall Inquiry maintained that drivers would always be looking at signals before the AWS sounded, in their study of driver eye movements Merat *et al.* (2002) found that on a significant proportion of signals the contrary was true. As predicted by quantitative attention models, attention was elsewhere and was attracted by the sound of the AWS.

*Driving Without Attention*

There is considerable interest in psychological research and among safety professionals in the concepts of 'driving without attention' (DWAM), and 'looking-without-seeing'. These notions may have relevance to the AWS context, although no specific literature related to train driving has been identified.

Allen (1965, p. 332)) wrote that:

> In modern multi-aspect signalling there is a tendency to shorten the signal sections … When traffic is dense through multi-aspect territory, a train will often come upon a quick succession of double or single yellow aspects. At every one the driver will get an AWS warning, which he has to cancel. In such conditions, the risk of frequent canceling action degenerating into a reflex action that loses its full meaning for the driver is a real one.

In their extensive review of human factors in road traffic safety, Dewar and Olsen (2002) consider the possible reasons for the large number of driving accidents associated with car driver inattention, 'highway hypnosis' and 'DWAM'. They refer to the state as being associated with monotonous, uneventful driving, where a lack of novelty promotes automatic responses.

*Expectancy*

One of the key benefits of route knowledge is that it allows the driver to prepare in advance for the route ahead. With good route knowledge, the driver is able to plan ahead, to prepare for events before they occur, and to quickly understand and interpret information. That is, it generates expectancy. In the great majority of situations, expectancy and the ability to plan ahead is effective; indeed, without it, it would not be possible for humans to perform many of the complex tasks they manage. On occasions, however, expectancy can lead to pre-planned behaviours being inappropriate, or can lead to incorrect interpretation of information.

*Trust in Automation*

A growing body of research has been investigating the role that trust and confidence in technology plays in influencing the way people interact with technology-based systems.

If a system is regarded as untrustworthy then it will tend not to be used correctly. Likewise, if an unreliable system is trusted then it will tend to be used even though it malfunctions. Despite the importance of trust in technology, it remains relatively ill-defined and only partly understood psychologically.

The dynamics of trust in technology have been dealt with by Lee and Moray (1992) and Parasuraman and Riley (1997) among others in connection with supervisory control and automated systems. They define the dynamics of trust around predictability, dependability and faith. In simplistic terms, trust is established by consistent and desirable behaviour. This relies on the observability of system behaviour, as well as the individual actively sampling it. As trust is gained, the focus shifts away from observing specific behaviours towards assessing the global disposition of the device (it is said to be 'dependable'). Over time, trust serves to reduce the effort expended in checking the system to ensure it is performing as expected.

In our questionnaire survey of 277 drivers conducted as part of the AWS study (McLeod *et al.*, 2003c), a substantial proportion reported issues suggesting drivers know they cannot always trust an AWS warning.

*Simple Heuristics and Recognition-primed Decision Making*

Perhaps most intriguing, is recent work by Gigerenzer and Todd (1999) on what they term 'Simple Heuristics'. Gigerenzer and Todd take the view that complex human performance in the real world cannot possibly involve the rational cognitive strategies assumed by much experimental psychology. Understanding decisions in the real world requires different, more psychologically plausible explanations. The work on simple heuristics focuses on what they term 'fast and frugal heuristics – simple rules for making decisions with realistic mental resources'.

The notion of simple heuristics is similar to the well-established concept of recognition-primed decision making (see for example, Klein, 1989, 2003). Instead of emphasising heuristics and rules, Klein's model of decision making is bound in the ability of humans to pattern match. Repeated experiences are unconsciously linked together to form a pattern. Links between features in past and current situations enable typical features of situations to be extracted, and typical responses generated (Klein, 1989). This model is similar to the well-known GEMS model of human error (Reason, 1990).

Whether or not 'simple heuristics' and/or 'recognition-primed decisions' are important in determining train driver performance remains to be seen. It might seem surprising if, in some form, they were not.

## Summary of Psychological Thinking

So how does consideration of these wide variety of psychological constructs help in understanding train driver performance? Two general observations seem relevant.

The first is that, however desperately the human factors professional might seek tools and analysis methods that can be applied systematically across application domains, the fact is that the basis of human performance is exceedingly complex. Generic information processing 'models' of the human operator can certainly provide extremely useful generalisations. They have value in systems engineering, and in seeking to predict asymptotic limits of human performance in complex systems. Even relatively simple models can have great value as engineering tools in providing structure and drawing attention to human capabilities.

However, as a means of understanding why specific individuals, on a particular day in a particular set of circumstances behaved (or failed to behave) in a particular way, or predicting how an individual might behave in a given set of unexpected circumstances, such models have very little to offer.

The second general observation is that if such an understanding of what happened in an incident, or what is likely to happen, is important, it is essential to identify and understand the characteristics of the context and situation at the time the behaviour of interest occurred or is expected to occur.

## A Situational Model of Driver Performance with AWS

A principle conclusion of the study was that driver performance with AWS can only be understood in terms of the context and situation at the time the system is intended to influence driver behaviour. From this perspective, existing approaches to modeling driver performance appear relatively superficial. Developing a comprehensive model of the psychology of driver behaviour would be a significant task. However, based on the material reviewed, and drawing on insight from discussion with drivers, a tentative framework that might underpin the assessment of risk associated with Extended AWS was developed.

The objective was to provide a framework to help understand the possible state of cognition of a particular driver between the point at which the AWS signal is detected and the driver's subsequent behaviour. The emphasis on time reflects the fact that many of the circumstances and situational factors involved are inherently time-dependent, or time-limited. These include:

- the role of time in controlling the safe passage of a train (such as the time for which signals are visible, and the time involved in accelerating or braking);
- events or features which are only true at particular locations on a track;
- expectations or beliefs a driver might hold because of the way in which events have evolved over time prior to an event (due to the spatial relationship between signals and speed restrictions, say, or because of the relative speed and distance between trains further down the track).

Figure 8.1 illustrates a 'situational model' of driver behaviour.

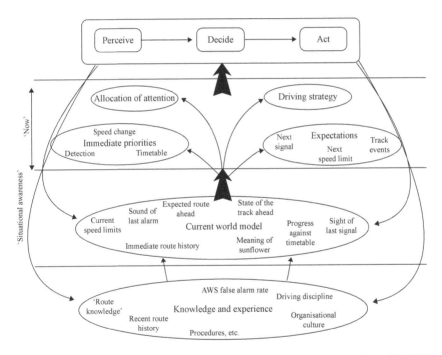

**Figure 8.1    Situational model of driver performance in interacting with AWS**

A number of points of introduction are needed to put this framework into context:

- the situational model has been developed as a framework to direct attention towards those situational factors which seem to be important in understanding driver behaviour;
- the model emphasises the state of cognition at a specific time ('now') in a specific situation and a specific context. the emphasis is therefore on the immediate

situation facing the driver, and the immediately preceding history (what is shown on the framework as 'the now'). Longer term knowledge and experience is clearly important. However, the emphasis is on understanding the behaviour of a particular individual at a particular moment;

- the situational model is aligned against a standard information processing model (perceive–decide–act) of human performance. This seeks to contextualise the model of the human as an information processor in terms of the situational factors that determine actual performance in a real-world, real-time, environment. This emphasises that any analysis of driver performance using an information processing model needs to take proper account of situational issues;

- the model can be no more than a 'snapshot'. Details of what matters at any moment, and how factors interact, will vary over time, particularly as the 'state' of the driver changes. For example, different states of attention and cognitive control could be particularly powerful in mediating how elements of the framework might interact at any time. Momentary distractions might also disrupt interactions between elements of the framework.

Briefly, the framework incorporates the following assumptions (each of these is represented as an element on the figure):

- driver performance is based on a foundation of knowledge and experience derived over a relatively long time from training and general experience. As well as explicit knowledge of the rules, procedures, and disciplines governing professional driving, this knowledge base includes recent experience of the route being driven (for example the next signal has been at caution on every approach over the past fortnight). It will also include such things as the typical false alarm rate associated with AWS signals;

- the driver has a mental representation of the current state of the world as it affects the current driving activity. This includes elements such as the current speed limit, state of the track (are trackside workers in the vicinity) and significant features in the route over a psychologically relevant timescale ahead. It will also hold awareness of the immediate history (the last six caution signals have cleared before the train reached them because of a queue of trains ahead). Much of this world model will be held subconsciously;

- two elements are particularly important in determining the driver's cognitive state at any specific moment ('Now'). These are the *immediate priority* and the *expectation* of what the world will be like in the next few moments. The immediate priority might be to reduce speed, or to ensure every individual in a track-side party can be seen. Expectation might include the location of the next signal, the point at which a speed limit will come into force, the expectation that an unsuppressed AWS magnet is about to cause an alarm that can be ignored, or the expectancy that the signal will change its state while it is still in view;

- both immediate priorities and expectations are continually changing and updating as the train progresses along the route;

- the driver directs attentional resources based on current priorities and expectations. Responses to events in the outside world are determined by a driving strategy which can vary from moment-to-moment, within reasonably broad constraints.

The nature of the driving strategy, and especially how they vary, will depend to an extent on the drivers training, experience and confidence.

We have found that the situational model of driver behaviour provided an extremely useful framework for assessing the risk of driver unreliability with AWS. Consideration of the model led to 18 situational factors being identified which might influence the driver's state of cognition at the time AWS warnings are encountered (that is, 'Now' on the model). These 18 factors provided the basis for an assessment of the likely risk associated with 20 scenarios involving encounters with AWS (see McLeod *et al.*, 2003b for a summary of the assessment).

## Conclusion

The study reported here sought to understand the nature of driver reliability with the Automatic Warning System, and to estimate the likely change in driver reliability if the use of AWS is further extended. Considerations from the psychological research base identified a reasonably large number of mechanisms and constructs that might be important in understanding moment-to-moment driver performance. Most important among these appear to be:

- the importance of expectancy and route knowledge;
- the driver's mental representation of the situation and how the understanding of immediate priorities and objectives are maintained and controlled; and
- trust in the system, specifically, the extent to which drivers' experience might cause them to expect a certain level of 'false' alarms or incorrect signals.

Both the situational model of driver behaviour, and the method used to assess driver reliability with AWS might have wider application in understanding safety issues associated with driver performance.

## Notes

1   Note that this approach is concerned with the analysis of cognitive work: it is not cognitive task analysis.

## References

Allen, G.F. (1965), *British Rail after Beeching*, London: Ian Allen.
Dewar, R.E. and Olsen, P. (2002), *Human Factors in Traffic Safety*, Overland Park, KS: Criterion Press.
Dore, D.M. (1998), 'The Pro-active Automatic Warning System (PAWS)', Report to the New Measures Working Group.
Gibson, J.J. (1979), *The Ecological Approach to Visual Perception*, Boston: Houghton Mifflin.

Gigerenzer, G. and Todd, P.M. (1999), *Simple Heuristics that Make us Smart*, Oxford: Oxford University Press.

Hall, S. (1999), *Hidden Dangers: Railway Safety in the Era of Privatisation*, Shepperton, Surrey: Ian Allen.

Hollnagel, E. (1998), *Cognitive Reliability and Error Analysis Method*, London: Elsevier.

Hutchins, E. (1995), *Cognition in the Wild*, Cambridge, MA: MIT Press.

Klein, G. ((1989), 'Recognition-primed Decisions', *Advances in Man-Machine Systems Research*, 5: 47–92.

Klein, G. (2003), *Intuition at Work*, New York: Currency Doubleday.

Lee, J. and Moray, N. (1992), 'Trust, Control Strategies and Allocation of Function in Human-machine Systems', *Ergonomics*, 35 (10): 1243–70.

McLeod, R.W., Walker, G. and Mills, A. (2003a), 'Assessing the Human Factors Risks in Extending the Use of AWS', paper presented to the 1st European Conference on Rail Human Factors, St Williams College, York, 13–15 October.

McLeod, R.W., Walker, G. and Moray, N. (2003b), 'Extended AWS Study: Review of the Knowledge Base', B/C271/FD.8, Nickleby HFE Ltd.

McLeod, R.W., Walker, G. Moray, N. and Love, G. (2003c), 'Project Summary Report: Driver Reliability with Extended AWS', B/C271/FD.5, Nickleby HFE Ltd.

Merat, N., Mills, A., Bradshaw, M., Everatt, J. and Groeger, J. (2002), 'Allocation of Attention among Train Drivers', in P.T. McCabe (ed.), *Contemporary Ergonomics 2002*, London: Taylor and Francis.

Parasuraman, R. and Riley, V. (1997), 'Humans and Automation: Use, Misuse, Disuse, Abuse', *Human Factors*, 39 (2): 230–53.

Rasmussen, J., Pedersen, A.-M. and Goodstein, L. (1995), *Cognitive Engineering: Concepts and Applications*, New York: Wiley.

Reason, J. (1990), *Human Error*, Cambridge: Cambridge University Press.

Sarter, N.B. and Woods, D.D. (1997), 'Team Play with a Powerful and Independent Agent: Operational Experiences and Automation Surprises on the Airbus A-320', *Human Factors*, 39 (4): 553–69.

Vaughan, A. (2000), *Tracks to Disaster*, Shepperton, Surrey: Ian Allen.

Vicente, K.J. (1999), *Cognitive Work Analysis: Towards Safe, Productive and Healthy Computer Based Work*, Marwah, NJ: Lawrence Erlbaum.

# PART 3
# DRIVING – VISION AND VISUAL STRATEGIES

# Chapter 9

# A Methodology to Investigate Train Driver Visual Strategies

N. Brook-Carter, A.M. Parkes and Ann Mills

## Introduction

Following a number of recent incidents on the railways, which have been attributed to errors in human performance, an increased effort has been focused on conducting research into understanding the behaviour and actions of the train driver. This has included focusing on visual behaviour and performance.

Train drivers are required to monitor the dynamic scene visually, both outside and inside the train cab. Poor performance on this visual task may lead to errors, such as signals passed at danger (SPADs). It is therefore important to understand the visual strategies train drivers employ when monitoring and searching the visual scene for key items, such as signals to inform on rail infrastructure design and driver training.

Currently within the UK, only one pilot study has been carried out investigating train drivers' visual behaviour and collecting data on driver monitoring of the visual environment (Groeger *et al.*, 2001). A larger set of data on train drivers' visual behaviour is required to provide a greater understanding of the strategies adopted, such as the visual cues from route knowledge and the relevance or irrelevance of trackside information.

In light of this need, TRL (Transport Research Laboratory) has produced a methodology for the assessment of train driver visual strategies, on behalf of the Rail Safety and Standards Board. This methodology will be used to conduct a series of trials, in which train drivers will drive in-service trains wearing a state of the art eye tracking system.

The corneal dark-eye tracking system chosen for these trials will track the visual search and scanning patterns of train drivers. Data collected will include the duration and frequency of glances made towards different aspects of the visual scene. In addition, the train drivers will be interviewed after driving the routes, to determine the thought processes behind the visual strategies they use. Effort has been made to produce a safe method of data collection that can provide useful and accurate data. The eye tracker and verbal protocol data are not without limitations and these need to be taken into consideration when analysing the data and making inferences from the results.

An expanse of data will be collected using the eye tracking equipment. An important aspect of the methodology is how these data are analysed to interpret train driver visual strategies. It is hoped that the data will be used to indicate both general eye movement behaviour and higher level visual strategies adopted (search and confirmation) by train drivers.

Results can then be used by the railway industry to improve driver training and driver interfaces. The results will also provide objective data with which to inform minimum signal reading times. Finally, the trials will be designed to answer a number of key research questions.

## Methodology

*Procedure*

The basic procedure to be adopted during the trials will be:

- fitting the eye tracker to a driver in a station office;
- calibrating the eye tracker in a station office;
- boarding and driving an in-service train;
- conducting re-calibrations at station stops when appropriate;
- carrying out a verbal protocol on video footage of the route following the drive.

*Eye Tracker*

The eye tracker to be used during this project is the VisionTrak, supplied by Polhemus. The VisionTrak is mounted on a headband that can be adjusted to all head sizes (from a 2-year-old to a full grown adult). It weighs approximately 170g. It uses a dichroic mirror to reflect the eye-image onto a highly sensitive head-mounted video camera. The scene camera is mounted below the driver line-of-sight.

The tracking technique used by the system is a 120Hz pupil and corneal dark-eye tracker. As the eye tracker uses dark-eye tracking rather than bright eye, it is less affected by changes in lighting conditions, such as glare or darkness when driving through tunnels.

**Figure 9.1    The eye tracker**

The eye tracker automatically tracks point-of-regard and the correlation of the raw eye position to the precise position on the scene, in real time. The image being viewed by the participant is identified by crosshairs on the recorded video footage. The participant has complete freedom of movement. The head mounted system with the head tracker allows the participant full six degrees of freedom of movement, and the participant is not limited in any way. The eye tracker works with most glasses and contact lenses. The eye tracker headband is connected to eye tracking equipment by a long cable and the equipment to be carried on the trains is portable and can be fitted into a small bag.

*Calibration*

Calibration of the equipment will be carried out in the office before boarding the train. This involves a five point calibration. The advantages of calibration at the station are the ability to calibrate with fixed points and the reduced time pressure under which the calibration takes place.

*Re-calibration*

Further calibrations will be recorded whilst on the train. They will be conducted by asking the participant to look at five specified points in the train cab. Practice re-calibrations have been found to take no longer than two minutes.

*Questionnaires*

Questionnaires will be administered to participants at various stages of the trials. At the beginning of the trials participants will be given a driver profile questionnaire asking details relating to age and experience of train driving. At the end of the study participants will be given a debrief questionnaire asking questions relating to the comfort of the eye tracking equipment, any visual impairment suffered due to wearing the eye tracker, the time it took to get used to the eye tracker and any further comments.

*Route*

Various different routes are expected to be driven during the trials. Each will contain a number of different train driving scenarios such as driving through tunnels and approaching complex signals.

*Participants*

The level of experience will differ between the various participants and a possible option will be to investigate driving experience as an independent variable.

*The Drive*

Prior to the study participants will be asked to read an information sheet containing instructions relating to the nature of the trial.

A maximum of two people will accompany the driver in the train cab; one investigator and a DSM (Driving Standards Manager). Once the eye tracking equipment is secured and the train driver is happy to proceed, the investigator will sit on the other side of the train cab and will not communicate with the driver. The other investigator and the scheduled train driver (on in-service trains) will ride in the train carriages.

The only communication between the driver and the investigator will be in a situation where the driver wishes to discontinue involvement within the trial or during re-calibration at station stops.

*Verbal Protocol*

In the first stage of the verbal protocol, drivers will be shown a video clip of a section of the route they have just driven.

The first set of video clips will be taken from video recorded prior to the trial on a fixed camera and will not therefore have crosshatches indicating the drivers' direction of gaze, as would be shown on the recordings made with the eye tracker.

The drivers will be asked to carry out a verbal protocol. Whilst watching the clips the drivers will be asked to imagine that they are driving the train and to talk through what visual behaviour and strategies they think they would adopt at each event. This will involve describing what they believe they are looking at and why they have adopted this particular visual behaviour.

If the drivers experienced difficulties in carrying out verbal protocols and only provided brief verbalisations, the experimenter will prompt the driver with probe questions such as: '*What are you looking at here?*' or '*Why are you focusing on that particular feature?*'. Whilst the driver is verbalising what he believed his visual behaviour would be the second experimenter will take notes on the drivers comments.

The second stage of the interview involves the driver watching a set of video clips of the same 'events', but taken from the actual drive just carried out.

Whilst watching the clips the driver will be asked to talk through his/her visual behaviour and strategies. This set of video data shows a crosshatch indicating the direction of the drivers' gaze and the stimuli on which the driver appeared to be focusing. The drivers will be asked to describe what they were looking at and why they believe they were adopting this particular visual behaviour.

If the drivers experience difficulties in describing their visual behaviour as shown on the video and only provide brief verbalisations, the experimenter will prompt the driver with probe questions such as: '*What were you thinking about here?*' or '*Why do you think your gaze was focused in that particular direction?*'. Whilst the driver is verbalising his visual behaviour the second experimenter will take notes on the driver's comments.

*Debrief*

At the end of each trial drive, participants receives a debriefing. This includes an explanation of the issues under investigation during the study. At this point, participants have the opportunity to ask the assessors any remaining questions they have regarding the research project or any concerns they might have relating to the use of the data collected. Participants are reassured that they were not being tested on their driving performance or driving behaviour during these trials and that their names would not be linked to any of the data.

## Eye Tracker Accuracy

The accuracy of the eye tracker can only be determined by asking a participant to look at specific points in the visual scene and analysing whether the cross hair on the video is positioned over these points. Experience to date using the eye tracker has found that when a good calibration has been conducted during which the participant remains still and looks at points two metres away, the accuracy provided by the eye tracker is high.

The advantage of re-calibrations is that as the eye tracker slips or is knocked, this can be accounted for rather than losing the data that follows. Re-calibrations can be successfully carried out in under two minutes and are therefore unlikely to disrupt the train timetable.

## Recognising the Limitations of the Eye Tracker

It should be noted that although the eye tracker provides a rich source of data relating to the visual behaviour of the participant, it does also possess a number of limitations.

Firstly, the eye tracker tracks the fixation point of the participant. However, this does not always directly relate to the direction of the participants attention. The participant may in fact be attending to stimuli in the peripheral or extra foveal vision. Alternatively the participant may be attending to auditory inputs. When analysing eye tracking data, it should be noted that the assumption is made that the fixation point is being attended to by the participant.

Another limitation of the equipment is that the participant may be looking at visual stimuli outside the view provided by the scene camera. In such situations the data are lost.

The image provided by the scene camera is not as high a quality as that provided by the human eye. Therefore aspects of the scene which may be detected in the distance may be visible to the participant before they are visible to the experimenter analysing the data. It may therefore appear that stimuli in the distance are detected later than they are in reality.

Further, the scene camera does not record the scene well in bright light and stimuli cannot be distinguished.

## The Data Set

The eye tracker will provide a vast quantity of video data. Video data will be available from each route (lasting approximately 30–40 minutes) showing a crosshair on the scene footage indicating where the train driver's foveal vision was directed.

Coordinates reflecting the position in the visual scene in which the foveal vision was directed will also be available.

This is an extensive amount of data, which need to be analysed by an observer frame by frame. As a consequence it would not be realistic to try to analyse each 30 minute route driven, particularly when the plan is to involve 20–40 drivers within each trial. The analysis therefore needs to be carefully planned in order to maximise the value of the information extracted.

## Analysing the Eye Tracker Data

It has been noted that the eye tracker provides vast amounts of data. Therefore, the eye tracking data need to be broken down into manageable chunks for analysis. For example, the visual behaviour in different visual environments, such as during signal approaches, at station throats, in tunnels, and in complex or simple environments, will need to be considered separately. Specific scenarios or aspects of the visual scene need to be chosen for analysis and might include:

- approaching a cantilever signal;
- approaching a gantry signal;
- approaching a tunnel mounted signal;
- approaching a complex signal;
- driving round a bend;
- driving on a straight;
- approaching a station throat;
- coming out of a tunnel;
- driving at high speeds;
- driving at low speeds.

Having separated the data into manageable chunks we then need to consider how they will be analysed. The dependent variables to be analysed initially will include:

- glance durations;
- percentage of time fixating on signals and signs;
- percentage of time fixating on in-cab systems;
- percentage of time fixating on other stimuli.

This analysis will indicate to us the basic visual behaviour of the drivers in different situations.

This basic analysis will provide an indication of where drivers look: however, it will not provide us with the more detailed strategies used by train drivers. Other methods might be employed which can more directly infer visual strategies and are currently under consideration.

One of these methods is link analysis. Link analysis uses algorithms to identify the frequency in which one aspect of the visual scene is viewed directly before or after another aspect of the visual scene and therefore indicates the links between stimuli, as well as the typical scan patterns that might be used. For example, if signals are frequently fixated on immediately following a fixation on signage, then a strong link would be observed between those two stimuli. This type of analysis will also help to identify the visual cues or 'passing points' which train drivers commonly use.

If it is possible to apply link analysis to the data, it would allow us to investigate whether there is an identifiable order in which aspects of the visual scene are searched or scanned in the various scenarios. Similarly, it might allow us to investigate whether there is a different order between experienced and inexperienced drivers. In car driving research the visual search of more experienced drivers is strongly related to their expectation of the location of relevant stimuli. Luoma (1986) describes how drivers appear to develop a mental model over time of where critical information can be obtained. This model is thought to be based on the experience of scenarios and the build up of a generalised schema of situations that leads to deeply ingrained expectancies. Particular visual scan patterns are based on this model in experienced drivers. This results in more predictable scan patterns in experienced drivers compared to novice car drivers. It might be expected therefore that from the train driver data a more defined and predictable link analysis is produced for experienced train drivers.

Another method of investigating the strategies adopted by train drivers would be to plot scatter graphs indicating the position of the fixations within the scene. These plots can be created using the coordinates generated by the eye tracker software. A disadvantage of this method is that it does not account for the change in scene due to the dynamic nature of the data, and in particular the change in scene due to head movements. However, it would provide an indication of the extent to which the visual scene is searched during different scenarios and by different drivers. For example, during complex situations scan patterns might be expected to be more focused on one or two areas within the visual scene, whereas during simple drives in rural areas, scan patterns might be expected to cover a larger area of the visual scene.

Similarly, scan patterns of inexperienced and experienced drivers might differ. In car driving research novice drivers have been found to search a smaller area of the visual scene that is close to the car. In comparison, experienced drivers focus on an area much further ahead of the car and their scan pattern is more centrally located as they rely more heavily on peripheral vision to identify hazards (Mourant and Rockwell, 1970; Zell, 1969). It might be hypothesised that train drivers of different experience will exhibit different scan patterns.

## Recognising the Limitations of the Verbal Protocol

The aim of the verbal protocol is to provide information relating to the cognitive processes of the train drivers. The verbal protocol requires introspection and verbalisation, and the train drivers' ability to do this depends on the level of consciousness they have regarding their behaviour, and will be influenced by their understanding of the trial requirements.

Nisbett and Wilson (1977) have argued that such introspection is almost worthless, as people are generally unaware of the processes influencing their behaviour, particularly behaviour dependent upon a practised skill. They claim that when trying to explain their behaviour, people apply or generate theories rather than remembering the actual cognitive processes. Further, as Hannigan and Parkes (1988) point out, the data collected will be a function of both the participant's ability to verbalise cognitive experiences and the degree to which the participant tailors the verbalisation to meet the expectancies and understanding of the investigator.

Despite findings to support the arguments of Nisbett and Wilson, verbal protocols can identify those processes which people are consciously aware of. Ericsson and Simon (1984) found that participants are more likely to provide accurate introspection when asked to describe what they are attending to or thinking about, than when required to interpret a situation or their own thought process.

Further, as Bainbridge (1974) had asserted, whilst a protocol does not give complete, or necessarily reliable, data on the participant's thoughts, it is a source of much interesting information which cannot be obtained in other ways.

Ericsson and Simon (1984) stated that the information less likely to be reported by verbal protocol are the cues that allow people to recognise stimuli. This should be taken into account when analysing the verbal protocols, in that the train drivers may well be unaware of a number of the visual cues or 'passing points' which they frequently use.

It is important to note the limitations of verbal protocols. The data obtained cannot be assumed to be the whole picture with regards to the underlying thought processes, strategies and tactical decisions relating to the train drivers' behaviour, but rather provide an indirect view of the underlying mental process.

In sum, during verbal protocol analysis, the train drivers are expected to present a rationalisation based on their own internal model of train driving, which will not completely mirror the actual cognitive processes underlying the visual behaviour, but will provide a valuable interpretation of events.

## Analysing the Verbal Protocol Data

Both the video sequence and the eye tracker recording containing a cross hair will be useful. During the video clip participants will be able to consider what they would normally look at and try to explain why. The eye tracker data will then act as an additional prompt or a reinforcer, where drivers can describe why they would look at different aspects in the visual scene.

The verbal protocol data will be recorded by an observer and these data can then be categorised in order to investigate whether train drivers provide similar reasons for their visual behaviour and adopt similar strategies during each driving scenario under investigation.

## Research Questions

One of the aims of this project is to answer a number of research questions with both the eye tracking data and the data provided by the verbal protocols. During a focus group several research questions have been identified as high priority and these include:

- how do different visual environments, such as rural and urban scenes, affect the types of visual strategies used?
- how do train speeds affect the types of visual strategies used?
- where do drivers look the majority of the time?
- how do changes in the infrastructure affect the visual strategies of expert drivers?
- are passing points consistently used?
- how can visual strategy data be used to inform signal reading times (in particular, GK/RT0037 'Signal Positioning and Visibility')?
- what are the differences between drivers in the visual strategies they use?
- what visual strategies do experienced drivers adopt?
- what aspects of the visual scene are used as components of driver route knowledge?
- what constitutes route knowledge and are specific 'passing points' used?
- what cues do drivers use?

**Hypotheses**

There are a number of hypotheses which might be investigated given the nature of the data being collected using this methodology. The hypotheses outlined here are based on findings from previous research into the visual behaviour of car and train drivers.

- experienced train drivers will use more advanced visual strategies than novice train drivers;
- experienced train drivers will use different strategies depending on the complexity and type of visual environment with which they are faced;
- train driver fixation patterns will follow the geometry of the track on a curve but not on the straight;
- the visual strategies adopted by train drivers in scenes of varying complexity are likely to differ;
- the number of fixations made by train drivers and the durations of these fixations will be longer when travelling at lower speeds;
- train drivers will actively search the most likely locations for visual information and will scan areas in which they have strong expectations that relevant visual information will be present;
- train driver visual strategies will be affected by a number of factors including:
  - the signal aspect;
  - the signal mounting;
  - the previous aspect passed;
  - the sounding of AWS.

It will be interesting to see which if any of these hypotheses are proven by the studies to come.

## Conclusion

A methodology has been produced to obtain a large scale set of data on train drivers visual behaviour and the thought processes behind this behaviour. This methodology will be used to conduct a series of trials, in which train drivers will drive in-service trains wearing a state of the art eye tracking system.

The corneal dark-eye tracking system chosen for these trials will track the visual search and scanning patterns of train drivers. Data collected will include the duration and frequency of glances made towards different aspects of the visual scene. In addition, the train drivers will be interviewed after driving the routes, to try and understand the thought processes behind the visual strategies they use.

The limitations of the eye tracking system need to be taken into account when considering results from the eye tracker data.

It is hoped that the data can be analysed both in terms of train drivers basic visual behaviour, such as glance durations and percentage of time spent fixating on certain aspects of the visual scene, and also higher level aspects of visual behaviour such as scan patterns and visual cues.

The verbal protocols will supply additional data that will provide an insight into some of the train drivers' cognitive processes and the strategies they use or motivations they have, in other words the reasoning behind why they look at what they look at.

The aim of the application of this methodology will be to answer a number of key research questions relating specifically to driver route knowledge, the cues in the visual scene which drivers use and how different environments and driving speeds affect drivers' visual behaviour

## References

Bainbridge, L. (1974), 'Analysis of Verbal Protocols from a Process Control Task. A Summary of the Cognitive Processes of Operators Controlling the Electricity Supply to Electric-arc Steel-making Furnaces', in E. Edwards and F.P. Lees (eds), *The Human Operator in Process Control*, London: Taylor and Francis Ltd: 146–58.

Ericsson, K.A. and Simon, H.A. (1984), *Protocol Analysis*, Cambridge, MA: MIT Press.

Groeger, J.A., Bradshaw, M.F., Everatt, J., Merat, N. and Field, D. (2001), 'Pilot Study of Train Drivers' Eye-movements', University of Surrey.

Hannigan, S. and Parkes, A. (1988), 'Critical Incident Driver Task Analysis', in A. Gale (ed.), *Vision in Vehicles*, Proceedings of the Second International Conference on Vision in Vehicles, Nottingham, September 1985, Elsevier: North-Holland.

Luoma, J. (1986), *The Acquisition of Visual Information by the Driver: Interaction of Relevant and Irrelevant Information*, Helsinki, Finland: Liikenneturva-Central Organisation for Traffic Safety.

Mourant, R.R. and Rockwell, T.H. (1970), 'Visual Information-seeking of Novice Drivers', International Automobile Safety Conference Compendium, New York: Society of Automotive Engineers.

Nisbett, R.E. and Wilson, T.D. (1977), 'Telling More Than We Can Know: Verbal Reports on Mental Processes', *Psychological Review*, 84: 231–59.

Zell, J.K. (1969), 'Driver Eye Movement as a Function of Driving Experience', Report No. IE–16, Columbus, OH: Driving Research Laboratory, Department of Industrial Engineering, The Ohio State University

Chapter 10

# A Structured Framework for Integrating Human Factors Design Principles into Railway Vehicle Cab Sightlines Specification

Paul Traub and William Lukau

## Introduction

This study investigated the optimum processes for addressing the requirements for railway vehicle driving cab sightlines and developed a coherent and measurable framework for specifying/auditing cab sightlines for all types of railway vehicle.

The framework was developed using a staged analysis process that included:

*   examining the gaps and inconsistencies in the requirements of existing Railway Group Standards with reference to good general design principles and human factors practices;
*   reviewing current best practice for defining requirements for field of vision in other safety-critical industries (including aviation and shipping).

The resulting framework provides a model for the development of Railway Group Standards for integrating human factors principles into the design and compliance testing of cab sightlines and signal sighting. A number of cab vision plots for different classes of trains were conducted as part of this activity. This enabled development of a new set of performance measurements for signal sighting that aligned with the cab sightline requirements (for the whole range of signal types specified in GM/RT2161).

The benefits of the framework are the establishment of a coherent link between cab sightline requirements and identified signal-sighting issues in order to specify driver's viewing angles in terms of degrees. This approach used a design eye point, also known as design eye position (DEP) rather than the current criteria within Railway Group Standards of distance to the signal from a reference cube. This study concluded that DEP provides a more accurate measure of sightlines (internal and external vision) that can be easily tested during design and compliance auditing. The novel framework that evolved from the study ensures that cab ergonomic issues can be addressed in a holistic manner rather than examining cab sightlines in isolation.

It is expected that, when properly applied, the framework will promote and facilitate greater integration of the human factors requirements into cab design and signal sighting criteria. This will ensure that drivers are provided with optimum lines of sight to effectively monitor both in-cab displays, controls and instruments, and also the outside operating environments and their associated requirements while ensuring at the same time the promotion of a more safe and comfortable driving operation.

This will, in addition, promote and facilitate compliance to the requirements of the relevant Railway Group Standards, thereby making driving a train more comfortable for the driver and safer for everyone using the railway.

## Objectives

The objectives of this study were to review and recommend minimum performance requirements relating to driver cab sightlines for external viewing of signals and associated rail infrastructure so as to:

- promote proper consideration, better management and coordination of cab design issues in relation to sightlines;
- support an effective and continuous development of the standard and ensure the delivery of adequate and appropriate data within GM/RT2161;
- provide performance recommendations (using theoretical and experimental evidence) to supplement those already in GM/RT2161.

The study established a framework for designing and assessing driver's cab sightlines. The resulting framework provides guidance on the implications and best practice applicable to the industry. This was done via:

- review of gaps, omissions and inconsistencies in current Railway Group Standards;
- developing and implementing criteria from best practice from other industries.
- identification of key technical requirements and measures;
- conducting a number of vision plot assessments on different classes of train with disparate cab structures;
- proposal of a cab sightline framework to address the gaps within GM/RT2161 and those of GK/RT0037.

## Review of GM/RT2161 and GK/RT0037

GM/RT2161 identifies important issues that have relevance in the establishment of the minimum requirements that would give an adequate field of view for the whole range of potential drivers. However, the standard falls short of linking such issues to the actual measurements or definition of cab sightline specifications.

The review concluded that it would be better if the standard utilised a seat reference point and a design eye point as the basis for cab sightlines rather than the current reference cube/seat adjustment criteria. The current approach provides huge variations in cab sightline visibility for different drivers and is very difficult to measure or ensure compliance with it. In addition, no guidance was provided for the application of specific anthropometric measures that are necessary to ensure an effective accommodation of the driver population (for example sitting eye and standing eye height ranges for male and female drivers).

The provision of the external view of the driver is referred to as a conditional option, and limited to considerations of visibility of train formation. These are related to the external view along the train and the external views forwards and backwards along each side of the train formation. However, the external view requirements should be linked more closely to requirements of signals and should be based on the field of view (FOV) rather than being solely distance based. The following summarises the main issues identified in reviewing Railway Group Standards GM/RT2161 and GK/RT0037:

- both Railway Group Standards fell short of linking the issues within them to the actual measurement or definition of cab sightlines;
- no reference was given of the actual profiling of the driver population, (and the need to specify driver population anthropometric ranges) as physical dimensions have changed significantly over time and are not typical from the current data that are utilised;
- no specific anthropometric dimensions are provided from which sightlines requirements should be established (such as sitting eye height and functional reach);
- no indication was made of the internal vision requirements of the driver with regard to the viewing angles and visibility of primary controls/instruments from the driving position and possible trade-offs against external vision requirements;
- lack of guidance for addressing obstructions when defining required viewing;
- lack of relevant requirements for the position, dimension and location of platform-mounted visual aids or mirrors from the driver's eye, or the positioning of in-cab CCTV with reference to the driver's internal field of vision;
- no consideration of the variability in driving cab designs/role of the vehicle;
- no indication of how the driver's eye level has been or would be derived for specific railway vehicle types;
- no consideration of the angle of view to subtend to the eye at design eye point;
- functional reach was not explicitly addressed.

## Review of Actual Cab Sightline Practice from Other Industries

A range of industry standards were reviewed from the following domains:

- shipping;
- military and commercial aviation.

The requirements for field of view for the shipping industry were found to be fairly high level (such as there shall be an adequate view) and those that were more detailed were generally not applicable or transferable to GM/RT2161. The key disadvantages of the shipping standards were as follows:

- no differentiations between standing and seated workstations;
- no consideration of anthropometry (for example male or female);
- no definition of workstation geometry to measure field of view;
- vague requirements that are difficult to quantify;
- no consideration of head and eye movement;
- how field of view is measured is unclear.

The most useful elements were that requirements were different for workstations in different operating conditions. This approach was recommended to be applied to GM/RT2161 in terms of generic classes of trains or their roles (such as shunting, high speed, local line and freight).

The review of standards from other industries found that there is an inconsistent approach to the measurement and specification of sightlines. The shipping industry is vague and non-prescriptive whilst military and commercial aviation is comprehensive and prescriptive in approach. Given that the main objectives of the aviation industry are to ensure collision avoidance and optimise weapon aiming their approach is structured, highly relevant and applicable to cab sightlines. The aviation approach is also aircraft role specific which was recommended for extrapolation to make cab type specific guidance for GM/RT2161.

Many of the approaches from the non-rail standards have direct application to GM/RT2161 and a list of the recommended requirements to be transferred to GM/RT2161 included:

- internal lighting;
- minimisation of internal reflections/surfaces;
- use of polarised and tinted windows;
- anthropometric considerations;
- seated vs standing workstations;
- definition and use of design eye point;
- trade-offs between internal and external vision requirements;
- requirements for specific scenarios/train classes/role;
- use of ambinocular vision;
- wiper sweep/rain repellent;
- field of view requirements in degrees (not based upon signal distance).

**Vision Plot Assessments**

The following diverse range of train cabs were assessed in terms of their field of view:

- Class 390 – High Speed Electric Multiple Unit (long nose and curved front);
- Class 92 – Conventional flat front freight locomotive;
- Class 319 – Conventional Electric Multiple Unit, where the driver sits to one side rather than the middle, with intermediate gangway.

The vision plots indicated from how far away the following signals could be seen for each of the train types with respect to:

- signal with a red aspect at the maximum permitted height of 6.3m above the rail;
- ground position light signals;
- side positioned signals.

The vision plots were conducted *in situ* (in a yard or depot). The conditions that were examined for the vision plots included the main and auxiliary driving position. The vision plots indicated the closest position of 'typically' positioned signals for each train cab. Key issues such as blind spots and restrictions to sightlines caused by cab structures were also quantified. Tables 10.1 and 10.2 summarise the results of a number of vision plots in terms of degree of viewing angles offered.

In order to position the DEP accurately it was not enough purely to take the sitting eye height from the seat reference point. What was included was the diameter of the head as this changes the eye position. The size of the head forward of the back rest is taken from the lowest point of eye socket to the back of the head, horizontally. By taking into account head size, the eye position is moved forward slightly, thus improving the sighrline for the purposes of external vision. This has the effect of placing the driver closer to the windscreen and hence improving the external view.

The results from the study indicated some interesting trade-offs. For high level signals, the best cab sightline for the 5th percentile female was from Class 390. However, for the 95th percentile male the best cab sightline for high positioned signals is that of the Class 319. In terms of low-level signals the Class 390 showed the worst cab sightlines across all scenarios. This was due to the driver not being to get as close to the windscreen as for the other classes. Conversely, the Class 319 had the best cab sightlines for low positioned signals for all scenarios, despite having the poorest overall FOV of all the classes tested. This can be attributed to the Class 319 having the largest vertical windscreen dimensions and the fact that the driver is closer to the windscreen.

In terms of high-level signals the Class 390 offered the best cab sightlines. This is unusual as the Class 390 has a 'long nose' and the other locomotives examined were flat fronted where the vision would have been expected to be better. The reason for this result is attributed to the tilted windscreen which, effectively increases the driver's upward line of sight and therefore improves upward vision. It is hypothesised that the rake of the windscreen was not intended to improve cab sightlines for high level signals but to meet crash worthiness/impact protection and aerodynamic requirements. However, given the function of the Class 390 is that of a high-speed passenger train the requirements to observe high level signals will be greater than for low-level signals. The results of the vision plots (for example Figure 10.1) have found that four key factors influence cab sightlines:

- how close the driver is to the windscreen (the closer the better);
- how high the driver sits (higher is better for low-level signals and vice versa);
- the rake of the windscreen (greater rake enhances line of sight for high level signals;
- vertical and horizontal dimensions of the windscreen (the larger the better).

**Table 10.1 Summary of results for high level signals**

| Cab survey | 5th percentile female seat fully back/lowered | 5th percentile female seat fully forward/lowered | 95th percentile male seat fully forward/raised | 95th percentile male seat fully back/raised | 5th percentile female seat mid position (forward and back) | 95th percentile male seat mid position (forward and back) |
|---|---|---|---|---|---|---|
| Class 390 | 3.8 | 2.6 | 9.9 | 13.5 | 4.5 | 8.6 |
| Class 92 | 4.9 | 4.0 | 10.0 | 12.0 | 6.5 | 12.0 |
| Class 319 | 4.9 | 3.9 | 7.5 | 9.3 | N/A | N/A |

**Table 10.2 Summary of results for low level signals**

| Cab survey | 5th percentile female seat fully back/lowered | 5th percentile female seat fully forward/lowered | 95th percentile male seat fully forward/raised | 95th percentile male seat fully back/raised | 5th percentile female seat mid position (forward and back) | 95th percentile male seat mid position (forward and back) |
|---|---|---|---|---|---|---|
| Class 390 | 49.8 | 44.3 | 6.7 | 8.3 | 38.5 | 15.9 |
| Class 92 | 22.2 | 18.4 | 5.0 | 5.8 | 7.6 | 5.1 |
| Class 319 | 7.8 | 6.5 | 3.0 | 3.6 | N/A | N/A |

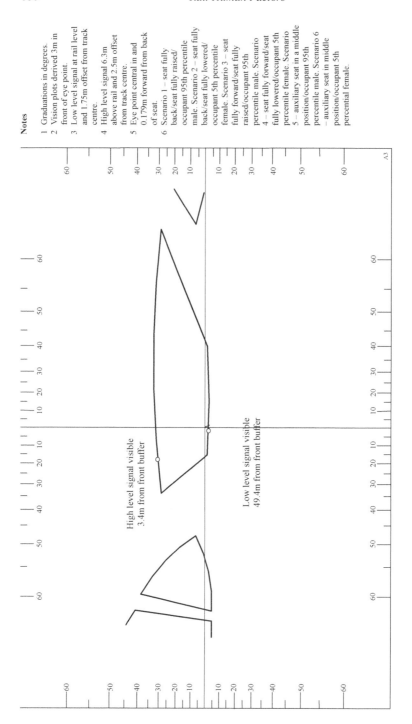

**Notes**

1  Graduations in degrees.
2  Vision plots derived 3m in front of eye point.
3  Low level signal at rail level and 1.75m offset from track centre.
4  High level signal 6.3m above rail and 2.5m offset from track centre.
5  Eye point central in and 0.179m forward from back of seat.
6  Scenario 1 – seat fully back/seat fully raised/ occupant 95th percentile male. Scenario 2 – seat fully back/seat fully lowered/ occupant 5th percentile female. Scenario 3 – seat fully forward/seat fully raised/occupant 95th percentile male. Scenario 4 – seat fully forward/seat fully lowered/occupant 5th percentile female. Scenario 5 – auxiliary seat in a middle position/occupant 95th percentile male. Scenario 6 – auxiliary seat in middle position/occupant 5th percentile female.

High level signal visible 3.4m from front buffer

Low level signal visible 49.4m from front buffer

**Figure 10.1  Example vision plot of a Class 390 – High Speed Electric Multiple Unit (long nose and curved front)**

The factor with the greatest influence is how close the driver can get to the windscreen. The optimum method would be to have a standing workstation to allow the driver to get as close as possible to the windscreen. This would need to be traded off against requirements for:

- crashworthiness/impact protection;
- aerodynamics;
- solar gain;
- fatigue from standing;
- visibility of other in-cab controls, displays and indications.

There is also a trade-off between observing high-level and low-level signals. In terms of addressing this trade-off the role of the train needs to be considered.

### Cab Sightlines and Human Factors Integration: Development of a Structured Framework for Cab Sightlines Specification

This study advocates that an effective definition of cab sightlines' operational requirements stems from a better accommodation of the driver within the cab workstation in terms of reach and other anthropometric considerations, instrument visibility and control, as measured from the design eye point and seat reference point. Design eye point was found to be the key starting point within the performance measurement framework. The design eye position is the midpoint of the design eye line from which all driver workstation dimensions are related and referenced. This is a fixed point in space (rather than the current cube that is cited in GM/RT2161) from which all other requirements for driver workstations commence. It should be the starting point for defining all cab requirements for vision, reach and anthropometric considerations.

The advantage of the DEP approach is that precise, accurate measures and definitions of vision can be given from a set point. The framework provides a structured process and identifies key issues that can be addressed in a prescriptive manner to adequately set out minimum operational viewing requirements for ensuring adequate viewing angles and performance measures. The framework is depicted in Figure 10.2.

The framework specifies key user requirements in terms of what needs to be addressed and considered as part of the specification and measure of cab sightlines. This includes consideration of the role and function of the vehicle. It also identifies a set of influencing factors that moderate and can improve signal visibility.

By utilising this framework precise requirements for the following can be provided:

- primary and secondary controls vision requirements (internal vision);
- horizontal and vertical vision requirements in degrees;
- functional reach envelopes (for primary and secondary controls);
- anthropometrics ranges for seat adjustability and clearance;
- auxiliary driving position vision and reach;
- functional reach;
- range of seat adjustability in terms of forward/aft and up and down.

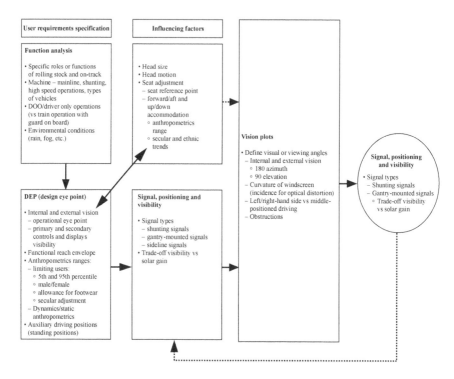

**Figure 10.2  Cab sightlines framework**

The following classification scheme was recommended to define the visibility areas in relation to the layout of in-cab instruments in terms of vision zones:

*Vision Zone 1* This zone shall include the area that shall be visible to the target driver population (as defined with reference to operational requirements)when located at the appropriate design eye position without head motion. Controls/ displays used in this zone shall include those frequently used during driving operation.

*Vision Zone 2* This zone shall include the area that shall be visible to the  target driver population (as defined with reference to operational requirements) when located at the appropriate design eye position with limited head motion and with average stretch of the shoulder and neck muscles.

*Vision Zone 3* This zone shall include the area that shall be visible to the target driver population (as defined with reference to operational requirements) when located at the appropriate design eye position with the maximum practicable upper body movement(such as leaning forward and sideways to gain better visibility).

Similar to the specification for internal vision requirements, the following definitions were recommended for GM/RT2161 in order to specify reach zones:

---

*Reach Zone 1* This zone shall include the area that shall be functionally reached and actuated by the target driver population (as defined with reference to operational requirements) when located at the appropriate design eye position without stretch of arm or shoulder muscles. Controls used in this zone shall include those frequently used during driving operation.

*Reach Zone 2* This zone shall include the area that shall be functionally reached and actuated by the target driver population (as defined with reference to operational requirements) when located at the appropriate design eye position with some extent of stretch of the shoulder and arm muscles.

*Reach Zone 3* This zone shall include the area that shall be functionally reached and actuated by the target driver population (as defined with reference to operational requirements) when located at the appropriate design eye position with the arm fully stretched.

---

To effectively address the above requirement for ensuring an adequate specification of cab sightlines the anthropometric ranges of the actual driver population need to be defined by the railway authority. The minimum scenario ranges to be addressed as part of design and assessment are as follows:

*Main and Auxiliary Driving Position*

- 5th percentile female sitting eye height seat fully back and fully lowered;
- 5th percentile female sitting eye height seat fully forward and fully lowered;
- 95th percentile male sitting eye height seat fully forward and fully raised;
- 95th percentile male sitting eye height seat fully back and fully raised.

**Practical Considerations for Railway Vehicle Cab Sightlines Implementation**

There is a need for a better mapping of driver visual needs with reference to internal and external vision requirements and the role and function of the vehicle. The design eye point must be the starting point from which all cab requirements for vision, accommodation of the driver within the cab workstation, reach, and anthropometric considerations are to be determined.

In order to properly accommodate the associated operational needs for driving a train the actual requirements for drivers' functional reach to effectively operate controls for train operation and carry out auxiliary tasks need first to be addressed. Secondly, the location of most critical and important in-cab controls and displays, external signals or information signs are then envisioned within the normal line of sight so that drivers can see them without the need for frequent eye or head movement from the normal line of sight.

The vision plots should define the appropriate visual/viewing angles and address issues of internal and external vision. The vision plots should indicate where the driving position is in relation to the vision plot. This may include the following positions:

- left/right hand side workstations (main driving position);
- left/right hand side workstations (auxiliary driving position);
- central driving positions;
- standing driving position.

One fundamental weakness that has been identified in GM/RT2161 was that it had been assumed that design eye point is uniform for all types of railway vehicles and for all drivers. This is despite the known variability of physical dimensions of body size and diversity in cab design of the extant railway vehicles.

Vision plots should determine the closest distance from where low-level (ground-mounted), side-position and gantry-mounted signals can be seen from the DEP. The vision plots make clear which position (seated, standing or auxiliary workstation) is being utilized for the plot and clearly indicate the DEP.

Under this framework, the visibility of each type of signal (shunting, gantry-mounted, side-positioned or ground signals) can be ascertained. Vision plots can then indicate whether signals can be seen at the DEP and at what distance these respective signals become visible from the DEP.

**Conclusion**

This framework constitutes a structured approach for assessing and specifying cab sightlines. In doing so, it is assumed that the improved standard will incorporate and address an essential dimension for drivers' reliability of signals sighting, visibility and observance. The key issues that the framework addresses are as follows:

- use of a driver DEP;
- how close the driver DEP can get to the windscreen;
- the range of seat adjustment in height and forward and aft positions;
- inclusion of the internal vision and functional reach requirements;
- trade-offs between the visibility of ground signals and requirements for high-level signals;
- forward, horizontal and vertical vision requirements (external vision) for direct view of infrastructure, signal sighting and viewing angles.

Application of this framework to GM/RT2161 will ensure that worst case scenarios of cab sightline issues are addressed. In addition, precise and accurate vision data can be provided to ensure that internal and external vision requirements are acceptable.

The role of the rolling stock must be considered. For example, for freight trains the emphasis should be upon low level signal sightlines. Conversely, for high-speed passenger trains the emphasis should be upon high level signals.

## Recommendations

This investigation into cab sightlines has resulted in the following recommendations that will consolidate work conducted to date:

1   vision plots are recommended for additional classes of trains including new build, to ascertain cab sightlines' problems. This should include an HST (high speed train), as visibility of a high positioned signal was raised as an issue in the Cullen Report;
2   cab sightline requirements should be specific to the train role and therefore be made more explicit by train function. This work should be linked to the further vision plots;
3   the study has found that standing positions enhance driver FOV as they allow the driver to get closer to the windscreen and this enhances the external FOV. Further work should be conducted that examines the trade-off between improved external FOV and the operability/fatigue effects of standing as opposed to seated workstations;
4   a separate study should be conducted that examines the cost benefits of standing workstation visibility against crashworthiness protection;
5   the framework for specifying and measuring cab sightlines should be expanded to address cab ergonomic issues such as reach, clearance and access to primary controls.

## References

Advanced Technology to Optimize Manpower on Board Ships (ATOMOS), 'Design Requirements for Ship Control Centre (Bridge)', Task 2309.

Defence Procurement Agency (1988), 'Design and Airworthiness Requirements for Service Aircraft', UK Defence Standard 00970, Vol. 2, Book 1.

Defence Procurement Agency (1988), 'Optical Transparent Components – Requirements for Satisfactory Vision-Recommended Methods for the Determination of Optical Properties', Defence Standard 00970, Part 1, Section 4, Leaflet 73.

Federal Aviation Authority (1993), 'Pilot Compartment View Design Considerations', Advisory Circular 25.773.–1.

International Association of Classification Societies (IACS), Standard for Bridge Design, Equipment and Arrangement.

Jankovich, J. (1972), *Human Factors Survey of Locomotive Cabs*, FRA–OPP–73–1, US Department of Transportation, Federal Railroad Administration.

LRMHA – Project Proposal 1381 for Investigation into Cab Sightlines as specified in:
    LRMHA (2003), 'Investigation into Cab Sightlines', Stage 1 Report, 14 October 2002
    LRMHA (2003), 'Investigation into Cab Sightlines', Stage 2 Report, 7 October 2002
    LRMHA (2003), 'Investigation into Cab Sightlines', Stage 3 Report, 17 March 2002
    LRMHA (2003), 'Investigation into Cab Sightlines', Stage 4 Report, 17 March 2002

Military Standard 850B, Aircrew Station Vision Requirements for Military Aircraft, http://iac.dtic.mil/hsiac/Std_Hdbk.htm#850B.

Military Standard 1333B, Aircrew Station Geometry for Military Aircraft, http://iac.dtic.mil/hsiac/Std_Hdbk.htm#850B.

Ministry of Defence (1996), *Human Factors for Designers of Equipment*, Part 7, Issue 2, 'Visual Displays', Defence Standard 00–25.

Ministry of Defence (1997), *Human Factors for Designers of Equipment*, Part 6, Issue 2, 'Vision and Lighting', Defence Standard 00–25.

Society of Automotive Engineers (1989), *Pilot Visibility from the Flightdeck*, ARP4101/2.

Railway Group Standard (1995), 'Requirements for Driving Cabs of Railway Vehicles', GM/RT2161, Issue 1, August, London: Railtrack plc.

Railway Group Standard (2001), 'Signal Positioning and Visibility', GK/RT0037, Issue 4, October, London: Railtrack plc.

Railway Safety (2002), 'Remit for an Investigation into the Cab Sightlines specified in GM/RT2161 Requirements for Driving Cabs of Railway Vehicles', Control Project No. 01/113, 18 July.

Robinson, J. (1978), *Locomotive Cab Design Development – Vol. IV Recommended Design*, Report No. FRA/ORD–76–275.4, US Department of Transportation, Federal Railroad Administration.

Robinson, J., Piccione, D., and Larners, G. (1976), *Locomotive Cab Design Development – Vol. 1 Analysis of Locomotive Cab Environment and Development of Cab Design Alternatives*, Report No. FRA/ORD–76–275.1, US Department of Transportation, Federal Railroad Administration.

# PART 4
# DRIVING – DEVICES IN THE CAB

Chapter 11

# Assessing the Human Factors Risks in Extending the Use of AWS

Ronald W. McLeod, Guy H. Walker and Ann Mills

## Introduction

The project reported in this chapter was conducted on behalf of the Rail Safety and Standards Board, and formed part of the SPAD reduction and mitigation research theme. It sought to assess the human factors risks associated with extending the use of the in-cab Automatic Warning System (AWS). The term 'extended AWS' refers to any situation where AWS is used other than to warn of the state of upcoming signals. This includes uses for permanent, temporary and emergency speed restrictions, certain level crossings, and, potentially, multi-SPAD signals. The chapter summarises the work performed in the study. It considers new areas of psychological investigation believed to be important for driver related research, the methods used to gather and analyse industry experience, and concludes by examining the risk of drivers failing to behave appropriately to AWS warnings.

## The Problem

AWS was originally intended to support semaphore signalling, and only later became extensively used within three and four aspect colour light signalling. However, it only provides warnings that discriminate between two signal states: clear (green, indicated by a bell or simulated chime, at 1200 Hz) and caution (red, single or double yellow, indicated by a steady alarm or horn sound at 800 Hz).

AWS is intended as a supporting system. The driver overrides the automatic application of the brakes by acknowledging the audible warning (horn) and retains manual control of the train. Cancelling a caution (horn) initiates a visual reminder in the form of the 'sunflower' on the driver's desk, indicating that the last signal passed was showing caution (or even stop). The maintenance of safety therefore depends, fundamentally, on the reliability of the driver.

Describing the 1936 Strowger-Hudd system adopted by BR as AWS, Vaughan (2000) writes that: 'No-one ever thought that the driver would ignore the warning. It was utterly taken for granted that a driver would always take notice of the signal' (p. 12). In connection with the crash at Watford Junction in 1996, Stanley Hall (1999), another very experienced railwayman turned author states: 'It seems inconceivable that a driver can acknowledge receiving two warnings and yet take no action to apply the brake to stop the train at the red signal, yet it happens' (p. 97). This paradox provided the background and context for the study; despite what, to all everyday experience, appears to be an extremely powerful attention-getting device, combined with a highly visible visual reminder, there is a significant, and recurring risk of AWS failing to prevent experienced train drivers passing signals at danger.

There were three main study objectives. The first was to investigate whether extended AWS could, or already has, led to a change in the risk of driver unreliability, potentially leading to signals passed at danger (SPADs). The second objective was to understand the circumstances in which further extending the use of AWS might reduce driver reliability. Finally, the study was required to provide a basis for predicting situations in which driver reliability with AWS might be significantly reduced. The overall aim was to provide a basis for understanding the implications of further extending the use of AWS to, for example, forewarn train drivers of approaching signals that have a history of multiple SPADs. (The study was specifically not required to consider the desirability or feasibility of any technical changes to current AWS equipment, such as additional tones or indications.)

**Industry Data**

Field research sought to capture the extensive operational experience available within the industry, and to gather detailed insights on the possible influences that might lead to driver errors with (or despite) AWS. Full details of this survey of industry are reported in Walker and MacLeod (2003). The information was gathered using focus group methods, cab-rides, a questionnaire survey of 277 drivers, a sample of OTMR (on-train-monitoring and recording) data, and incident data from SMIS and CIRAS. The results of this exercise have been validated by industry stakeholders.

*Extended AWS and SPAD Risk*

In general, drivers expressed very positive views towards AWS as an aid to safe driving. Drivers report that the obvious limitations of the system (for example the simple two state warnings) are largely overcome through the use of defensive driving techniques. Drivers also express confidence with the way in which AWS is currently used, and suggested that they would consider the network less safe if AWS was restricted solely to signal aspects (that is, no extended AWS at all). Despite this, there are clearly many occasions where drivers experience doubt, confusion, or uncertainty about the meaning of AWS warnings associated with extended usage.

In our questionnaire survey of 277 drivers, more than three-quarters reported having experienced an AWS magnet that they were not expecting. Eighteen per cent of these unexpected magnets were due to extended AWS usage for temporary speed restrictions (TSR). Similarly, 16 per cent of drivers report having experienced confusion about the meaning of an AWS warning and 34 per cent of these sites were for extended AWS. When asked to suggest what factors are likely to increase risk with AWS usage, emergency speed restrictions (ESR), temporary speed restrictions (TSR), and permanent speed restrictions (PSR) (all extended uses) were placed in first, second, and third places respectively. The findings from this survey clearly indicate that extended AWS has and does potentially increase the risk of SPADs.

*AWS and Driver Behaviour*

Results from the driver questionnaire survey show that a significant number of drivers (155 out of 277, or 56 per cent) report having 'automatically' cancelled an AWS warning on one or more occasions. As Figure 11.1 shows, the great majority state that they have cancelled AWS automatically between one and six times (that is, a few times). Interestingly, a very small percentage (2 per cent) report that they experience this on a daily basis. Once again, extended AWS issues (for example, 'When the relevance of the horn for a TSR seems a little less important than coming up to restrictive signals') accounted for around 20 per cent of the reasons given by drivers for automatic AWS cancellations. Personal factors, the quantity of warnings, and signalling issues (such as prolonged running on cautions) also feature prominently.

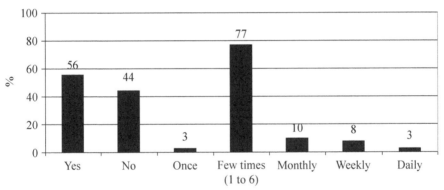

Have you cancelled AWS automatically and how often?

**Figure 11.1   Responses to question 4, 'Have you ever experienced cancelling AWS automatically?'**

A small sample of OTMR data was made available to the study and, although limited to the extent that it merely provides a snapshot of a specific day, driver, driving style and route, it does offer some objective indication of how drivers react to AWS. (A larger scale analysis was beyond the scope and resources of the project.) Figure 11.2 illustrates that on seven out of 11 occasions where an AWS horn indication was received, the driver was operating the cancellation button *before* the horn had started to sound. This seems to suggest that the driver's response was anticipatory and in response to sight of the AWS track magnet ahead. On occasions where the driver receives the horn *and then responds* the average response time calculated from the OTMR data was 0.60 seconds (minimum time is 0.49 seconds, maximum time is

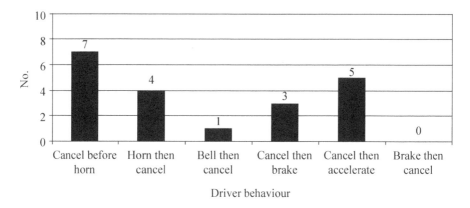

**Figure 11.2   Driver behaviour in response to AWS horn**

0.89 seconds). The speed of this response time again suggests that the cancellation behaviour is highly learnt, and/or anticipatory in nature.

The OTMR data confirm that there is no single, fixed behavioural response expected of a driver when in receipt of an AWS warning. For three of the AWS cancellation events illustrated in Figure 11.2 above, the immediate behaviour was to begin braking, but on five out of 11 occasions the AWS cancellation was followed or accompanied by *accelerating* the train. Although seemingly counter to the explicit purpose of AWS (which is to prompt the driver to the possible need to slow down) this behaviour is for the most part entirely appropriate and legitimate. Another interesting aspect of the data is that there are no instances of the AWS horn occurring *after* the driver has initiated a braking manoeuvre. That is, situations where the driver had sighted the signal and has started to respond in advance of the AWS activation. Behaviour of this sort would be expected to be reasonably frequent. It is not clear why it was not seen in this sample of OTMR data.

*Mode Errors*

Many factors specific to the driver, the class of rolling stock involved, the nature of the movement, and the situation at the time the warning occurs will determine how and when an individual driver reacts to an AWS warning. As Woods (2003) notes, this situation 'shifts to a reliance on human adaptability to make [AWS] work (in principle people can keep track of the context) ... but sometimes they will misinterpret in circumstances where it matters'. This misinterpretation is defined as a mode error, with the result that the likelihood of a SPAD is significantly increased. To reinforce this point, SMIS data (covering the period 1996 to 2002) define the most common cause of a SPAD as being where a driver fails to respond to a cautionary signal. In these situations the driver correctly cancels the AWS warning, but misinterprets the circumstances (that is, commits a mode error) and does not react with appropriate action to brake the train. The industry data presented above provide insight into why, and possibly when such situations may arise.

## Assessment of Driver Reliability with Extended AWS

There was no requirement in the AWS study to attempt to quantify the absolute level of risk – in terms of the likelihood of occurrence – associated with extended AWS. However, in common with established safety management practice, the assessment sought to establish whether the level of risk associated with extended uses of AWS is 'as low as is reasonably practical' (ALARP).

### Analysis Method

The approach used to assess the risk of driver unreliability associated with extended AWS was based on consideration of the likely psychological mechanisms and constructs involved in influencing driver behaviour. This included a 'situational model' of train driver performance. The background thinking is summarised in McLeod *et al.* (2003). The approach emphasises the need to try to understand the driver's possible psychological state *at the time* at which warnings are current. The analysis method involved the following steps:

1   a set of 20 AWS usage scenarios were developed. These were based largely on discussions with drivers, and are thought to be representative of many situations encountered across the UK rail network. These scenarios are summarised in Table 11.1;
2   each scenario was presented graphically along a timeline (the time dimension of the evolution of risk with AWS has emerged as possibly the key situational factor);
3   by consideration of the way in which each scenario develops over time, situational factors considered likely to be 'active' at each point in the scenario were identified and mapped onto the graphical presentation. These were drawn from a set of 18 potential AWS specific situational factors generated from information gathered in the industry survey;
4   a very simple influence model was used to explicitly represent the way in which the situational factors might affect driver reliability with AWS. This model is encapsulated in the following two simple assumptions. First, if any of the situational factors are identified as being likely to be present at the time that an AWS warning is active, there will be an increased risk of the driver making a mode error. Second, the more factors present at the time an AWS warning is active, the greater the risk of driver unreliability. No attempt was made to account for interactions between the situational factors, or cumulative effects if multiple factors might be active.

### Analysis Example

An example of an analysis is shown in Figure 11.3. The scenario develops from left to right. The critical point exists between points A and B. Here the driver has accelerated to 60mph and as usual does not need to take any action for the PSR (at point 6) because trains cannot reach the previous line speed of 90mph. From point 6 onwards the active sunflower is ambiguous because it refers to the *un*usual cautionary

**Table 11.1  Brief description of the 20 AWS usage scenarios**

| Scenario | Brief description |
|---|---|
| 1 | Incremental increases in line speed which do not reflect actual train acceleration. PSR and attendant AWS redundant, and in close proximity to signal. |
| 2 | Driver already braking for signal when AWS received for PSR. Driver responds to line speed increases instead of AWS indications for cautionary signal. |
| 3 | Driver is braking appropriately in response to AWS for TSR, but not appropriately for upcoming danger signal. TSR commencement is beyond signal at danger. |
| 4 | Four aspect signalling interspersed with PSR's. Driver is distracted by guard communication while receiving AWS warnings in quick succession for signal and PSR. |
| 5 | Interrupted view of facing signals on bi-directional line, driver expects certain aspect(s) to be displayed. AWS does not discriminate between signals at caution or danger. |
| 6 | Multiple AWS warnings due to trains bunching in rear of ESR. Repetitive AWS cancellations with no action required as train speed is already low. |
| 7 | Closely spaced AWS magnets for PSR and signal. Driver receives horn but sunflower remains black, or, receives continuous horn activation requiring one cancellation not two. |
| 8 | Sunflower left activated after departing from AWS-fitted route onto an unfitted route. TSR in place with AWS, sunflower remains active for TSR and last fitted signal. |
| 9 | Driver receives AWS at distant signal. Driver subsequently departs from station with sunflower active, proceeds through successive stop signals, but last in sequence is showing danger. |
| 10 | Driver reads-through to signal on wrong line. Correct AWS indication is assumed to be a Code 2 fault. |
| 11 | Single line working with unsuppressed magnets. Initial reaction to AWS warnings is that they are redundant even though they may apply to an unexpected stop aspect. |
| 12 | TSR encountered by freight driver on 125mph line. Sunflower already active due to cautionary signals. Driver forgets TSR within lengthy 125mph braking distance. |
| 13 | AWS magnet missing. Driver expecting warning, none received, braking occurs late. |
| 14 | AWS magnet buried by track ballast. Driver using defensive driving techniques, cannot site magnet which leads to late braking. |
| 15 | Signal located in tunnel. AWS coincides with having to sound horn and other train approaching. Poorly sited sunflower combined with AWS gap area. |
| 16 | Driver passes signal at danger. TPWS activates, driver does not notice TPWS light. Assumes problem is an unsolicited AWS brake demand, resets and departs. |
| 17 | Competing priorities (view of signal and responding to AWS versus safety of track workers). |
| 18 | Confusing TSR layout. Driver cancelling AWS just to keep the brakes off due to insufficient time to attribute individual warnings to specific signage. |
| 19 | Driver braking for station when AWS received for ABCL crossing. Distractions during station stop, driver attributes sunflower to next signal not to level crossing. |
| 20 | Change in meaning of AWS referent. AWS refers to spate board, spate board then missing so AWS warning is a Code 8 fault, same AWS then changed to active TSR. |

signal aspect encountered at point 5, and the *usual* redundant PSR encountered at point 6. Between points A and B six situational factors are considered as potentially active. Using the simple influence model, this combination of factors clearly offers the potential for the driver to make a 'mode error', by confusing the meaning of the sunflower and not applying the brake in time for the upcoming signal at danger.

In this case there are a number of actions that could potentially be taken which would align the scenario with ALARP principles. These include reducing the line speed in order to reflect the accelerative abilities of trains using the route, and thus doing away with a reduction in speed that necessitates an AWS installation. Alternatively, the PSR could be enforced using the existing practice of approach release signalling, or using TPWS in a new way to provide protection against over-speeding without the need for an overt AWS warning. These possible solutions would mean that the sunflower refers to only one source, therefore reducing ambiguity, as well as ensuring that route characteristics (or driveability) matches appropriate driver behaviour more closely.

*Overall Analysis Results*

Figure 11.4 presents a summary of the number of times that each of the 18 situational factors apply within all the 20 AWS usage scenarios. Four situational factors are identified most frequently, these are: route expectancy (expectations about the route ahead), uncertainty (over what a warning refers to), ambiguous AWS referents (where the sunflower refers to more than one thing), and no action needed (in response to a warning). Consideration of these scenarios suggests that there are a number of actions that are both reasonable and practical that could potentially be taken to reduce the risk to a level which is ALARP.

Potential interventions that might help reduce the risk of driver unreliability with extended AWS include:

- ensuring that the AWS referent is clear and unambiguous;
- ensuring that the implementation of TSR/ESRs follows standardised practice;
- designing line speed profiles that approximate to the accelerative and speed capabilities of stock using the route (to avoid redundant drops in line speed and AWS);
- providing incremental drops in line speed instead of large decrements that require an AWS warning (also consider incremental drops in speed in advance of an ESR to lessen the effects of bunching traffic);
- in some cases consider imposing a different ESR speed to that normally required for running under single yellow aspects (this would help break the routine of AWS cancellation followed by no action needed);
- employing TPWS (or TPWS+) as the primary safety device at PSRs instead of AWS;
- considering the acceptable spacing distance for signals and static AWS warnings for PSRs;
- controlling the speed of approaching trains with approach release signals;
- considering the use of reminder signs (for example for active speed restrictions) at critical decision points in a situation;

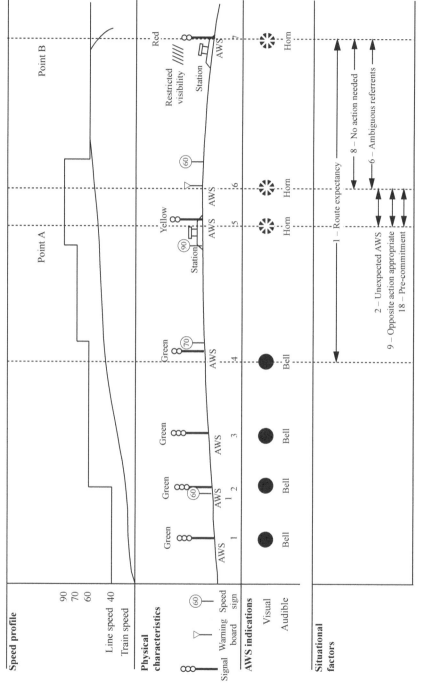

**Figure 11.3  Graphical representation of scenario 1 of 20**

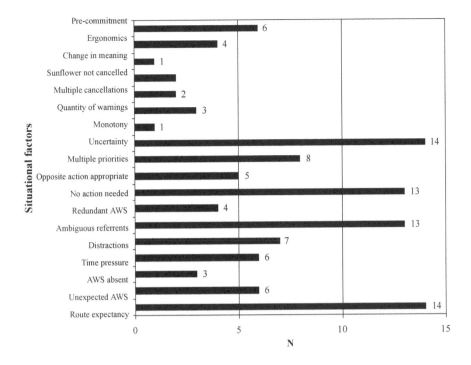

**Figure 11.4 Summary of situational factors associated with the 20 scenarios**

- considering the position of TSR/ESR termination boards in relation to signals (to ensure that the mode change from multiple to single AWS referents does not cause confusion);
- reconsidering the use of AWS at spate boards.

## Conclusions

The study set out to understand and assess the risks of driver unreliability associated with Extended AWS. There were three objectives.

*Objective 1: Has Extended AWS Increased Risk?*

The answer to the first objective is clearly 'Yes'. The evidence strongly suggests that extended uses of AWS should be assumed to have increased the risk of driver unreliability. There are four principal reasons for this:

- the same AWS warning can refer to any of a number of different states. Other than the driver's own awareness and route knowledge, there is nothing that discriminates between the different sources;

- multiple AWS warnings, with different sources, are frequently current simultaneously. There is therefore a reliance on the driver's memory not only to know how many warnings are current, but what they refer to;
- the sunflower is inherently ambiguous: it does not discriminate between the source of the last AWS warning, but simply indicates whether the last warning was a horn or a bell;
- because of the variety of causes, extended AWS has many more implications for driver behaviour than was originally the case.

These four factors potentially increase the probability of the driver misinterpreting the circumstances and making a mode error.

### Objective 2: The Circumstances Leading to Driver Unreliability

A key premise underlying the study has been that driver performance with AWS can only be understood in terms of the context and situation at the time the driver encounters a warning. In particular, the events immediately preceding and coincident with the warning, and the driver's expectations about the route ahead, are critically important.

Because of this, the response to the second objective is less conclusive than for the first. In general, given the controls and procedures that are already in place, there are unlikely to be easily identifiable circumstances inherent in a particular AWS site that would tend to lead to driver unreliability. Essentially, it would be incorrect to view an AWS site itself as being inherently high or low risk (assuming it has been implemented and maintained in accordance with Group Standards). The conclusion from this study is that it would only be correct to consider the likelihood of a driver encountering an AWS warning under conditions which cause one or more of the situational factors influencing driver behaviour to be active.

### Objective 3: A Predictive Method

The third objective of the study was to provide a basis for predicting situations in which the risk of driver reliability with AWS is likely to be significantly reduced. The analysis method developed in the study and used to analyse the 20 AWS usage scenarios provides such a basis.

Further validation of the method is likely to be required to improve its predictive power and validity. In particular, further development is required to represent the combined effect when multiple situational factors are thought likely to be present simultaneously.

The method may be considered to have greater power as a retrospective analysis tool, for example in SPAD investigations; that is, in seeking to identify the factors that led to an incident which has already occurred. The method has significant potential to help avoid simplistic conclusions such as 'driver inattention', and to help identify what might have gone on 'in the driver's head' before an incident. As it stands, the method is thought to provide a reasonably structured approach to reviewing AWS sites with a view to identifying the potential for driver unreliability.

## References

Hall, S. (1999), *Hidden Dangers: Railway Safety in the Era of Privatisation*, Shepperton, Surrey: Ian Allen.

McLeod, R.W. and Walker, G.H. (2003), 'Assessment of Driver Reliability with Extended AWS', Nickleby HFE Ltd, August.

McLeod, R.W., Walker, G.H. and Moray, N. (2003), 'Extended AWS Study: Review of the Knowledge Base', Nickleby HFE Ltd, July.

Vaughan, A. (2000), *Tracks to Disaster*, Shepperton, Surrey: Ian Allen.

Walker, G.H. and McLeod, R.W. (2003), 'Extended AWS Study: Report on Industry Data', Nickleby HFE Ltd, July.

Woods, D.D. (2003), Personal communication, May.

Chapter 12

# Driver Vigilance Devices

Adam Whitlock, John Pethick and Ann Mills

## Introduction

There has been considerable interest in alertness, vigilance and fatigue in safety critical work environments, particularly where exacerbating factors such as shift work are present. These environments include civil aviation (flying), road (car and truck driving) and rail (train driving) where periods of intense activity occur within periods of monotonous work. These tasks often occur against a background of irregular work and rest schedules for the operator, which may also mean that sleep quality and duration are poor. These issues have given rise to a number of proposed devices for monitoring vigilance and alerting the operator to lapses of concentration.

This study concerns a review of i) devices currently used on railways, ii) novel systems to measure driver vigilance and iii) the appropriateness of transfer of technology from other transport industries.

Railway vigilance devices (for example the deadman's pedal and Automatic Warning System) measure frequency of actions taken by the driver to monitor for loss of driver function. Drivers are said to respond to these types of vigilance devices in an automatic manner as a result of driver habituation or prediction, which can significantly reduce the effectiveness of monitoring devices.

## Purpose of the Study

The study was conducted between March and August 2002. It set out to provide the Rail Safety and Standards Board (then Railway Safety) with a practical understanding of the underlying physiological characteristics that could be used to monitor and evaluate driver vigilance. This provided the foundation for assessing reliability, validity and suitability of devices for use in the UK rail environment.

There were five high-level objectives:

1   to conduct a comprehensive and systematic information search of vigilance devices, and methodologies and techniques that measure vigilance;
2   to develop i) user acceptance, ii) UK railway context and iii) scientific feasibility criteria that a device would have to meet for use on the UK rail network;
3   to evaluate candidate vigilance monitoring devices against the feasibility criteria, and recommend suitable devices for further validation;

4    to perform a high-level cost-benefit analysis of implementing a vigilance monitoring device;
5    to identify a high-level implementation plan and experimental characteristics for trialing a device.

## Vigilance, Train Operator Readiness and Train Driving

Porter (1992) identified key skills that train operators require to perform train driving tasks:

* the ability to remember and recall information;
* the ability to think ahead and evaluate factors that affect train performance;
* fast reaction time;
* precision control skills;
* the ability to maintain vigilance and concentration.

Current UK signalling, prior to Train Protection and Warning System (TPWS) and European Rail Train Management System (ERTMS), relies on the driver to monitor for, and respond to, danger and warning signals. The ability to maintain vigilance and concentration, in addition to good vision, are key driver skills. Routes are currently being updated with protection systems that will prevent or minimise the consequences of Signals Passed At Danger (SPADs). Even with train protection, and current selection and training policies and programmes, driver errors occur. Halliday and Porter (1996) provided a categorisation of the driver errors and their underlying causes for SPAD conditions. These included:

* failing to spot a danger signal in time to initiate the appropriate braking;
* failing to apply route knowledge;
* failing to spot caution signals;
* failing to spot trackside signs, boards and indicators;
* failing to spot changes in track gradient or curvature;
* responding to AWS system without knowing the type of signal aspect passed;
* spotting a danger signal but assuming that it will change.

The ability of a driver to execute driving skills depends on their level of arousal. Arousal is a general physiological state, ranging from coma, asleep, drowsy to alert. Different levels of arousal affect task performance in the form of an inverted U (Yerkes and Dodson, 1908, Figure 12.1), with optimum performance at an intermediate level. Furthermore, the optimum value of arousal is inversely related to the difficulty of the task.

Low levels of arousal can be caused by factors such as i) sleep loss, ii) physical fatigue, iii) time of day, iv) driving duration, v) perceptions of boredom, or vi) the consumption of drugs. Low levels of arousal can cause a decline in a person's ability to sustain attention and increase the probability of distraction. High levels of arousal can be caused by either personal or task related anxiety and have the effect of narrowing attention and neglecting tasks. These relationships are not straightforward.

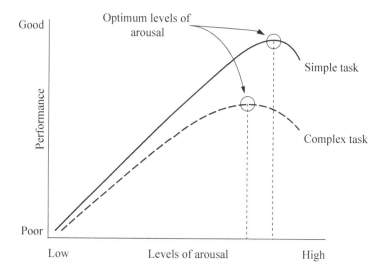

**Figure 12.1   Yerkes Dodson inverted U**

During periods of low arousal it is still possible to respond automatically, even though task awareness is lost. Boredom can lead to distraction even at normal arousal levels. Gibson (1999) in a review of Category A SPADs (1994–97) stated that 30 per cent of drivers attributed their SPAD to distractions from events inside and outside the cab.

The definitions of the terms vigilance, attention and arousal overlap. Vigilance has both physiological (arousal – Head, 1926) and psychological (monitoring, search and attention – Boff and Lincoln, 1988) components. According to Wickens (1992), attention relates to the allocation of mental resources that are available (capacity) and the mental operations that must be performed (demand).

Attention can be selective and directed to one or more sources depending on the perceived priority. It also involves the conscious processing of information. The results of inappropriate allocation of attention during task performance include incorrect task prioritisation or shedding critical information in preference to non-critical information.

A vigilance monitoring device should i) predict when a driver's level of arousal is sufficiently low to render them incapable of responding reliably to signals or trackside information and ii) identify when a driver is distracted from the driving task.

To rate candidate devices a concept of 'driver readiness' was developed, defined as 'the state at which a driver is ready to perceive, process and act on signal aspects and trackside signs'. This merges aspects of sustained attention, allocation of attention, and level of arousal into a concept appropriate for driving tasks. Current rail vigilance devices are not able to detect subtle changes in behaviour or physiology that may predict a reduced capacity to drive the train. The data collected indicate that driver readiness can be rated as follows:

1   the driver is incapacitated/asleep/dead;
2   the driver is alive, but drowsy or not sufficiently alert;
3   the driver is sufficiently alert, but is not attending to the train operating tasks;
4   the driver is sufficiently alert, but is inappropriately time-sharing between train operating tasks;
5   the driver is sufficiently alert and attending to the appropriate train operating tasks.

## Information Search

A search of devices, and of methodologies and techniques that measure driver readiness, was conducted across a range of industry sectors including rail, air, land, sea, and defence. The findings of the search were recorded in databases containing information on physiological indices and vigilance devices.

The types of physiological measures identified were:

*   eye movement or eye blink related – blink rate, eyelid droop/closure, pupil occlusion;
*   brain activity – electro encephalogram (EEG);
*   heart activity – electro cardiogram (ECG);
*   respiratory activity – breathing rate, depth, expired air;
*   body, head or skin movements, electrodermal activity (EDA).

The vigilance device database contains information regarding i) devices that are either currently available to procure, ii) in prototype or testing phase or iii) in design or concept phase.

Database sections and fields followed the study's categories for feasibility, namely i) user acceptance, ii) UK railway context, iii) scientific validity and iv) cost-benefit acceptance.

## Conclusions of the Information Search

Thirteen devices were identified. One of these employed a device provided by another company, and one had been taken off the market by its manufacturer. Eight of the 13 devices employed some form of eye blink or movement measure.

Heart and respiration measures were not represented. Physiological measures of task-based visual attention were not identified.

The technology is available to track and analyse eye movement reliably, but it is assumed that not enough is known about task specific search patterns and eye fixations for rail driving tasks to allow task-based visual attention monitoring.

Eleven devices were selected for paper-based feasibility assessment.

**Feasibility Assessment**

The key scientific acceptance criteria were developed from the literature on vigilance, attention, arousal and scientific processes as follows:

- published research information on the physiological measure and the impartiality of any research conducted on the measure;
- internal validity of i) experimental design, ii) use of subjects and iii) statistics, including:
  - criterion validity – related to the results of the experiment;
  - reliability – related to the consistency of the experiment;
  - sensitivity – related to the ability of the measure to consistently detect a potential error-producing physiological state across the target audience;
  - specificity – related to minimising false alarms;
- generalisability/external validity, including:
  - population validity – related to subject variability, and the application of experimental results to the wider population;
  - ecological validity – related to how representative the research was of train driving tasks.

For the purposes of the research, the scientific feasibility was assessed using literature findings. It was recommended that the literature-based assessment be substantiated with formal experimentation.

The criteria for i) user acceptability and ii) UK railway context were developed during working groups convened by the Rail Safety and Standards Board; including representatives from ASLEF, Train Operating Companies, the Rail Safety and Standards Board and the Health and Safety Executive (Railway Inspectorate). The scientific criteria were developed from basic principles of research design and analysis.

The key criteria for user acceptability are shown in Figure 12.2 and the key criteria for UK railway context are shown in Figure 12.3.

**Results of the Evaluation**

The results of the evaluation are summarised in Table 12.1.

The paper-based evaluation showed that only three devices met the user acceptability criteria of which only the Engine Driver Vigilance Telemetric Control System (EDVTCS) third generation device met all of the applicable criteria.

The requirement to minimise potential driver distraction will need to be examined in a trial of the EDVTCS device. The auditory characteristics of the EDVCTS will also have to be assessed for compliance against Network Rail's noise intensity standard. The EDVTCS has the greatest potential as it is already being used on a rail network. It may be comparatively easy to integrate EDVTCS with current rail safety devices, such as the driver safety device.

The effectiveness of the MicroNod Detection System (MINDStm) would have to be evaluated with respect to wearing head-dresses and hard hats in the cab, and to the height of cab ceilings.

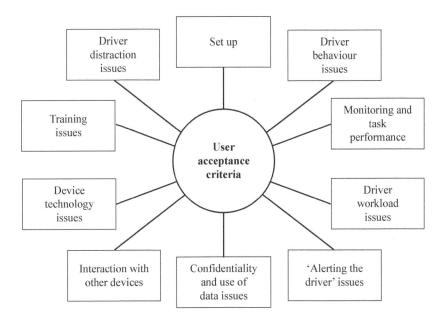

**Figure 12.2  Key criteria for user acceptability**

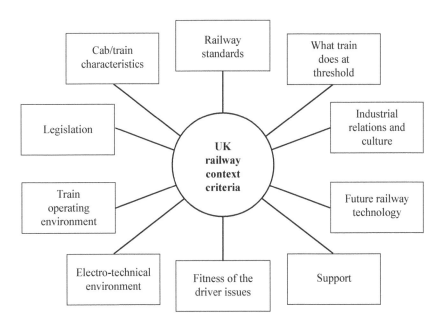

**Figure 12.3  Key criteria for the UK railway context**

The ETS-PC eye tracking system may also be a future suitable candidate, however, currently it does not incorporate an alert. Neither the MINDStm nor the ETS-PC predicts lapses in attention.

## Engine Driver Vigilance Telemetric Control System (EDVTCS)

The EDVTCS uses a watch-based wrist sensor to measure electrical activity from skin on the wrist (electrodermal activity, EDA). This electrical activity includes muscle stimulation in the arm, brain and heart activity. The wrist sensor transmits the activity to a computer unit. The computer unit uses bespoke algorithms to remove unwanted electrical activity and to determine electrodermal reaction (EDR), a derivative of EDA that is sensitive to reductions in arousal. The device is currently in use on the Russian railway for both long distance and suburban routes.

The literature provided by the EDVTCS manufacturer states that 6,500 test subjects were used over a 20 year research period to prove the technology. Experiments have been conducted to determine:

- methods to differentiate electrodermal reactions from other components of electrodermal activity;
- the differences in electrical resistance distribution over the surface of the hand;
- potential individual, gender or race differences in electrodermal activity;
- the effect of ambient temperature and limb movement on the measurement of electrodermal activity;
- the relationship between electrodermal reactions and changes in vigilance and alertness;
- the relationship between electrodermal reactions and task performance;
- the relationship between electrodermal reactions and other measures of alertness (such as EEG).

The research also indicated that the device is an accurate predictor of a person's transition from a state of relaxation to drowsy sleep, and indicates a reduction in vigilance related to task performance.

## Cost-benefit Analysis

The cost-benefit analysis gave an indication of safety benefits (in terms of reducing the number of SPADs and other permanent way incidents) over and above the financial investment. The data for the cost-benefit analysis were collected during workshop meetings, and during a meeting convened between the Rail Safety and Standards Board's Head of Standards Justification, the Rail Safety and Standards Board's Principal Human Factors Specialist, and Quintec. Aspects of the Rail Safety and Standards Board's cost-benefit analysis process (Blacker, 2002) were adopted by the study.

A device implementation lifecycle was suggested to assist with the identification of through life costs. Device dependent and device independent acquisition cost issues

**Table 12.1 Summary of device evaluation outcomes**

| | Device | Device company | Reason for non-compliance |
|---|---|---|---|
| **May be suitable** | *Rail mature* | | |
| | Engine Driver Vigilance Telemetric Control System (EDVTCS), 3rd generation | J-S Co. NEUROCOM | |
| | *Rail immature* | | |
| | ETS-PC eye tracking system | Applied Science Laboratories | |
| | MicroNod Detection System (MINDStm) | Advanced Safety Concepts, Inc (ASCI) | |
| | Eye tracking alertness monitor | Future of Technology and Health | Does not accommodate sunglasses Possibly invasive |
| | ABM drowsiness monitoring device (DMD) | Advanced Brain Monitoring Inc. | Not compatible with head wear or hard hats |
| **Unsuitable** | Photo driven alert system | Michael Myronko | Does not accommodate sunglasses |
| | Device for monitoring haul truck operator alertness | Australian Coal Association Research Programme (ACARP) | Potentially distracting |
| | Vehicle driver's anti-dozing aid (VDADA) | BRTRC Technology Research Corporation | Device not compatible with head wear or hard hats |
| | Copilot PERCLOS monitor | Driving Research Center | Only works at night |
| **More information required** | faceLAB™ 2.0 | Seeing Machines | |
| | Hypovigilance diagnosis module | EU Information Society Research Programme | |

and loss control benefits were identified during the workshop. Device specific costs were identified for the EDVTCS, based on an estimate on the number of cabs in service.

The safety benefit assessment assumed that a device will always prevent incidents (SPADs and others) where the root or underlying cause can be attributed to the driver being in a state of low arousal. Safety benefits of a new measure are found by:

- establishing relevant hazardous events given in the Risk Profile Bulletin (2001), and the hazard precursors (causes);
- evaluating the level of risk associated with the hazardous events and precursors;
- evaluating the measure's effect on the frequency and consequence of each hazard and its precursor, and converting the effects into equivalent fatalities.

Case studies can be used to assess how a new driver vigilance device may have prevented previous SPADs or minimised their effects.

It was possible to describe whole life cost for each stage of the device's life cycle. The majority of costs would be incurred during device assurance and procurement. Overall cost effectiveness may depend on whether the device is retrofitted to current rolling stock, or whether it is implemented only in future rolling stock. Safety benefits may differ according to the type of line or type of service (for example high speed).

It is anticipated that the safety benefits of TPWS will reduce the severity of the consequences of SPAD incidents by reducing the number of SPADs on signal overlaps. This reduces the potential benefits of a new vigilance device. Further research should be performed to determine the types of incidents that a new vigilance device would avoid, in addition to identifying safety benefits above those provided by TPWS. More detailed human error event precursors may be required in the Risk Profile Bulletin to identify fully the potential safety benefits of a new device.

### High-level Implementation Plan

The high-level implementation plan and required characteristics for experimental design were discussed and developed during a working group workshop convened by the Rail Safety and Standards Board. Once a device is selected there are levels of approval for safety critical devices that have to be met before acceptance for use on the railway:

- vehicle approval;
- technical approval for signals;
- legislative approval;
- non-technical aspects of approval.

Vehicle approval includes i) design, build and maintenance conformance and ii) engineering acceptance from the Rail Safety and Standards Board Vehicle Acceptance Board. Technical approval for signals includes route acceptance for the equipment, mainly to demonstrate that a device will not cause spurious signal aspects. Legislative approval includes approval under the Railways and Other

Transport Systems (Approval of Works, Plant and Equipment) Regulations 1994. Non-technical aspects of assurance include i) scientific proving, ii) training, iii) procedures, iv) management, and v) rules. The culmination of the approval process would be to trial the device on a part of the rail network.

## Conclusion

The safety benefits provided by the implementation of TPWS (and full train protection and in-cab signalling) may surpass the potential benefits of a vigilance device that monitors driver arousal or alertness. The study has identified the current availability of devices that provide reliable measures of arousal, but the prevalence of drowsiness related incidents would have to be investigated and quantified to ascertain any potential safety benefits. The evidence from this study also suggests that the extent of the assurance requirements for a device may dissuade the railway industry from investing in the implementation of such a device.

The following recommendations are made for investigation of the EDVTCS:

1   further evaluation of the EDVTCS device should be conducted in situ, to learn from current device users. This will assist in establishing 'real world' validity;
2   a device could be purchased for use in laboratory or simulator studies to confirm the scientific validity of the device, to understand more about the effects of alerting drowsy drivers, and to investigate the possible benefits of providing drivers with an indication of their arousal level;
3   an appropriate, representative and practical benchmark for arousal/alertness should be identified, either through literature search or by experimentation and trial. This should be aimed at providing a standard and rulebook entry for driver monitoring devices.

Overt indicators of visual attention, such as gaze direction and fixation time, could be used in the future to determine incidences of driver distraction. Typical patterns of gaze direction for track monitoring (between signals) and visual search (before signal is in view) could possibly be identified. The increasing sophistication of eye tracking technologies could lead to a greater understanding of eye movements in the context of train driving tasks which may lead to the production of an attention monitoring device. This could be combined with GPS technology and future signalling systems to ensure that drivers are sufficiently aroused and exhibiting the correct eye movements at the appropriate time for detecting and identifying oncoming signals.

## References

Blacker, C. (2002), 'Cost-benefit Analysis for Railway Group Standards, Task Procedure' (draft), Railway Group Standards, Standards Justification, Railway Safety, March.
Boff, K.R. and Lincoln, J.E. (1988), *Engineering Data Compendium: Human Perception and Performance*, Vols 1, 2 and 3, Wright-Patterson Air Force Base, OH: Harry G. Armstrong Aerospace Medical Research Laboratory.

Gibson, H. (1999), 'Special Topic Report: Human Factors Review of Category A SPADs', London: Railtrack Safety and Standards Directorate.

Halliday, M.W. and Porter, D.R. (1996), 'SPADRAM: Driver Distraction, Human Factors Assessment', Report No. SHE 952046, London: British Railway Board.

Head, H. (1926), 'The Conception of Nervous and Mental Energy, Ii. Vigilance: A Physiological State of the Nervous System', *British Journal of Psychology*, 14: 126–47, reprinted in D.R. Davies and G.S. Tune (eds) (1970), *Human Vigilance Performance*, London: Staples Press.

Porter, D. (1992), *A Detailed Task Analysis of Four Types of Train Driving*, Issue 1, Cheshire: SRD.

Railway Safety (2001), 'Profile of Safety Risk on Railtrack Plc. Controlled Infrastructure', Report No. SP–RSK–3.1.3.11.

Wickens, C.D. (ed.) (1992), *Engineering Psychology and Human Performance*, 2nd edn, New York: HarperCollins.

Yerkes, R.M. and Dodson, J.D. (1908), 'The Relation of Strength of Stimulus to Rapidity of Habit Formation', *Journal of Comparative Neuropsychology*, 18: 459–82, reprinted in C.D. Wickens (ed.) (1992), *Engineering Psychology and Human Performance*, 2nd edn, New York: HarperCollins.

Chapter 13

# Human Factors Issues Raised by the Proposed Introduction of GSM Radio Telecommunications into the UK Rail Environment

Mark Young and James Jenkinson

## Introduction

GSM mobile phones are already used in many parts of the UK rail industry. GSM–R is a railway-specific mobile phone system, based on a mandatory European standard that is being introduced in the UK. To support safe and effective implementation of the new system, Rail Safety and Standards Board commissioned a study to identify human factors and equipment issues that may affect the introduction and acceptance of the new technology. This chapter outlines the objectives of the study and the methods used. Key results, including those likely to influence safety, are discussed together with some possible implications of the new approach.

For a number of years, British trains have been fitted with 'ship-to-shore' radio communications that enable a train driver to contact the signaller or line controller to report incidents or to seek advice and assistance. The need for such systems arose, in part, from railway accidents and incidents, where poor quality of communications between drivers and those controlling the rail system were found to be a contributory factor or even a cause.

Two voice radio systems are in use, the National Railway Network (NRN) and the Cab-Secure Radio (CSR). The latter system is also used to improve train control. In recent years, telephone technology has developed markedly, particularly with the introduction of mobile telephone technology, which is now so much a part of everyday life. Public GSM phones are already in use in many rail situations. However, a railway-specific version of the commercial GSM system, known as GSM–R (Global System for Mobiles – Railway), has now been developed. The system requirements are defined in two comprehensive documents (EIRENE, 2000 and 2001), the EIRENE Specifications.

A number of current railway development projects will introduce GSM–R facilities to the UK. In the longer term, GSM–R will be fitted nationally. As part of upgrading the West Coast Main Line, GSM–R will be used for voice communication and improved train control facilities. Certain trains will be fitted early with handheld GSM–R radios in association with improved train detection systems, before the majority of stock is fully equipped with the new system. Similarly, the West Coast Main Line fitment will precede fitting of other lines. These early systems may establish *de facto* standards for some users, setting user expectations and norms.

The Rail Safety and Standard Board (RSSB) is responsible for developing standards to support safe operation of Britain's national rail network. RSSB standards embrace a wide range of technical and operational subjects, including railway telecommunications. In 2002, RSSB initiated a study to investigate ergonomic factors that might influence the introduction of GSM to the British railway environment. This chapter describes a study carried out on behalf of RSSB by one of the authors.[1] The specific objectives of the study were:

- to establish what GSM–R equipment is currently available and to assess it against UK needs;
- to analyse existing published research concerning the effects of using mobile phones in road, rail and air transport and relate this to the UK railway domain;
- to obtain the views of a number of UK users of existing railway radio communications systems;
- to provide an input to new Railway Group Standards for the introduction of GSM–R on a national basis.

## GSM–R Equipment

Although a number of companies are producing COTS (commercial off-the-shelf) GSM–R equipment, the market is not yet mature. At the present time, there are four product approaches regarding the train cab radio user interface:

- a DIN sized COTS 'portrait' style interface unit permanently mounted in the train cab;
- a COTS handheld portable mounted in a purpose designed cradle suitable for mounting in train cabs. This method of mounting allows the unit to be fixed during normal train operations, but de-mounted and used outside the vehicle when required;
- a UK-designed 'landscape'-oriented, DIN sized unit designed as a permanently mounted direct physical replacement for NRN/CSR.
- handheld portable units.

Each product approach carries advantages and disadvantages. The handheld portable is well suited to general railway use. However, when the handheld unit is mounted in the purpose-designed plate for in-cab use, it exhibits a number of features that may make it difficult to use and error prone, for example long reach required, small display and key pad size, etc.).

UK manufacturers are producing project-specific equipment to be fitted to the modernised West Coast Main Line. The radio head unit (the part containing the user interface) is being designed as a like-for-like physical replacement for existing train radios. As such, the design has inherited a number of physical and logical constraints. Some of these will provide continuity with existing equipment, which users may find useful. However, by imposing such constraints, the opportunity to establish an ergonomically sound design is excluded and the resulting interface may not be optimal. The West Coast Main Line GSM–R equipment may be in service in some trains before the corresponding national GSM–R programme has provided equipment to other trains. Thus the West Coast design may establish a *de facto* standard in users' minds, against which any other cab radio design will be judged.

The West Coast line is defined as part of the new Trans-European Network of high-speed lines and international operability is a key consideration in the project. Trains fitted with GSM–R equipment may run on a variety of lines. When trains operating normally on these lines become fitted with a national standard GSM–R user interface, there may be the potential for confusion and human error, if different radio interfaces are in concurrent use. A national implementation plan being developed by RSSB and Network Rail will address these issues.

At present, there is no standard design for signaller GSM–R interfaces. This study revealed the need for a sound signaller GSM–R interface design that meets both safety and operational needs. The design needs to take into account the differing traffic patterns in different locations throughout the network and the consequent differences in emphasis on various features. These include effective means of managing broadcast calls and the presentation of trains-in-area and incoming call lists. The EIRENE standard does not prescribe such features, but equally does not preclude their provision. The design of these facilities has both safety and operational implications. The conclusion must be that the design of the interface should be based on sound task analyses of signaller tasks under all operational modes.

### Literature Survey

*Road Transport Research*

The influence of mobile phones on driving and related tasks has been well researched over a lengthy period (1969 to date). Various distinct approaches can be identified, including: on-road studies, test track studies, high-fidelity simulator studies, low-fidelity simulator studies, laboratory experiments, epidemiological and related studies, human-machine interface studies and heavy goods vehicle studies.

On-road studies are often claimed to offer the highest degree of validity when compared with simulator and other laboratory studies. In general, results from these show that driver reaction time and situation awareness are degraded when phones are used during driving. Subjective assessments by drivers indicate that phone use represents a significant mental task. Particular concerns that emerge are due to the effects of manual dialling. The use of text messages incurs a high potential level of visual distraction and absorption of attention. Test track studies are argued to overcome the limitations imposed in true on-road experiments by allowing the experimenter to impose a greater degree of control over experimental conditions. The results of the test track studies generally support the findings of on-road studies.

High-fidelity and low-fidelity simulator studies offer an attractive approach to investigating driver behaviour under controlled conditions. Laboratory investigations also make a useful contribution to the overall body of research. A variety of performance effects have been investigated using these aproaches. Whilst laboratory experiments often focus on particular issues, the overall trend of the results of the studies confirms the findings from road-based and simulator based research.

Certain epidemiological studies have considered the effects of mobile phone use. These have investigated relationships and associations between continuous human behaviour (mobile phone ownership and use) and real-life events (accidents). As yet, these studies do not prove any causal relationships.

The trend towards higher degrees of automation in road vehicles and the associated development of more sophisticated user interfaces has led to the need to examine driver behaviour in greater detail. Mobile phone technology forms only one part of this trend. In a modern vehicle, there may be several other competing sources of driver distraction present. An emerging trend is for higher amounts of integration in information displays. Some information has emerged from research studies to allow the relative effects of competing user interface designs to be compared and evaluated.

Several studies consider methods and tools for the study of phone use whilst driving. The task of car driving appears to have been modelled in some detail. Studies have sought to confirm the validity of previous models of human information processing, such as those of Wickens (1984), or to establish alternative theories. Such models may bring improved understanding of the mechanisms of driver error and degraded response. They may be extendable to represent the heavy goods vehicle-driving task, but they cannot be readily related to the task of the train driver. Hankey *et al.* (1996) appear to have produced a set of task models that could be adapted to railway use.

An obvious question that arises is 'How readily can research on phone use in cars be translated to the rail domain?' A limited amount of research has been carried out with drivers of HGVs and trucks. It may be argued that truck driving is more like train driving than car driving. As such there may be a greater degree of perceived relevance from these studies. The vehicle dynamic characteristics are similar to a degree, and the training, work pattern and ethos of the commercial vehicle driver are arguably similar to those of the train driver. In general, the results of research in the HGV sector generally support findings from studies of car drivers.

Of particular significance to the UK is the report of the Stewart Committee on the safety of mobile phones (2000). Although this government-sponsored group looked primarily at radiation safety, it also considered hazards from phone use in vehicles. The report draws firm conclusions and makes firm recommendations. Similarly, the recent government sponsored report by RoSPA (2002) has clear findings. Both studies conclude that mobile phone use whilst driving a road vehicle is unacceptably hazardous.

In most studies, significant driving performance effects have been found. Again, driver distraction effects, affecting the tactical level of driving, are of more concern than the effects on 'operational' driving skills. Comparisons between handheld phones and so-called hands-free phones show little safety benefit in the latter type. In order to have a less detrimental effect, the phone must be truly hands-free, meaning in practice that a call may be received with zero or minimal operator manual action and a call initiated with no more than a single key press action. BSI Draft Document DD235 provides a basis for the safe design of in-vehicle user information system interfaces and makes firm, clear recommendations about HMI design, device location, etc. (Hannigan and Herring, 1990).

*International Rail Cross-border Operations*

The report of a research project entitled HUSARE, funded by the European Commission under the 4th Framework Programme identifies, *inter alia*, high levels of communication between rail staff as an essential component in effective cross-border operations. There are benefits in having a common user interface across European boundaries. Further problems could arise if rail staff operating outside their home country were unsure about the responsibilities of the various parties controlling the infrastructure, for example signaller, zone controller, dispatcher, etc.

Work on behalf of SNCF developed a computer-based training system to enable drivers of TGV trains to operate a new radio control unit whilst driving on French, British or Belgian railway networks. The training used multimedia techniques, including video images and speech recognition to simulate realistic rail operating scenarios. In addition to providing a structured user training facility, the system allowed human factors issues to be tested and evaluated in a systematic manner, particularly for emergency situations, which are not easily simulated in the real world. Study topics included error-resistant message protocols and the use of icons to overcome language problems.

*Aviation Research*

Voice communications form an integral part of the pilot's task. Increases in the volume of air traffic and increased pressure to improve aircraft handling times at airports have shown that voice communications channels can become a limiting factor in safe and efficient aircraft operations. Digital communications technology is increasingly used to provide aircraft crews with information necessary to plan and execute flight plans and aircraft operations, providing information through visual displays and auditory displays, thus utilising an additional human information channel. The development of mobile device fax and text messaging may represent the beginning of a similar trend toward multimodal information displays in train cabs. Experience with digital data links in aircraft may provide some insights into possible future issues that may arise with in-cab information displays. The proven value of aircraft party line communications provokes the question 'Should a party line system be introduced for railway operations?' Also, based on aviation studies, there appear to be strong arguments in favour of more experiential training, through which users can develop the appropriate confidence in their own communications technique and command of the necessary language.

**User Opinion Surveys**

To ensure that future GSM–R user opinions were fully explored, a number of face-to-face interviews were conducted with railway stakeholders. The survey embraced system functionality, operability, and consistency of user interface design, migration, emergency response procedures, training, and user documentation. The user groups involved included:

- signallers and signaller supervisors;
- train drivers, driver managers and train managers (guards);
- train preparers (shunters);
- electrical power supply control room operators.

Various operational situations were studied, including high-density/high-speed passenger and freight traffic routes, low traffic density routes, DC Electrified lines, train preparation facilities, mixed radio system areas and mobile user facilities.

A high degree of cooperation and openness from users was found. Discussions with train driver representatives quickly established that train drivers are no longer accessible in large numbers at booking-on points, mess rooms etc. Drivers typically receive booking-on instructions by mobile phone, report for duty individually at specified times, prepare trains and complete specific planned journeys. Current work schedules mean the drivers are tightly scheduled and do not have periods when interviews could take place. A decision was therefore taken to access driver views based on a limited number of cab ride interviews together with more in-depth discussions with driver managers. In addition, the interviews that were carried out with signalling staff included discussions of driver tasks.

*Current Radio System Uses*

At present, calls from drivers to signallers typically concern reports of lineside events, incidents and problems. A further major use is where trains that come to a stand at a signal set to danger announce their presence to the signaller. In some situations, drivers call the signaller in order to obtain information on the likely duration of their wait at the signal, and possibly to obtain information on the reasons for any delay, even when local circumstances make the reason obvious (for example a train visible in the next section). This type of call can occupy a significant proportion of a signaller's resources. There is scope to rationalise these arrangements, possibly on a national basis. A facility is available within GSM–R for a train to announce its presence at a stop signal using a unique key or key code and for the signaller to acknowledge the incoming call using a variety of pre-defined responses. These could include pre-defined text such as 'Signaller busy – Continue to wait', 'Traffic ahead – wait', 'Problems – call signaller', etc.

Typical outgoing calls from signallers to trains include advice to drivers of general traffic problems, advice of lineside incidents, or advice of adverse railhead conditions. In NRN areas, such calls cannot be practically made directly to individual or groups of trains. Where justified by operational conditions, such calls must be made via Zone Controllers. In CSR areas, signallers can make such calls. Again, with the existing facilities, these tasks can occupy significant amounts of signaller resources. There is wide agreement amongst users that the existing CSR facilities for group broadcast of recorded messages are wholly inadequate. Broadcasts interfere with the normal voice channels; message time is too short; and repeating the message several times is time and attention consuming. All CSR users wish these facilities to be improved.

*System Functionality*

All users expressed the view that a new radio system should have improved functionality and reliability rather than simply being a direct replacement for NRN and CSR. A new system should meet long-term railway communications needs. Most users believed that a national radio network should provide standardised facilities that operate in a standardised way. Differences between existing radio systems in local areas compounded training and usability problems and might have safety implications in some circumstances. Users believed that GSM–R should achieve similar or better customer service levels than those experienced for public GSM services.

A core of users expressed strong views that train radio systems should be kept exclusively for purposes of maintaining line safety and that operational communications should be kept separate. This view contrasts with that of other users who could envisage a more open and integrated use of the GSM–R system.

Improved facilities for drivers to announce their arrival at a stop signal would be useful. More importantly, major improvements in the facilities to allow signallers to record and broadcast pre-defined messages to trains are essential. Many signallers argue for more selective tools to stop trains. The existing 'Stop All Trains' facility is too coarse. It should be possible for a signaller rapidly to select particular trains to receive a stop message.

A number of mandated requirements identified in the EIRENE Specification are prohibited on UK rail networks. An example of this is the requirement for drivers to be able to make broadcast and point to multi-point calls, which would contravene the UK requirement that states drivers shall not be able to communicate with other train drivers. Difficulties may also arise with the logging on procedures. The EIRENE specification does not support location dependent addressing, which requires trains to log on using alphanumeric train running numbers. This shortcoming will require signallers to perform a proxy log on.

In contrast to established aviation practice, many users expressed concern about the use of party line communications. Most felt that on balance, this type of communication could be unsafe. Rail experience had shown the need to clearly establish the identity of called and calling parties. Users were able to cite examples where overheard conversations or failures to use established communications protocols had led to a 'near miss' situation. Several users cited examples of problems with 'Push-to-talk facilities' on handsets. It is clear that the quality of equipment having this feature is important. Clearly, the introduction of party line working would require significant changes to existing railway communications protocols.

*Equipment Design*

Several users felt that the provision of handheld radios to train drivers would be an advantage in allowing communications when the driver was out of the cab. However, users felt that the small physical size of a handheld unit could cause problems if it was used as the cab radio on a mounting plate. Cab radio equipment must be suitable for worst-case conditions. Cab vibration levels could be high. Some drivers wear gloves. Designs should provide larger keys and display than the existing NRN and CSR equipment.

Shunters were opposed to the use of an integrated radio system for shunting operations. They stressed the need to establish clear unambiguous contact between the members of a shunting group. Handheld radio pairs on a dedicated frequency provided this assurance. Particular problems might be caused by adjacent shunting operations. Shunters also did not like the small physical size of the example handheld radio that was presented for discussion. They felt it was too small, unsuitable for gloved use and overly complex compared to existing established designs.

Most signaller-users felt that a computer-based interface would be widely acceptable. However, any display design should fully meet operational user needs. This implies the need for increased task analysis of signaller tasks in the design of such interfaces, including infrequent and high–consequence activities. Whilst a detailed display specification could be produced from such an analysis, it would be better to establish an interactive development facility with which a number of experienced and novice signallers could interact as part of the development process. Several users reported feeling that the existing NRN and CSR radio systems were presented to them as *faits accomplis*. They would clearly like to have been involved in development of the existing systems. The nature of GSM–R development may produce a similar situation.

*Training*

Almost all interviewees commented on the required quantity and quality of training in the use of radio systems. Many users acquired their existing knowledge through 'sitting with Nellie', rather than through formal training. Whilst all users felt confident in the use of basic radio communications facilities, some felt they, and others, might have problems in some infrequent, high stress situations, such as railway fatalities. Users all felt that the training for a new radio system should be improved to include better documentation, issued on a personal basis, dedicated professional training staff (rather than experienced users) and more facilities and time to allow trainees to actually practice communications skills.

**Study Conclusions**

*Hazards and Risks*

There is clear evidence that using a mobile phone while driving a moving vehicle on roads is hazardous. However, the degree of risk that pertains is highly situation-dependent, and also dependent on individual driver characteristics. There is some evidence of gender-specific performance differences. Under light demand conditions, where driver boredom may be present, phone use may possibly bring some enhancement to a driver's arousal state. Under the majority of road driving conditions, phone use produces negative performance effects. Direction control is impaired. This is more relevant to road vehicles, since lateral control of a rail vehicle is provided by the infrastructure. Speed control is also impaired, and drivers may drive more slowly whilst engaged in conversation. Again, this phenomenon would affect a train driver differently from a car driver, due to the vehicle inertial characteristics and the different methods of applying power and brake control.

The most significant research finding is that phone use diverts and consumes mental resources away from the primary task of driving. In experiments of all kinds, participants have been found to spend less time looking at the view ahead, more time looking at, searching for and manipulating the phone device, and have significantly poorer situation awareness when using a phone whilst driving. These results are likely to transfer more or less directly to the train-driving situation. The main concern is that a driver of a moving train, whilst using a phone, might fail to see, fail to interpret correctly or fail to act upon lineside information (that is, signals) or track and vehicle conditions. The concept of an acceptable eyes off the road time (EORT) has been established for car studies. It may be useful for the railway industry to develop an equivalent 'eyes off the rail' metric. Some data exist that may assist with this, but the likelihood is that experiments using a high-fidelity train simulator might be necessary to provide the necessary data.

*Legislative Issues*

Many countries already restrict or ban the use of mobile phones in a moving car, especially for public service drivers. In many countries, employers who provide employees with mobile phones increasingly restrict phone use while driving. Much interest has been caused by recent UK research which concluded that mobile phone use is more dangerous than driving under the influence of alcohol. The UK government has now introduced significant restrictions on phone use in a moving vehicle.

In contrast, radio telephones have been safely used in train cabs for a number of years. Available data would suggest that such devices do not directly contribute to railway accidents. Current operational practices that restrict radio use may be the foundation of this safe situation. However, whilst there appears to be no evidence the train radios are unsafe, there is no evidence that they are positively safe. In the light of changing public attitudes, it may be desirable for the rail industry to seek more positive data in this area. The research indicates that at the very least, phone use when a vehicle is moving should be constrained or restricted to ensure safety under all operating and emergency conditions, particularly where increased functionality is provided. Rules and procedures have a part to play in this, but there are features, such as restricted presentation of text messages, that need to be achieved using technological solutions.

*Integration Issues*

Hitherto, in-cab radios have been present as an isolated piece of technology amidst the plethora of disparate systems, indications and controls that go to make up a train driver interface. As technology matures, system functionality increases and additional automation is applied to train operations, the driver workspace is becoming equipped with more and more in-cab information systems. Clear parallels can be seen in the development of other technologies, including aircraft, ships, and process plants, etc. Lessons learned in those industries may benefit the rail sector. There is a clear need to assess in-cab radio as an integral part of the driver interface, rather than as an isolated system. As European integration progresses, the likelihood of cross-border rail operations increases. Existing UK experience with CTRL is relevant. The drive towards common operating procedures and practices may be an important factor.

Driver and signaller training systems are well specified and well regulated under existing railway standards and guidance. These systems include training in radio use and communications protocols and practices. However, the study indicated the need for adequate technical user training in use of the new equipment. Research indicates that training should also provide users with the skills and confidence to communicate effectively under all circumstances, including emergencies and critical situations. This implies much more practice and tutor feedback than is common in many training systems. A further concern arises with track-side workers, where there may be cultural communications issues. There is a need to inculcate in all radio users, a high level of discipline in radio use, message formulation and message delivery.

*Cab Fitment Issues*

Existing cab radios are fitted in a variety of positions relative to the driver's sight and reach envelopes. There are strong arguments for locating the radio in a position that allows it to be used with minimal impact on the driver's focus of visual attention outside the cab (that is, straight ahead for most of the time). Similarly, if a driver needs to operate a control to receive or make a call whilst on the move, the radio controls should be readily and safely accessible without undue driver body movement. This is true for purpose designed radio interfaces, but it is more so for COTS equipment such as hand-portable radios mounted on cradles. In the latter case, screen visibility and push button selection criteria may dictate the best location.

Notwithstanding the above, many existing train cab designs will present a series of constraints when new or replacement equipment is being considered. Cab radio equipment design needs to take vibration into account, requiring larger controls and displays. Rather than adopting a purely pragmatic approach, where a home for radio equipment is found expediently, a more systematic approach would be to develop ergonomic guidelines for such installations, based upon sound criteria. The existence of such guidelines could better inform the development of any new designs of GSM–R equipment for both the UK and the general EU market.

*Usability Issues*

The practical constraints of running an operational railway network often mean there is a need to compromise from what would otherwise be 'ideal' human factors designs. At times, even some of the human factors principles would appear to be in conflict when attempting to apply them in a practical setting. For instance, the introduction of GSM–R presents a prime opportunity to completely redesign the radio interface to optimise usability. The disadvantage with this is that a new head unit might initially cause problems for drivers who have grown accustomed to existing systems over a number of years – it becomes a trade-off between flexibility and compatibility. Similarly, the increased functionality associated with GSM–R means that some functions may necessarily be relegated to embedded menus, in order to reduce display clutter. Therefore, while there are certain usability principles associated with cab radios that should be upheld, others need more detailed investigation or discussion.

In addition to the design and functionality issues identified in the user opinion survey, the literature review also revealed some findings on the usability of in-car devices. General good practice of functional grouping, clear labelling, and display size and readability were all applicable (Hannigan and Herring, 1990); Johnson and Van Vianen, 1992), and ought to be a given in the design of any new GSM–R interfaces. Certain specific studies on in-car telephone tasks are also pertinent to the rail environment. A driving simulator experiment found that dialling a number proved to have the worst effect on driving performance (in terms of lane deviation), although voice dialling was preferred by users and had less of an effect than manual dialling (Serafin *et al.*, 1993). In another study, a 'satellite' design, in which the phone controls were mounted on the steering column and the display located in the driver's line of sight, was compared against a traditional interface (Parkes and Ward, 2001). The satellite design resulted in lower workload, greater ease of use, and reduced eyes-off-road time.

Whilst mandating the general good practice guidelines in the new GSM–R Railway Group Standard has been relatively straightforward, adapting some of the specific findings is more problematic. For example, voice dialling will probably be too unreliable in a train cab where the ambient noise is much higher than in a car. Employing a satellite-type design would face stiff competition for real estate on the driver's desk, and is unlikely to be accepted for an interface that is not viewed as a primary control or safety system. To challenge the latter point would mean reviewing the standard on train cab design.

The signaller's GSM–R interface is rather less constrained, and in many ways has not yet been standardised. Early human factors input here could potentially have a much greater impact than in the train cab. One can envisage an integrated system for signalling and telecommunications, which exploits the enhanced features of GSM–R to identify train movements and contact drivers. For instance, the entire signaller's panel could be a touch-screen, such that a call could be placed simply by pointing to the relevant train on the display. Needless to say, such a revolutionary design would entail a large investment in signalling control centres, but it does represent an ideal to work towards.

In sum, translating the results of human factors research into requirements for Railway Group Standards is not a simple process. Maintaining a high-level, generic set of principles allows for a level of interpretation in design, which can help to avoid apparent contradictions as well as accommodating practical constraints. The next challenge is to manage the transition from existing systems to GSM–R. Again, it is anticipated that compromises will have to be made between the ideal human factors solution (a 'dual-mode' unit with a single interface for seamless transition between old and new networks) and the realistic options (dual fitment of two different pieces of kit during the transition period). The results of the literature review and user survey reported here will help RSSB make the human factors decisions to ensure a successful roll-out of GSM–R.

**Note**

1 The study was carried out when the author (JJ) was with Systems Engineering and Assessment Ltd.

## References

European Integrated Railway Radio Enhanced Network (EIRENE) (2000), *Functional Requirements Specification*, Issue 5.0, Ref. MDA029D009, 20 December.

European Integrated Railway Radio Enhanced Network (EIRENE) (2001), *System Requirements Specification*, Issue 13.0, Ref. ITA078D018.

Hankey, J.M., Dingus, T.A., Hanowski, R.J., Wierwille, W.W., Monk, C.A. and Moyer, M.J. (1996), 'The Development of a Design Evaluation Tool and Model of Attention Demand', retrieved 20 June 2005 from http://www-nrd.nhtsa.dot.gov/departments/nrd-13/driver-distraction/PDF/8.PDF.

Hannigan, S. and Herring, V. (1990), 'Human Factors Implications of Mobile Data Services', *Proceedings of the 13th International Symposium on Human Factors in Telecommunications*, 10–14 September 1990: 489–93.

Johnson, G. and van Vianen, E.P.G. (1992), 'Comparative Evaluation of Basic Car Radio Controls', in E.J. Lovesey (ed.), *Contemporary Ergonomics*, London: Taylor and Francis: 456–62.

Parkes, A.M. and Ward, N. (2001), 'Case Study: A Safety and Usability Evaluation of Two Different Car Phone Designs', *International Journal of Vehicle Design*, 26 (1): 12–29.

Royal Society for the Prevention of Accidents (RoSPA) (2002), 'The Risks of Using a Mobile Phone While Driving', retrieved 20 June 2005 from http://www.rospa.com/roadsafety/info/mobile_phone_report.pdf.

Stewart, W. (2000), 'Mobile Phones and Health. Report of the Independent Expert Group on Mobile Phones', retrieved on 20 June 2005 from http://dddc.org.uk/lpi/inquirydocuments/core%20 documents/CD041StewartReport.pdf.

Serafin, C., Wen, C., Paelke, G. and Green, P. (1993), 'Car Phone Usability: A Human Factors Laboratory Test', *Proceedings of the Human Factors and Ergonomics Society 37th Annual Meeting*, Santa Monica, CA: Human Factors and Ergonomics Society: 220–24.

# PART 5
# DRIVING – SIGNS, SIGNALS AND SPADS

# Chapter 14

# Driver Detection and Recognition of Lineside Signals and Signs at Different Approach Speeds

Guangyan Li, W. Ian Hamilton, Ged Morrisroe and Theresa Clarke

## Background

A study was carried out using simulation based on desktop virtual reality (VR) to investigate driver responses to lineside signals and signs at various approach speeds. The objectives of the study were:

- to find out whether train speed would significantly affect signal/sign reading;
- to examine at which point certain types of signs or signals could be detected or recognised; and
- to determine a speed cut-off level above which certain types of signs or signals are no longer recognisable or detectable.

Fifty-seven train drivers from 12 train operating companies (TOCs) participated in the trials. Twenty different types of lineside signs and ten types of signals were tested under six different approach speeds ranging from 100km/h to 350km/h (62–218mph). Driver performance measures were 'time remaining to the signal/sign' at the point of detection or recognition, and reading error rate. The results showed a significant influence of train speed on driver responses to lineside signals/signs and demonstrated a non-linear relationship between driver responses to signals/signs and approach speed. This has been used to estimate a maximum approach speed limit within which a specific signal or sign can be correctly detected or recognised. The implications of the findings of the study are discussed.

## Introduction

Following a preliminary study of driver recognition of lineside signals and signs at different train speeds using a simple computer simulation (Li *et al.*, 2003), a further study was carried out to validate and develop these findings in a more realistic VR environment.

The scientific justification for this study has been documented in report (HEL/ RT/01651/RT1, 2002) to Network Rail. This revealed that most of the research that has been carried out about the design and evaluation of traffic signs has investigated different driving situations for road, as opposed to rail transport. Little is known about the effects of train speed on driver responses to different types of lineside signals and signs. For example, how much information can be correctly identified while travelling at different speeds, and what are the speed 'cut-off' points beyond which the presentation of the current signage is no longer effective? The scientific literature does not provide a satisfactory answer to these questions, and consequently the present study set out to explore these issues.

**Experimental Studies**

The test paradigm was a VR representation of the UK track environment between Handforth and Manchester South. The total length of the track is 15km and in the simulation the train runs in both 'up' and 'down' directions.

*Controlled Variables*

The controlled factors (independent variables) were approach speed (six levels) and the type of lineside signs/signals (30 types). The six speed levels were: 100km/h (62mph), 150km/h (93mph), 200km/h (124mph), 250km/h (155mph), 300km/h (186mph) and 350km/h (218mph). Each speed was tested independently in a separate trial session.

Table 14.1 shows the lineside signs (20 types) and signals (10 types) tested. All signs were based on GE/RT8043 (Rail Group Standards, 2002a) and all signals or indicators unless noted were based on GK/RT0031 (Rail Group Standards, 2002b). Table 14.2 shows the trial session arrangement.

*Signal/Sign Positions*

Both signs and signals were positioned at least 2.08m from the track, measured from the nearest face of posts to the centre line of the track (Cope, 1993). Colour light signals were placed 3.3m above the rail, measured to the centre of the red aspect and signs were placed 2.5m above the rail, measured to the centre of signboard (GK/RT0037). The signals and signs were separated from each other so that the participants would have to respond to only one of them at a time. Driver eye level was set at 2.75m above rail level (GK/RT0037).

*Driver Performance Measures*

Driver responses (dependent variables) were measured either in terms of detection or recognition of signs/signals on approach.

For sign/signal recognition, drivers pressed the spacebar as soon as they could recognise the meaning of the sign or the aspect of a colour light signal, and the time remaining from this point to the sign/signal was automatically recorded. The sign/signal recognition was tested under six speeds. Following each key press, the drivers verbalised what sign/signal they had just seen and their answers were recorded by the experimenter using a checklist.

**Table 14.1  Lineside signs and signals tested in the trials**

| ID Signs | Name | Specification |
|---|---|---|
| 1 | Whistle board | Diameter: 750mm; letter height: 300mm |
| 2 | Commencement of cab-signalling | 500 × 500mm; letter height: 140mm |
| 3 | Commencement of special AWS working | 800 × 800mm (each side) |
| 4 | Countdown marker | W × H: 700 × 1200mm |
| 5 | Neutral section warning board | 600 × 600mm; black background |
| 6 | Neutral section indication board | 600 × 600mm |
| 7 | Emergency indicator | $W_1$(bottom) × $W_2$(top) × H: 650 × 450 × 1070mm |
| 8 | Permissible speed indicator | 900 in diameter; letter height: 370mm |
| 9 | Temporary speed restriction indicator | W × H: 390 × 780mm; letter height: 230mm |
| 10 | Permissible speed warning indicator | W × H: 480 × 700 mm (*Ref. p. 41 of GE/ RT8043*) |
| 11 | Temporary speed restriction warning board | W × H(base): 560 × 200mm; W × H(top): 390 × 780mm; total height: 980mm |
| 12 | Permissible speed indicator – limited clearance applications | W × H: 450 × 650mm |
| 13 | Permissible speed warning indicator | W × H: 480 × 700mm (*Ref. p. 41 of GE/ RT8043*) |
| 14 | National radio network channel change | W × H: 450 × 850mm |
| 15 | Cab secure radio channel change | W × H: 450 × 850mm |
| 16 | SPAD indicator and identification plate | W × H: 310 × 500mm |
| 17 | Sandite and low adhesion markers | W × H: 700 × 1200mm |
| 18 | Coasting board | W × H: 250 × 400mm; R=20mm |
| 28 | EPS advance warning indicator (limited clearance) | Overall W × H: 450 × 450mm (miniature warning and commencement board) |
| 29 | EPS speed commencement board (limited clearance) | Overall W × H: 450 × 450mm (miniature warning and commencement board) |
| **Signals** | | |
| 19 | Banner repeater on or off | Theatre box : 600 × 600mm |
| 20 | Semaphore signal on or off | W × H: 1060 × 260mm |
| 20a | Semaphore distant signal on | W × H: 1060 × 260mm |
| 21 | Colour light signal – green | Standard four-aspect design |
| 22 | Colour light signal – double yellow | Standard four-aspect design |
| 23 | Colour light signal – single yellow | Standard four-aspect design |
| 24 | Colour light signal – red | Standard four-aspect design |
| 25 | Flashing double yellow | Standard four-aspect design |
| 26 | Flashing single yellow | Standard four-aspect design |
| 27 | Junction indicator (45° to the left or right) | Current standard design |

**Table 14.2 Trial session arrangement**

| Trial session | Approach speed | | Route | Note |
|---|---|---|---|---|
| | *km/h* | *mph* | | |
| Session 1 | 100 | 62 | 1 | Random sign/signal sequence |
| Session 2 | 150 | 93 | 2 | Random sign/signal sequence |
| Session 3 | 200 | 124 | 1 | Random sign/signal sequence |
| Session 4 | 250 | 155 | 2 | Random sign/signal sequence |
| Session 5 | 300 | 186 | 1 | Random sign/signal sequence |
| Session 6 | 350 | 218 | 2 | Random sign/signal sequence |
| Session 7 | 100–300 | 62–186 | 2 | Signals in normal sequence, with varying speed between each set of four signals. |

**Note**

Route 1 – Manchester south to Handforth; Route 2: Handforth to Manchester south.

For sign/signal detection, drivers pressed the return key as soon as they could see a signboard, without needing to read its details. For a colour light signal, this means that the participants do not need to tell whether it is a double yellow or a single yellow. Detection was tested under three speeds (100, 200 and 300km/h). Driver detection and recognition of signals/signs were tested in separate trial sessions.

*Procedure and Participants*

On their arrival, the drivers were briefed about the purpose of the study and what they would be required to do during the trials. They were then asked to complete a participant record and consent form. The tests were conducted individually, one driver at a time.

Two types of participant training exercises were given. First, 30 colour printed copies of signs/signals were randomly presented to each driver who was asked to recognise/describe each item. This was to familiarise the drivers with the names and features of each sign/signal and to enable the experimenter to understand some particular words a driver might use when naming a certain type of sign. The second part of the training was performed on the VR simulator using a specially designed training session. At least two VR training sessions (one for detection, one for recognition) were given to each participant and the whole training lasted approximately 10–15 minutes. At the end of the training, the participants were asked to confirm whether they felt ready to start the formal testing.

The driver then started the formal trial sessions with sign/signal recognition or detection at different speeds in a randomised fashion. It took approximately 60 minutes for each participant to complete the trials and the participants took a short break in the middle of the trials with refreshments.

A desktop PC (with 18″ screen) was used to run the VR and data capture. The trials were conducted at different locations in order to get a reasonable representation of the UK driver population.

Fifty-seven participants (55 males, two females) from 12 TOCs took part in the trials. Their average age was 41.5 years (SD=8.27, range=30–63) and their average train driving experience was 14.3 years (SD=10.62, range=0–43).

## Results

### *Signal/Sign Recognition and Detection at Various Speeds*

The trial data of the 57 participants' performance measures were analysed using ANOVA on SPSS. The results show that approach speed had a significant influence on sign/signal recognition and detection (both at $p<0.0001$ level). There were also significant variations between different types of signage with respect to their influence on recognition and detection ($p<0.0001$).

Figures 14.1 to 14.4 show the time remaining (mean values) at the point when signals/signs were recognised or detected corresponding to a particular approach speed (see Table 14.1 for sign/signal IDs). IDs in brackets indicate that the computer-recorded data for that sign/signal were unavailable due to technical reasons, but the driver recognition errors were still recorded by the experimenter. The dotted line represents recognition error rate in reference to the second axis on the right of the chart.

Further analysis using Tukey's HSD test confirmed that there were significant differences between speed levels 100-300km/h with respect to their impact on signage recognition and detection ($p\leq0.0001$), but there was no difference between 300–350km/h. Figure 14.5 shows how speed affects signal recognition and detection.

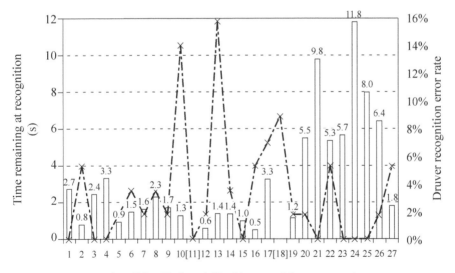

Signal/sign ID (overlaid with recognition error rate)

**Figure 14.1  Recognition of signals/signs at 100km/h (62mph), overlaid with error rate**

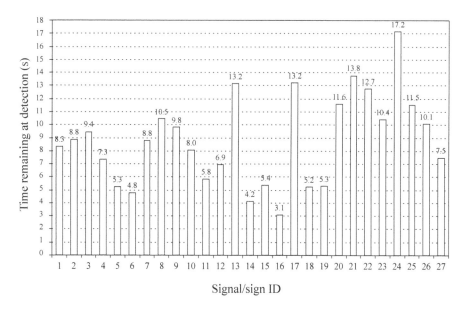

**Figure 14.2   Detection of lineside signals/signs at 100km/h (62mph)**

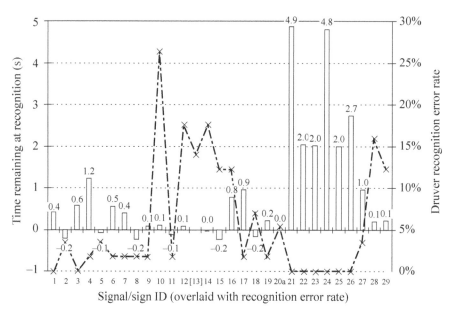

**Figure 14.3   Recognition of signals/signs at 200km/h (124mph), overlaid with error rate**

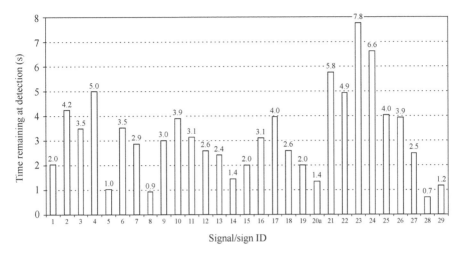

**Figure 14.4  Detection of signals/signs at 200km/h (124mph)**

Statistical analysis (paired-samples t-test) confirmed that detection (dotted lines in Figure 14.5) was made significantly sooner than recognition (solid lines) ($p<0.0001$). This was consistent for both signs and signals across the 100–300km/h speed range.

Multiple comparisons also show that symbolic signs are relatively easier to recognise than numeric signs, and the detectability of most lineside signs is similar regardless of approach speed, except for the radio channel change markers and SPAD indicator which are more difficult to recognise or detect than the remaining signs tested. For colour light signals, their detectability is similar, but the readability of yellow signals is not as good as green or red as double yellow aspects are sometimes mistaken by some drivers for single yellow. Red aspects were recognised/detected significantly sooner than any other aspect.

*Reading Error*

Approach speed had a significant effect on signage reading errors ($p \leq 0.0001$). Further analysis revealed that speed affected recognition error mainly for signs ($p \leq 0.0001$), but not for colour light signals ($p>0.05$); with significantly more sign reading errors occurring as the speed went beyond 200km/h (that is, at or above 250km/h). Speed also affected reading errors of the 'non-colour light signals', including banner repeater, semaphore signals and route indicator ($p \leq 0.01$).

*Mathematical Modelling of Signal/Sign Recognition And Detection*

Regression was performed with the driver performance data to develop relationships between time remaining at signal/sign recognition or detection and approach speed. This produced a 'best line of fit' with a general equation $V = a(b^t)$, where 'V' is approach speed (km/h); 't' is time remaining at the point of recognition or detection

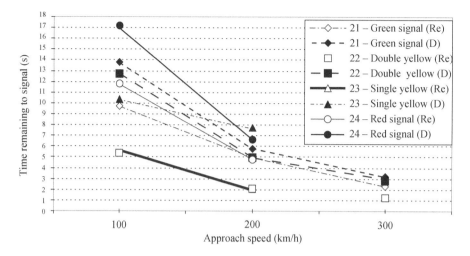

**Figure 14.5  Effect of approach speed on signal recognition (Re) and detection (D)**

(seconds); 'a' and 'b' are constants which can be calculated from the trial data. Table 14.3 summarises these equations for different types of signs and signals.

## Discussion

### *Simulated VR and the Real World Environment*

This study has demonstrated a dynamic relationship between driver responses to lineside signals/signs and train speed. Using the mathematical modelling technique, a critical train speed can be estimated, beyond which the corresponding sign or signal may not be detectable or recognisable within a given time span. However, it should be understood that these results were based on driver performance in a VR environment, in which a signal/sign's visibility is limited by the smallest pixel size possible. An initial calibration effort suggested that, given other conditions are constant, a sign's detectability or readability is between 1.3–3.2 times better in the real world than it is in the VR (depending on the type of sign) (HEL/RT/02785, 2003). Therefore, it can be assumed that if a sign or signal can be detected or recognised at a certain speed with a certain amount of time remaining in VR, it should be detectable or readable in the real world with more time remaining.

It should also be understood that the simulated environment is simpler than what the drivers see and do in the real world, in which, for example, the information can be cluttered, the signs/signals can be covered with dirt or partially blocked by trees, the windscreen of the train may be smudged, or there may be glare from the sun. The drivers also need to perform other tasks in addition to reacting to signals or signs. Therefore, the real speed limit could be lower if signage is to be identified correctly.

**Table 14.3 Mathematical modeling of V-t relationship for recognition and detection**

| Type of signage | Signage ID | Models of 'best line of fit' | Examples |
|---|---|---|---|
| All symbolic signs | No. 1, 2, 3, 4, 5, 6, 7, 17, 18 | *Recognition*: $V_{R\text{-}SS}=226.1(0.83^t)$ | If t=2s, V=156 If t=4s, V=107 |
| | | *Detection*: $V_{D\text{-}SS}=273.1(0.9^t)$ | If t=2s, V=221 If t=4s, V=179 |
| All numeric/text signs | No. 8, 9, 10, 11, 12, 13 | *Recognition*: $V_{R\text{-}NS}=213(0.88^t)$ | If t=2s, V=165 If t=4s, V=128 |
| | | *Detection*: $V_{D\text{-}NS}=263.7(0.92^t)$ | If t=2s, V=223 If t=4s, V=189 |
| All signs tested | No. 1–18 | *Recognition*: $V_{R\text{-}AS}=217.8(0.85^t)$ | If t=2s, V=157 If t=4s, V=114 |
| | | *Detection*: $V_{D\text{-}AS}=256.9(0.92^t)$ | If t=2s, V=217 If t=4s, V=184 |
| All colour light signals* | No. 21–24 | *Recognition*: $V_{R\text{-}Sig}=292.8(0.92^t)$ | If t=4s, V=210 If t=8s, V=150 |
| | | *Detection*: $V_{D\text{-}Sig}=310.7(0.93^t)$ | If t=4s, V=232 If t=8s, V=174 |
| Single yellow signal* | No. 23 | *Recognition*: $V_{R\text{-}SY}=287.5(0.89^t)$ | If t=4s, V=180 If t=8s, V=113 |
| | | *Detection*: $V_{D\text{-}SY}=321.1(0.9^t)$ | If t=4s, V=211 If t=8s, V=138 |
| Red signal* | No. 24 | *Recognition*: $V_{R\text{-}Red}=413.8(0.9^t)$ | If t=4s, V=271 If t=8s, V=178 |
| | | *Detection*: $V_{D\text{-}Red}=453.4(0.9^t)$ | If t=4s, V=297 If t=8s, V=195 |

\* Based on driver responses to signals approached in normal sequence.

*'Time Remaining' Measure and Visual Reaction Time*

In this study, a sign or signal's visibility was represented by a time remaining measure, recorded from the point when the signage was detected or recognised to when it was reached. This measure is different from the conventional concept of visual reaction time (RT) which is defined as 'the interval between the onset of the stimulus and the response under the condition that the subject has been instructed to respond as rapidly as possible' (Teichner, 1954). RT can be represented as RT=PT+MT, where PT is perception time (the time between stimulus onset to the moment of stimulus detection), and MT is motor time (the time needed for the activation of specific muscle groups) (Lupp *et al.*, 1978). Whist MT is assumed to be independent of stimulus variables and is largely age-related (Era *et al.*, 1986), PT has been proved to be affected by stimulus intensity and by parameters of peripheral stimulation such as wavelength, retinal locus, stimulus size and duration (for example, Mansfield, 1973; Tolhurst, 1975).

   The conventional concept of RT is based on assumption that the onset of a stimulus is known to the experimenter, so that the RT can be measured from this point. In the present study, however, the signals/signs were gradually brought into a driver's view and the theoretical onset of these stimuli would depend on many factors such as their design parameters, environmental conditions and driver individual factors. A similar problem has been encountered in studies of driver RTs to road hazards and one of the reported methods to overcome this problem is to 'hide' the visual target (obstacle) behind a hill so that the vehicle position from which the top of the obstacle is just visible over the hillcrest can be fixed (Olson and Sivak, 1986). Using the time remaining measure to represent the detectability/readability of the signals or signs in the present study appears to be a practical approach which is in line with the current railway standards (GE/RT8043; GK/RT0031) in which a signal's (or sign's) sighting quality is specified as the minimum time remaining on approach.

*Detection and Recognition*

Drivers commented during the trials that, in real life, they often do not need to recognise the signs before reaching them and the signs are used as reminders, in combination with their route knowledge. Practically, therefore, it may be reasonable to assume that a sign's functionality can be largely achieved from the point of its detection. However, there are occasions where route knowledge may not work such as driving in poor visibility (for example, in fog or at night) or for new drivers.

*Viewing Object in Motion*

Approach speed has been found to significantly affect driver detection or recognition of lineside signs and signals. One of the possible explanations for this is dynamic visual acuity which is the ability to distinguish moving objects in fine detail and has shown to have the strongest and most consistent relationship with driving performance (Burg, 1967). However, further analysis of Burg's data by Hills and Burg (1977) revealed that only the older drivers (over 54 years) showed a consistent relationship between their dynamic visual acuity and accident rate. In addition, dynamic visual acuity is assessed by moving the visual target across the visual field, but in the present study the visual targets are located within a relatively small visual field and are rapidly brought closer to the driver.

Viewing two or more objects which are moving sideways across the visual field may also create a problem known as 'motion parallax'. One example of this is when several signals come into a driver's view and they appear to be moving across the visual field at different speeds as the train travels along a curved line. This has been a concern for SPADs especially as trains enter or leave stations. Further discussion of this topic is beyond the scope of the present study, but more relevant information can be found in published sources (for example, Adelson and Movshon, 1982; Mestre and Masson, 1997).

Another possible factor is the time available to view the targets. North (1993) estimated that a person can transmit up to 10 'bits' of visually displayed information per second and concluded that it is not the input of the visual system that limits visual performance because the human sensory system has the capacity to transmit millions of bits per second, but the processing, decision making and motor output.

Visual accommodation, the ability of the eyes to focus over a considerable range, may provide another explanation. The adjustment of the eye from one sighting distance to another takes time, which can be considerable even for relatively 'young' people. For example, Krueger (1980) showed that it took about two seconds for a 41 year-old subject to refocus from infinity to a 25cm viewing distance, whereas a 28 year-old person spent only 0.8 seconds. However, Krueger's (1980) study only examined accommodation ability between fixed distances, information is still limited regarding the focus/refocus ability of human eyes with continuously and rapidly changing target distances.

## Conclusion

This study has identified a dynamic relationship between driver detection or recognition of lineside signs/signals and approach speed. The maximum speed limit relating to each type of lineside signs or signals can be estimated, beyond which the corresponding signage may not be detectable or correctly identifiable within a given time limit. However, due to the differences between the VR environment and the real-world, further work is still required to convert the VR findings into their real-world equivalents. Therefore, the results presented in this chapter should be regarded as being indicative rather than conclusive.

The present study also suggested that driver responses to railway signs or signals at a fast approaching speed appear to be complex, which requires not only the capacity of the visual system, but also different levels of information processing and decision-making, as well as psychomotor abilities and skills. This may have implied that, in the design of a certain type of lineside sign or signal to be viewed within a time limit on approach, care should be taken to consider this requirement together with the line speed of the region to ensure that the specification is practicable.

## Acknowledgements

The completion of this work would not have been possible without the help and technical support of many people to whom we owe our grateful thanks. In particular to John Robinson, Bob Muffett, Julie Fowler, Anthony Slamen and Philippa Murphy at the Ergonomics Group of Network Rail. Thank you also to John Wilson for his comments on the work and the manuscripts. Grateful thanks are also due to the Train Operating Companies for making the drivers available.

## References

Adelson, E.H. and Movshon, J.A. (1982), 'Phenomenal Coherence of Moving Visual Patterns', *Nature*, 300: 523–5.
Burg, A. (1967), 'The Relationship between Vision Test Scores and Driving Record: General Findings', Report 62–74, Los Angeles: University of California.
Cope, G.H. (ed.) (1993), *British Railway Track – Design, Construction and Maintenance*, London: The Permanent Way Institution.
Era, P., Jokela, J. and Heikkinen, E. (1986), 'Reaction and Movement Times in Men of Different Ages: A Population Study', *Perceptual and Motor Skills*, 63: 111–30.
HEL/RT/01651/RT1 (2002a), 'Driver Recognition of Lineside Signals and Signs during the Operation of High-speed Trains – Scientific Basis for the Study, Technical Report prepared by G. Li and authorised by W.I. Hamilton for Network Rail.
HEL/RT/01651/RT2 (2002b), 'Driver Recognition of Lineside Signals and Signs at Different Train Speed – Findings of Preliminary Phase-1 Trials', Technical Report prepared by G. Li and authorised by W.I. Hamilton for Network Rail.
HEL/RT/02785 (2003), 'Driver Recognition and Detection of Lineside Signs and Signals at Different Approach Speeds – Findings of Signal/sign Trials in a Virtual Reality Environment', Technical Report prepared by G. Li and authorised by W.I. Hamilton for Network Rail.
Hills, B.L. and Burg, A. (1977), 'A Re-analysis of California Driver Vision Data: General Findings', Report No. LR768, Crowthorne, UK: Transport and Road Research Laboratory.
Krueger, H. (1980), 'Ophthalmological Aspects of Work with display workstation', in A. Mital (ed.), *Advances in Industrial Ergonomics and Safety I*, London: Taylor and Francis: 47–55.
Li, G., Hamilton, W.I. and Clarke, T. (2003), 'Driver Recognition of Railway Signs at Different Speeds – A Preliminary Study', in P.T. McCabe (ed.), *Contemporary Ergonomics*, London: Taylor and Francis: 373–8.
Lupp, U., Hauske, G. and Wolf, W. (1978), 'Different Systems for the Visual Detection of High and Low Spatial Frequencies', *Photographic Science and Engineering*, 22: 80–84.
Mansfield, R.J.W. (1973), 'Latency Functions in Human Vision', *Vision Research*, 13: 2219–34.
Mestre, D.R. and Masson, G.S. (1997), 'Ocular Responses to Motion Parallax Stimuli: The Role of Perceptual and Attentional Factors', *Vision Research*, 37 (12): 1627–41.
North, R.V. (1993), *Work and the Eye*, Oxford: Butterworth Heinemann.
Olson, P.L. and Sivak, M. (1986), 'Perception–Response Time to Unexpected Roadway Hazards', *Human Factors*, 28 (1): 91–6.

Railway Group Standard (1999), 'Train Driving', GO/RT3251, Issue Three, December, London: Railtrack plc.

Railway Group Standard (2000), 'Lineside Signal Spacing', GK/RT0034, Issue Four, December, London: Railtrack plc.

Railway Group Standard (2001), 'Signal Positioning and Visibility', GK/RT0037, Issue Four, October, London: Railtrack plc.

Railway Group Standard (2002a), 'Railway Signs Required for Safety', GE/RT8043, Issue One, June, London: Railtrack plc.

Railway Group Standard (2002b), 'Lineside Signals and Indicator', GK/RT0031, Issue Four, February, London: Railtrack plc.

Railway Group Standard (2002c), 'Provision of Lineside Signals', GK/RT0032, Issue Two, February, London: Railtrack plc.

Teichner, W.H. (1954), Recent Studies of Simple Reaction Time', *Psychological Bulletin*, 51: 128–49.

Tolhurst, D.J. (1975), 'Reaction Time in the Detection of Gratings by Human Observers: A Probalistic Mechanism', *Vision Research*, 15: 1143–9.

# Chapter 15

# Signal Sighting – Development of a Framework for Managing Conflicting Requirements

Nicola Stapley, Tidi Wisawayodhin and Ann Mills

## Introduction

Signal sighting is controlled by Railway Group Standards (principally GK/RT0037), but in practice the activity still relies heavily on professional judgement on the part of the Signal Sighting Committee (SSC) members, as standards cannot cater for the contextual uniqueness of particular signal arrangements. It is necessary to consider a number of factors in relation to the location, operational needs and mandatory requirements for any given signal. These factors can often lead to conflicting recommendations that must be resolved to achieve the optimum arrangement for the positioning of the signal. The solution to the requirement for resolving conflicts in signal sighting lies in the development of an appropriate and explicit model of performance. The driver's task in signal reading is a dynamic behavioural process that has to balance competing demands for the driver's attention and take account of knowledge and experience. Performance of the signal reading task is prone to failure in a number of predictable ways through behavioural mechanisms that are susceptible to the influence of identifiable external hazards. This chapter reports on the development of a framework of guidance to assist signal sighting committees resolve conflicts to achieve the optimum signal arrangement. The work was undertaken through an iterative consultation process with subject matter experts in order to capture their insight and experience within the framework and gain user acceptance of the framework's development. The framework can be used to identify the order and strength of influence of sighting factors, enabling them to be prioritised in terms of their relative importance for successful signal reading. Signal sighting professionals can then be trained in the application of this framework and on how to use it to define the optimum arrangement for any signal. This work served to identify best practice human factors signal sighting principles and their conflicting requirements. Signal sighting performance shaping factors and mitigation measures to resolve the conflicts between principles were also identified.

**Aim**

The aim of the project was to develop a framework to aid signal sighting practitioners resolve the conflicting signal sighting requirements, in order to achieve an optimum signal arrangement. The key objectives for development of the framework required the following:

- identification of principles for good signal sighting based on evidence;
- investigation into the relative effectiveness of the principles;
- prioritisation of the signal sighting principles;
- identification of performance shaping factors (PSFs);
- identification of potential conflicting principles and solutions to the conflicts.

**Method**

In order to develop a usable and systematic tool for signal sighting practitioners, it was first necessary to identify the components that formed the signal sighting task, including the knowledge and expertise required and the effect of any given signal arrangement on the driving task.

*Subject Matter Expert Consultation*

The programme of work required continuous consultation with subject matter experts (SMEs). Workshops and visits to stakeholders took place over a six month period. In addition to the workshops, a representative from Human Engineering attended five Zonal SPADRAM meetings and four signal sighting committee meetings. A SPAD incident investigation was also attended.

*Signal Sighting Principles*

A list of good practice signal sighting principles was developed following an extensive literature review that considered the issues related to signal detection and driver responses to signals. The literature review included relevant Railway Group Standards, scientific research, rail-related studies, other transport-related studies and international studies relating to signal reading.

*Evidence of Effectiveness*

SME opinion regarding the use and effectiveness of each signal sighting principle was sought in order to support the scientific evidence identified during the literature review, and confirm the relevance of each principle identified.

Historical evidence regarding the effectiveness of the application of the signal sighting principles was explored through the use of *post hoc* data available within the industry. SMIS (safety management information system), all multiple SPAD CD-ROMS since the year 2000 and Zonal SPAD databases obtained from SPADRAM meetings were reviewed to source specific evidence for the application of the principles.

*Prioritisation of Principles*

In order to understand the effect of particular signal arrangements on the driver signal reading task, the signal sighting principles were classified according to the impact each had on the information processing stage within the driver's signal reading task. A representation of the stages involved in the driver's signal reading task is presented in Figure 15.1.

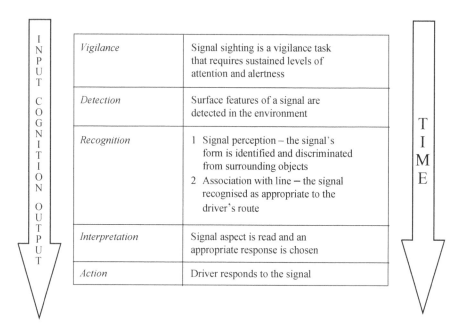

**Figure 15.1    Driver information processing stages during the signal reading task**

*Source*: Human Engineering, 2002a.

   The model provides a comprehensive representation of the driver information processing stages required throughout the signal reading task and whilst it cannot be said that this description of the signal-reading task is universal and unvarying, it does represent all the stages of information processing involved. It can therefore be considered an appropriate engineering approximation of the stages involved in the task and thus identifies the limitations and potential sources of failure.

   The stages involved in the driver signal-reading task are sequential, as shown in the model in Figure 15.1. The earlier cognitive processing stages must be completed first in order to move to the next stage, that is, a signal must first be detected in order to be perceived, associated with the correct line and then interpreted. Therefore, the allocation of each principle to its information processing stage also provided a means of prioritisation between principles, as it is obvious that if the driver cannot see the signal then other design parameters will be of limited or no value. Once it is established that the signal is visible, consideration moves on to perceptual influences that determine whether the signal has a recognisable form and can be associated with the correct line.

The model was used to ensure that the relevant information processing stage was allocated to each signal sighting principle. In order to further prioritise between principles within each stage, SMEs were asked to rate each principle on a scale of 1 to 5, from desirable to essential.

*Performance Shaping Factors*

A list of performance shaping factors (PSFs) that could affect the application and value of the principles was developed. The list comprised features that could be present at the signal and its location.

*Functional Allocation of Principles*

In order to enhance the usability of the framework, a range of signal types and functions was identified. Principles were allocated to each signal type and function as appropriate.

**Results**

*Data Collation*

Through consultation it was established that the term 'signal sighting principles' was already used by other signal sighting guidance documentation. In order to distinguish between the existing material and the framework under development it was decided that an alternative label – human factors (HF) signal sighting principles, be used.

Data from the literature review and SME consultation were collated before developing the tool for managing conflicting signal sighting requirements and the following factors were identified:

- 45 HF signal sighting principles;
- driver performance impact, for example, source of distraction;
- reason for inclusion of the HF signal sighting principle;
- performance shaping factors that impact on the application of the HF signal sighting principle;
- suggested mitigation measure for dealing with a particular PSF in association with the HF signal sighting principle;
- conditions of use for certain mitigation measures;
- validation of the mitigation measure;
- conflicting HF signal sighting principles;
- reason for conflict, that is, where the application of an HF signal sighting principle together with a PSF conflicted with another principle, or where the application of

an HF signal sighting principle with a PSF resulted in a mitigation measure that conflicted with another HF signal sighting principle;
- priority of the conflicting HF signal sighting principles;,
- reason for priority;
- suggested solution.

*Human Factors Signal Sighting Principles*

The initial list of signal sighting principles was rationalised. Any PSFs within the principles were removed, which resulted in a final list of 45 good practice statements regarding signal sighting.

*Evidence of Effectiveness*

SME opinion supported the use and effectiveness, and confirmed the relevance, of each HF signal sighting principle identified. However, the databases reviewed did not contain any specific records on the effectiveness of the application of SPAD mitigation measures.

*Prioritisation of Principles*

The information processing stages in the driver signal reading task provided a means of prioritising principles by processing stages. Each principle was allocated to the earliest possible information processing stage in the driver signal reading task. It was acknowledged that some principles could be allocated to more than one processing stage. SME rating of principles within stages provided a further level of prioritisation.

*Mode of Application*

SME opinions, obtained through the consultation and interview process, suggested that not all signal sighting principles were relevant for all circumstances. Some principles were considered essential for the sighting of newly commissioned signals, but were of limited use for existing signals. Similarly, a number of principles were fundamental to the consideration of existing signals, but not relevant to the sighting of newly commissioned signals.

   For example, driver duties whilst driving are a vital consideration for the signal sighting process following a SPAD. It is considered that drivers' duties such as NRN (national railway network) changeover, or arming ATP (automatic train protection), could potentially be a primary contributor to a signal being passed at danger. These duties, however, would not be taken into consideration for the sighting of newly commissioned signals.

   In addition, it is essential to consider the positioning of signals in relation to section gaps, OLE neutral sections, or converging junctions during the commissioning stage. However, these are of limited value to the sighting of existing signals, as the relocation of signals in complex areas could have a knock-on effect on other signals within the area.

It was therefore decided to structure the framework to continue the signal sighting response theme documented within the SPAD Hazard Checklist (Human Engineering, 2002), that is, proactive for new signal design and reactive for existing signal design.

It was agreed that the framework should be developed for the reactive mode of application, as a greater amount of information was readily available for existing signal design. It was also agreed that further investigation should be undertaken into the new signal design process before the proactive framework was developed.

*Performance Shaping Factors*

The PSFs were categorised into one of eight groups. A graphical representation is provided in Figure 15.2 below to show the key components of the HF signal sighting framework and their interrelationships, which is applicable to both the proactive and reactive modes of application.

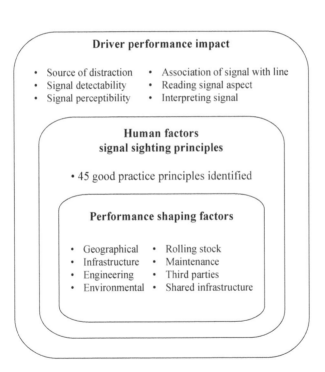

**Figure 15.2   Human factors signal sighting framework key components**

The existence of a PSF could directly affect the application, or value, of a principle for the signal sighting task. Therefore all PSFs at each signal location should be identified in order to determine whether the application of the principles could be affected. Mitigation measures could be applied to compensate for some PSFs, but their existence could still devalue the principle.

There were no direct conflicts between principles, but the existence of a PSF could result in a conflict between two principles. Similarly, the application of a mitigation measure could introduce a further PSF, which in turn could create a conflict between principles.

*Example of a Conflict*

An SSC may be faced with the situation where a group of adjacent signals cannot be clearly associated with their lines. One mitigation measure may be to introduce vertical staggering[1] of the adjacent signals to assist the driver in identifying the relevant signal with his/her line at an earlier stage of approach. The application of two HF principles, numbers 15 and 25, are affected by these PSFs (parallel lines and adjustment to signal height). The following paragraphs provide further explanation regarding the types of conflict that signal sighting practitioners encounter.

HF principle 15 states, 'Signal height along particular stretches of route should be consistent'. HF principle 25 states, 'Signals should be easily distinguishable from adjacent signals of the same group'. The signal sighting practitioner would have to decide whether to introduce vertical staggering, which would result in the signals being more distinguishable, but would introduce an inconsistent signal height along the route.

HF signal sighting principle 15 was allocated to an earlier stage in the driver's signal reading task (signal detectability) than principle 25 (association of signal with line). Therefore, HF principle 15 was allocated a higher priority in terms of its application than principle 25. However, data collected through the series of workshops identified that it would be more important in this example to apply 25 before 15, as breaking principle 25 would create a higher risk of a SPAD. Therefore, the signal height should be adjusted to improve visibility, provided that the signal height is below 5.1m, as misreading an adjacent signal represents a higher risk of a potential SPAD than an adjustment to the signal height.

PSFs can therefore devalue a HF principle by making it necessary to overrule the priority originally allocated to the principle in favour of a lower priority principle.

*Functional Allocation of Principles*

Two signal types and six signal functions were identified (Figure 15.3) and the properties of each principle were analysed to determine whether it was relevant to each combination of signal type and function. It should be noted that the level crossing indicator is applicable only to colour light signals.

In order to fulfil the aim of the work, the data were translated into a paper-based framework that guides the user through its use in order to identify and manage any conflicting signal sighting requirements and achieve an acceptable signal arrangement.

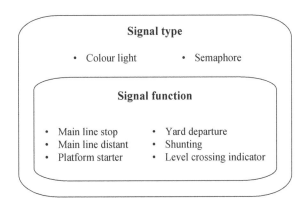

**Figure 15.3   Signal type and function categories**

## Framework Development and Structure

A user-friendly framework was developed in order that signal sighting practitioners would be able to identify the elements (PSFs) of any signal and its location, in a systematic and consistent manner, and be guided to the likely conflicts that might occur, together with their appropriate mitigation measures.

The paper-based framework for managing conflicting requirements was developed as a seven stage process. The structure is by mode of application, followed by signal type and function. A separate framework was developed for each signal type and function for the reactive mode of application.

The following paragraphs describe the contents of each section of the framework, that is, one section represents one signal type and function by its mode of application, for example, an existing, colour light shunting signal.

## Framework Contents

The contents of each section of the framework comprise the following documents, which have been tailored to the relevant signal type and function:

- list of relevant HF signal sighting principles – for reference purposes;
- header sheet – detailing the signal type, function and location;
- PSF questions – to prompt the signal sighting practitioner to consider which PSFs exist at the site that have an impact on the application of the HF signal sighting principles;
- framework checklist – to identify the relevant signal sighting principles that may be affected by PSFs at the site;
- principle number tick list – a basic aid to rationalise any duplication and provide a list of relevant principles in numerical order;
- mitigations and solutions – to provide solutions for unique combinations of PSFs and principles specific to the signal type and function.

**Framework Method of Use**

*Step 1 – List of Relevant Principles*

A list of the relevant principles is provided for reference purposes and contains details of the HF signal sighting principle number, its allocation to the information processing stage of the driver signal reading task, and the reason for inclusion of the principle. The list of principles is provided to give a general overview of those that are relevant to the signal type and function and offers the signal sighting practitioner an opportunity for familiarisation.

*Step 2 – Header Sheet*

A header sheet should be completed for each signal to be sighted, which details:

- signal number;
- location;
- signal box (or control point);
- line;
- direction;
- prepared by name and date;
- reviewed by, for example, signal sighting chairman name and date.

The framework checklist, principle number tick list, and mitigations and solutions pages constitute the main body of the framework and should be completed for each signal.

*Step 3 – PSF Questions*

The list of questions provided is to prompt the signal sighting practitioner to consider the PSFs that may be present at the site. Some questions may appear obvious, but others seek to encourage the signal sighting practitioner to think beyond what is obvious at the site, for example, the question related to a building being a PSF asks, 'If there are buildings on approach to the signal, is there any part of, or any future extensions to, the building that could obscure the signal?'

   The questions are presented in alphabetical order of PSFs. The signal sighting practitioner should indicate in each Yes/No box where a PSF exists, or will exist, at the signal location. When completed, the signal sighting practitioner should proceed to the framework checklist.

*Step 4 – Framework Checklist*

The framework checklist is also presented in alphabetical order by PSF and, when completed, will identify the HF signal sighting principles that may be affected by the presence of PSFs.

   The user should navigate to each framework checklist section containing the PSFs that were identified in the PSF questions and tick the Yes/No box for the appropriate

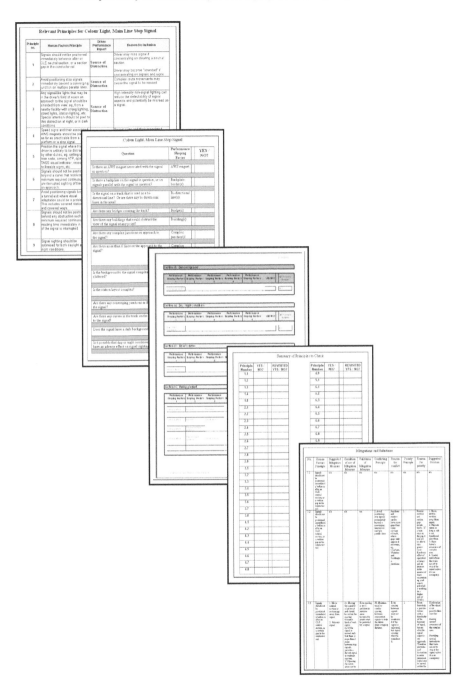

**Figure 15.4 Framework method of use**

combination of PSFs. The framework checklist has been designed to eliminate the need to look up unnecessary HF signal sighting principles.

*Step 5 –Principle Number Tick List*

The principle number tick list should be completed by ticking the relevant HF signal sighting principle numbers highlighted in the framework checklist. When complete, the principle number tick list can be used to direct the user to the appropriate HF signal sighting principles in the mitigations and solutions section.

*Step 6 – Mitigations and Solutions*

The mitigations and solutions are presented in numerical order according to the HF signal sighting principle number, which also represents the order of priority for application of the principles. Each line in the table applies to a unique combination of the HF signal sighting principle and PSFs. Where appropriate, conflicting principles are presented together with the reason for the conflict, the desired priority of application of the conflicting HF signal sighting principles, the reason for priority between conflicting principles and suggested solutions for managing the conflict.

The combination of one or more PSFs with an HF signal sighting principle may not result in a conflict with another principle, but may require the introduction of a mitigation measure in order for successful application of the principle. Mitigation measures to resolve the difficulties in application of the principles with certain PSFs are also presented in this section of the framework.

*Step 7 – Revisit*

Revisiting the framework is a vital part of the process. When the mitigation measures and solutions have been selected, it is necessary to go through the framework once again, as the application of mitigation measures may introduce additional PSFs. The combination of PSFs may therefore have changed from those originally identified.

Steps 3–6 should be repeated, to capture any new PSFs and their effect on the application of HF signal sighting principles, until no further PSFs are identified.

**Conclusions**

The aim of the project was achieved through development of the human factors signal sighting framework. The iterative development process adopted to generate and refine the framework has ensured that a wide range of stakeholders have been consulted and have contributed to its structure and content.

The framework enables signal sighting practitioners to approach all signal sighting work in a consistent manner, irrespective of the signal type and function. The framework will guide the user through the necessary stages, to identify the relevant performance shaping factors, any potential conflicts, and appropriate mitigation measures.

The interaction between signal sighting requirements, driver performance, performance shaping factors and the merits of different mitigation strategies have all been taken into account during the development of the framework. This information has been translated into a framework structure, according to signal type and function, and has created a systematic tool for signal sighting practitioners that will provide a more focussed approach to the effective positioning and arrangement of signals.


## The Way Forward

The framework should be tested through an iterative process of its use and feedback from users. The acceptability test programme should include measurements of the framework's usability, reliability and validity. The results from the test programme would allow the framework to be refined and user acceptance could be assured to maximise the value of this work to the railway community.

The framework forms the most comprehensive source of information regarding solutions to signal sighting arrangements and mitigation measures currently available to signal sighting practitioners. A study is currently being undertaken by Network Rail to integrate the various support tools available to the signal sighting practitioners which will include this framework.


## Note

1   Research in a VR environment has shown that staggering does not work in all circumstances (Human Engineering Limited, 2002b)


## References

Human Engineering Limited (2002a), 'Development of a Human Factors SPAD Hazard Checklis', Technical Report for Railtrack plc (part of the Network Rail Group), HEL/RT/02719/RT1, Issue 01.

Human Engineering Limited (2002b), 'Evaluation of the Signal Design Scheme for the Paddington Throat in a Virtual Reality Environment: Test Phase 2 of Human Factors Support To The Fibre Optic Signal Assessment', Technical Report for Great Western Zone, Railtrack plc (part of the Network Rail Group). HEL/RGWZ/01633/RT3, Issue 3.

Multiple SPAD Summary Briefing (2003), CD-ROM, April, Issue 13, London: Network Rail.

Network Rail Company Standard Specification (2003), 'Signal Sighting', RT/ES10157, draft issue, approximate issue date August.

Railway Group Standard (1996), 'Lineside Signs', GK/RT0033, Issue Three, July, London: Railtrack plc.

Railway Group Standard (1999), 'Train Driving', GO/RT3251, Issue Three, December, London: Railtrack plc.

Railway Group Standard (2000a), 'Lineside Signal Spacing', GK/RT0034, Issue Four, December, London: Railtrack plc.

Railway Group Standard (2000b), 'Signing of Permissible Speeds and Speed Restrictions', GK/RT0038, Issue 2, December, London: Railtrack plc.

Railway Group Standard (2001a), 'Lineside Signals and Indicators', GK/RT0031, Issue Four, February, London: Railtrack plc.

Railway Group Standard (2001b), 'Signal Position and Visibility', GK/RT0037, Issue Four, October, London: Railtrack plc.

Railway Group Standard (2002a), 'Provision of Lineside Signals', GK/RT0032, Issue Two, February, London: Railtrack plc.

Railway Group Standard (2002b), 'Signals Passed at Danger', GO/RT3252, Issue Four, September, London: Railtrack plc.

Reason, J. (1990), *Human Error*, Cambridge: Cambridge University Press.

Chapter 16

# Ergonomics Relating to the Migration of Lineside Signals to ETCS L2 Cab Signals

D.P. Rookmaaker, D.W. de Bruijn, C.E. Weeda, M.P.H.J. Koedijk and
J.L. Zwartenkot

## Background

This study focuses on the transition (migration) phase from existing lineside signalling systems (where the aspect of trackside signals provides safety-determinant information) to a cab signalling system under an ETCS (European Train Central System) Level 2 regime (where the ETCS display in the driver's cab provides safety-determinant information). During the transition period of several years, both protection systems may co-exist for some time (dual signalling period).

Concerns that this situation could be confusing to drivers led to INTERGO being requested to develop and test a specially designed signal aspect with extra attention value.

The aim is to minimise the likelihood of an erroneous reaction when extinguished lineside signals are observed. With active ETCS cab signalling, ETCS signals are designed to be displayed in close proximity to an extinguished lineside signal.

INTERGO have developed five signal variants. All have been subjected to laboratory examination to determine the extent to which they can be distinguished from a signal extinguished as a result of a failure and to what extent the ETCS signal itself can be distinguished from regular signal aspects and colours (red, yellow and green).

In a tachistoscope experiment (100msec exposure time), these variants were displayed in a dual-task situation and unfavourable viewing circumstances in day and night-time conditions. Thirty-four test subjects took part in the experiment. All were experienced drivers from the Dutch railways. Scores were given for correct observation of both an ETCS signal variant and an extinguished signal. Following the experiment, the test subjects were asked to give their subjective view of the variants under study. The results show that in the laboratory situation approximately 75 per cent of all signals were correctly observed. Statistically, an equivalent number of mistakes were made in the day and night-time situation. The ETCS subvariant H9 with a vertical row of white LEDs gave the best objective and subjective results. Its white line appears to embody a unique code leaving little room for error. The track between Utrecht and Amsterdam is currently being equipped with ETCS variant H9 signals.

## Introduction

The European Train Control System (ETCS) is gradually being introduced on the European railways. Under ETCS Level 2, lineside signals are to disappear and speed and safety information is to be displayed on a screen in the driver's cab (ETCS cab signalling). This study addresses the dual mode migration period with some trains running in ETCS Level 2 mode on certain routes while other trains still receive their speed and safety information in the 'traditional' manner on the same route through lineside signals.

When ETCS cab signalling is functional, all information that is relevant to the driver relating to permitted running speed is displayed on a screen inside the cab (Vorderegger *et al.*, 1998). Outside information via signals and signs etc. is not relevant to the driver. Lineside signals will therefore be extinguished. During 'traditional' running (that is, with cab signalling switched off) on the other hand, the driver has to act entirely as directed by the signs and signals placed along the track.

The lineside system of light signal aspects used in The Netherlands is extremely simple and is based on the following principles: red means stop, green means go, yellow means reduce speed.

The dual mode situation will prevail for a number of years. In this situation it will be important to avoid confusing drivers as a result of the unfamiliar conditions. Particularly since a signal extinguished because the ETCS cab signalling is switched on is identical in presentation to a defective/failed signal amongst normally functioning lineside signals. In the latter case, the driver must stop immediately; in the first case the extinguished lineside signal is of no importance to the driver. Confusion between these two situations is to be avoided by showing the driver a specially designed signal aspect with extra attention value.

Intergo was given the following remit:

1   develop proposals for variants of signal aspects to prevent confusion between intentionally extinguished signals (cab signalling mode) and unintentionally extinguished signals (failure);
2   test these proposals in order to make recommendations concerning the best lineside signal for the purpose of preventing confusion.

### Variants of Lineside Signal Aspects

*Premises and Considerations*

The following premises and considerations were applied to develop the special dual mode signal aspect:

1   the functionality of variants is in principle independent of visibility conditions (light/dark; good or bad visibility);
2   the message of the variants is always the same, that is, 'lineside signals not functioning, cab signalling determinant'. The extra signal aspect information

supplied in these conditions must make this clear to the driver at a glance at high speed;

3   the function of the variants consists in providing redundancy and avoiding (high-risk) cognitive confusion;

4   the added information always provides status and not speed information.

5   the light signal design was taken into account as far as possible during the work to develop variants;

6   existing Dutch railway signal aspects were taken into account when choosing the proposed variants. Variants that could lead to cognitive confusion with existing signal aspects, and thereby represented a high risk, were consistently avoided.

## Variants Studied

A large number of variants were developed in cooperation with the client. The variants could be classed according to additional colours, the addition of shapes or the addition of extra signs to the existing signal aspects (Turatto *et al.*, 2000).

Agreement was finally reached and a shortlist of signal aspect variants drawn up that were then subjected to further testing. The coding used is not logical in every case.

The five variants for high signals (H) given below were included in the ensuing study (Figure 16.1):

1.   variant H4, for which a fourth signal light emitting a blue light is mounted above the existing lights. This light is completely integrated into the background;

2.   variant H6, for which white LEDs are used to produce the image of a single cross;

3.   variant H7, for which the indication 'CAB SIGN' is introduced in a round white sign above the base signal;

4.   new variant H8, for which a white sign 'end of all prohibitions' is introduced in a round sign to be added above the signal;

5.   new variant H9, for which a white vertical line of white LEDs is shown adjacent to the extinguished signal lights.

Finally – H1 – the completely extinguished signal.

## Tachistoscopic Simulation Study

*Introduction*

Objective of the extra signal aspect variant to be selected:

•   it must be possible to distinguish between a signal extinguished as a result of ETCS cab signalling and a signal that has been completely extinguished erroneously. In addition it must be possible to distinguish the signal from the normal signal aspects red, yellow and green.

*Rail Human Factors*

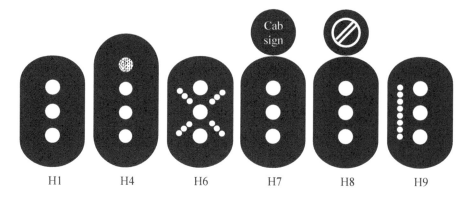

**Figure 16.1   The variants tested, including the zero variant H1**

It was decided to apply a test in a simulated environment for this study (Hughes *et al.*, 1986). The five signal aspect variants examined were presented in the study in conditions comparable from the point of view of perception psychology to unfavourable conditions in real life. These conditions are:

a) gaze not directed straight at signal.

Inside the eye (the retina) we distinguish between a central part (fovea) and a peripheral part (around the fovea). The fovea is a small central part of the retina with which an object can be regarded in detail with maximum attention.

Human visual perception is capable of recognising simple patterns and colours on the periphery of the eye/retina, the 'corners of the eye' or 'parafovea' (Fiorentini, 1989). This allows people to read a paper, for example, whilst walking through a busy street.

In driving practice the driver may need to be able to perceive objects on the periphery of the eye in the following conditions:

- the driver is concentrating on a hazardous situation on the line. The signal may then be perceived 'out of the corner of the eye';
- the driver is running on a four-track section and sees signals on the adjacent line. Seeing an extinguished signal (interpreted by the driver as 'failed'), even on a secondary line, may startle the driver into reacting. For this reason, the added signalling must provide immediate cognitive reinforcement in this kind of situation;

b) signal is visible for a very brief period only.

In driving practice the driver may be confronted with a signal suddenly appearing out of the mist. The driver is expecting a signal, but may at that moment be looking at the track. The time available for perception will clearly be short.

The signal aspect must therefore have such (simple) characteristics that their visual perception can be processed at a glance;

c) driver's attention diverted because of extra task.

The central part of the retina (the fovea) is directed at the point of focal attention. Studies have shown that the more difficult the perceptive task in this central part, the less peripheral perception functions. This fact is comparable in driving practice to a driver whose attention and eyes are directed principally at an acutely hazardous situation on the line, and who pays attention to the signals with 'half an eye' only.

The signals are therefore presented on the periphery as a secondary task with a primary foveal task consisting in the perception (and reading) of a cipher. The degree of difficulty of the primary task is varied by the number of symbols shown. The greater the number, the greater the uncertainty about the answer and the more difficult the task. A study (Williams, 1985) revealed that a choice of six letters was considerably more difficult than two letters. In this study the ciphers 0 through 5 can appear in the image.

In summary, the clarity (that is, visibility and distinguishability) of the signal aspects under study was tested by presenting them in unfavourable conditions:

- on the periphery of the field of vision;
- visible for a very short period;
- while the driver's attention is on reading a cipher in the central part of the retina.

*Object of the Study*

*Stimuli/signal aspects*

The main task involved the ciphers 0 through 5 shown simultaneously with the signal. The cipher was presented in the middle of the field of vision: the test subject directed his gaze to it by looking at a dark or light point. The signal aspects were presented peripherally, with a retinal eccentricity of approximately 10 degrees. This corresponded more or less to real conditions in that the images were presented from the right in the field of vision, since trains in The Netherlands in principle run on the right-hand track on double-track lines.

The screen was 38 by 28.5cm (1024 × 768 pixels). The signal on the screen was approximately 4cm[1] high. The ciphers on the screen were 0.5cm high. The distance between cipher and signal on the screen was 11.5cm. The viewing distance was 60cm.

The images were presented tachistoscopically for 100msec. The driver looked at a fixed point in the middle of the screen obeying the directions of the person in charge of testing. The driver himself was in control of producing the flash, which took place shortly after the space bar on the computer was pressed.

The experimental signal aspect was alternated with normal signal aspects (red, yellow, green) and twice by a completely extinguished signal (H1). Each signal was shown in the same place and each time a cipher was also shown. The order of presentation was random. The signals were presented in a day series and a night series, with a light or dark background.

Per signal aspect variant (H4, 6, 7, 8, 9), the following signal aspects were presented:

- 2 × red signal (shown in the experimental signal design, for example with sign above signal);
- 2 × yellow signal;
- 2 × green signal;
- 2 × 'ETCS' signal aspect variant H4, 6, 7, 8, 9);
- 2 × signal completely extinguished.

A total of ten images for each signal aspect variant were shown in sequence and the ten images of the next variant then followed without interruption. Day and night series were separated by a short pause and were identical in terms of their order of sequence. The set of ten images for a variant began each time with a normal signal aspect. An ETCS signal aspect variant or a completely extinguished image was therefore never presented at the start.

### Procedure

Each driver was asked to give a number of details. This was followed by an explanation of the objective and set-up of the experiment. A check for the right viewing distance from the screen was also made.

Prior to the test itself, each test subject was familiarised with all the ETCS signal aspect variants separately and foveally (= straight in front) in a flash presented via the screen. They were asked to describe what they saw. This gave a first impression of the capacity to recognise a hitherto unfamiliar signal aspect rapidly and whether at first sight it resembled an existing signal aspect.

The test subject was then allowed to look at all the signal aspect variants on paper at his leisure and to ask questions, if necessary. The drivers were told about the signal aspects they could expect (that is, red/yellow/green/ETCS variant/extinguished). Each test subject tested all five ETCS signal aspect variants. Each variant (H4, H6, H7, H8, H9) was presented tachistoscopically at random a number of times. Following this familiarisation, a few images were displayed to practise applying the procedure. When everything was clear, the experiment was started. Immediately following each presentation, the test subject read the cipher he had seen out loud. The test subject also identified the signal aspect shown. The experiment was performed in a meeting room without daylight openings. The night-time part was performed separately after a short break due to having to work in the dark. The light in the test room was extinguished.

After the test, the driver could state which signal variant he preferred and what his reasons were. The test session lasted around half an hour in all.

### Test subjects

Thirty-two test subjects took part in the study: 31 men and one woman, all were qualified drivers and all came from the region of Utrecht. Participation was on a voluntary basis. The test subjects varied in age between 27 and 57 (average: 41.3).

*Materials and means*

'Flash' images were presented via a Macintosh G3 computer (300 Mhz). The Sony screen in front of the driver was a 20 inch type Trinitron Multiscan 20SEII. The resolution set was $1024 \times 768$ pixels with an image refreshing frequency of 75 Hz. In the 100 msec interval used this enabled a good image to be produced on the screen.

The graphic card used was part of the monitor drive provided standard with the computer. The person in charge of testing had his own monitor with a separate graphic card.

The test subject started the reproduction of the flash image by pressing the space bar on the keyboard. The person in charge of testing used a mouse to control the experiment and to record the data. The recordings were made on a specially designed electronic form.

*Results*

*Data analysis*

All 32 test subjects answered ten series of ten items each, that is, a total of 100 items, bringing the complete number of reactions scored to 3200. The principal consideration was for correct and incorrect identifications. In total 75 per cent of the items presented were correctly identified. Then the errors/mistaken identifications were analysed and categorised. Only error types relevant to this study were taken into closer consideration.

Relevant types of error were defined as follows:

* failure to identify an ETCS signal aspect variant;
* failure to differentiate between two variants;
* failure to differentiate between an ETCS signal aspect variant and a completely extinguished signal;
* failure to differentiate between an ETCS signal aspect variant and a signal showing red, yellow or green;
* failure to identify a completely extinguished signal;
* failure to differentiate between an extinguished signal and a signal showing red, yellow or green.

The data analysis focused on two categories of errors:

Category 1: errors related to an incorrect reply when shown a specific ETCS signal aspect variant (H4, 6, 7, 8 of 9);
Category 2: errors related to an incorrect reply when shown an extinguished signal (H1).

*Errors in Category 1*

First of all, the mean correct scores for the different signal aspect variants were calculated (Table 16.1).

**Table 16.1  Mean scores for various signal aspect variants**

| Signal aspect variant | Mean |
| --- | --- |
| H4 | .66 |
| H6 | .91 |
| H7 | .84 |
| H8 | .87 |
| H9 | .97 |

Subsequently the data were used to perform a variant analysis. The hypothesis that the signal aspect variants did not differ from one another as to the number of cases of mistaken identification was rejected ($F(4, 635) = 15.2$, $p<.01$). A Bonferroni test was performed in order to establish which mean group values differed from one another. This showed that the mean score for variant H4 differed significantly from all the other variants (sign. $\leq.05$). The mean for variant H7 differed significantly from the mean for H9. The mean values for the remaining variants did not differ significantly from one another. No general effect from the day and night situation could be established ($F(1, 638) = .303$, $p = .58$). Neither could a significant interaction effect from day and night conditions be found ($F(5, 634) = .33$, $p = .56$).

*Errors in Category 2*

The mean correct scores for extinguished signals in experimental conditions were calculated (Table 16.2).

**Table 16.2  Mean scores for extinguished signals in experimental conditions**

| Variant | Mean |
| --- | --- |
| Extinguished under H4 | .73 |
| Extinguished under H6 | .73 |
| Extinguished under H7 | .74 |
| Extinguished under H8 | .73 |
| Extinguished under H9 | .77 |

These differences are not significant, ($F(4, 635)= .122$, $p = .97$), indicating that the extinguished signals were recognised in all the experimental conditions to approximately the same degree. No day and night effect was found ($F(1, 638) = .008$, $p = .92$) to exist and no significant interaction effect between the day and night situations was found ($F(5, 634 ) = .099$, $p = .99$).

*Errors in Categories 1 and 2 combined*

A variant analysis was also performed on the combined scores from both categories.

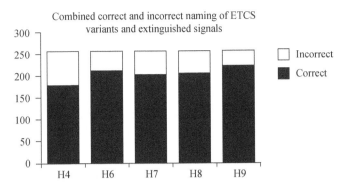

**Figure 16.2  Identification of ETCS signal aspect variant and extinguished signals**

This analysis also revealed that the hypothesis by which the experimental cases of ETCS signals did not differ from one another as to the number of mistaken cases of identification $(F(9,1270) = 7.93$, p<.01) could be rejected. A Bonferroni test again showed a significant difference between variant H4 and the other experimental conditions and the extinguished signals (sign. ≤.05). Variant H7 differed significantly from H9 and revealed a negative connection to all the other conditions, indicating that this variant had a non-significant lower mean correct score compared to H6 and H8. Variant H9 scored significantly $(p ≤ .05)$ well. No general effect from the day and night situation was established $(F(1,1278) = .049$, p = .82) and no significant interaction effect from the day and night conditions was found $(F(10, 1269) = .05$, p = .82).

*Subjective assessment*

When processing the subjective data, the observations made by all 32 test subjects were considered. In all, reactions were recorded on three occasions (immediately at the beginning of the session, after the variants had been presented on paper and on completion of the study).

The following reactions to each signal aspect variant were provided: preference/ clarity/confusing associations/lack of information content/too much similarity with existing signal aspects with a completely different significance. Suggestions were also made for other variants and for modifications to the presented variants. The reactions revealed a trend towards a positive assessment of signal aspect variants H7 and H9 and a negative assessment of variants H4 and H8; H6 was given both positive and negative assessments, seemingly arbitrarily.

**Discussion**

Two types of perception were crucial in the light of the remit:

- the correct perception of an ETCS signal aspect variant;
- the correct perception of an extinguished signal.

Approximately 75 per cent of all signal aspects were correctly perceived. Most errors were made with variant H4, the least errors with variant H9. Variants H7 and H8, both with a separate sign above the signal, scored equal. Variant H6, the design of which most resembles H9, has less errors than H7 and H8. For the perception of extinguished signals, the difference between the diverse variants is much less pronounced. The least errors were made with H9 (non-significant). Statistically, the same number of errors is made in the day situation as in the night situation.

H4 can be eliminated. H4 is characterised by a blue light. The drivers frequently mistook the blue light for a white one. This was evidenced both by the errors made and by the assessment afterwards. From the point of view of perception physiology, this can be explained by the fact that blue is a colour that is difficult for the eye to recognise. The blue H4 light was also often mistaken for a green light, and was assessed by many (26 drivers) afterwards as being difficult to distinguish. Another explanation for the number of errors with H4 may be that it is more difficult to determine which colour amongst the four colours of the four lights was seen than with three colours and three lights. In short, the poorer results obtained with H4 in the study would appear to have an ergonomic explanation that tallies with the subjective assessments.

Variants H7 and H8 come second to last in terms of correct perception. They differ considerably from H4, though not in a significant way. However, variant H7 is significantly worse in terms of errors than H9. H8 comes immediately after H7, in third position. With both variants, the drivers consider the most important disadvantage to be that a combination of several signs is not easily read. However, H7 is the one receiving the largest number of positive assessments from the drivers. Ergonomic aspects are that the sign is the most explicit, but the complex letter shape is less rapidly recognised than the white lines for instance. It should be noted in this connection that H7 was relatively often confused with H8, while the converse was not common.

H6 has second place in terms of correct perception. Relatively often it was said that the similarity with an existing (although flashing) white 'X' for goods trains in The Netherlands was confusing. It is indeed feasible that the cross could be mistaken in fog for a signal applying exclusively to goods trains. However, the significance of the cross is obvious and it is an easily recognisable symbol.

H9 comes first, with the least number of errors. The white line is seen as a unique code that does not elicit any associations with existing signals. H9 is clearest, even compared to H6, specifically where confusion in identifying the ETCS signal aspect variant against extinguished signals is concerned.

## Conclusions and Recommendations

It is recommended not to choose H4. The largest number of errors were made with this because the blue colour could not be readily seen. It is also recommended not to choose H7 or H8 because of the number of errors made, although H7 was often given a positive assessment by the test subjects.

In terms of perceptual errors, H9 is the best variant, receiving numerous votes of preference, whilst confusion with other signals is unlikely. In terms of errors, H6 is a good alternative, but it was judged by some drivers to cause confusion (similarity to the white flashing 'X' for goods trains). It is recommended to use H9. H6 could be an alternative.

The client has in fact chosen Variant H9, requesting information on the precise construction of the vertical white LED line (intensity of LEDs, positioning of the white line *vis-à-vis* signal lights, number of LEDs, etc.). Signal Aspect Variant H9 is still a provisional design and has been installed on the Utrecht – Amsterdam line, the first line in The Netherlands to be equipped with dual signalling.

## Note

1   H4: 43 mm; H6 and H9: 36 mm; H7 and H8: 50 mm.

## References

Fiorentini, A. (1989), 'Differences between Fovea and Parafovea in Visual Search Processes', *Vision Research*, 29 (9): 1153–64.

Turatto, M. and Giovanni, G. (2000), 'Color, Form and Luminance Capture Attention in Visual Search', *Vision Research*, 40: 1639–43.

Hughes, P.K. and Cole, B.L. (1986), 'Can the Conspicuity of Objects be Predicted from Laboratory Experiments?', *Ergonomics*, 29: 1097–111.

Vorderegger, J.R., Verhoef, L.W.M. and Rookmaaker, D.P. (1998), 'Een taak-gebaseerde interface voor Europese treinbestuurders' ('A Task-based Interface for European Train Drivers'), *Tijdschrift voor Ergonomie*, 1: 11–17.

Williams, L.J. (1985), 'Tunnel Vision Induced by a Foveal Load Manipulation', *Human Factors*, 27 (2): 221–7.

Chapter 17

# A Human Factors SPAD Checklist

Emma Lowe and Claire Turner

## Background

That there are human factors issues connected with the occurrence of SPADs (Signals Passed at Danger) is well known, but the understanding of these issues in the rail industry is sometimes limited and often based only on subjective opinion. An objective and consistent approach to the investigation of human factors causes of SPADs was therefore necessary to create a generic performance-based understanding of human factors, eliminate the need for subjective opinion, and pinpoint the root human factors causes of SPADs. The development of the Human Factors SPAD Hazard Checklist, outlined in this chapter, was intended to address this need. A model of driver behaviour and error, based on established cognitive theory, and a validated database of SPAD hazards, form the basis of this work. The final checklist has been developed for ease of use both proactively, by signal engineers and sighting committees, and reactively, to investigate the root causes of a SPAD incident and to help to prevent their future occurrence.

## Introduction

The general trend in the occurrence of SPADs (Signals Passed at Danger) has decreased over the last year. For example, 24 SPADs were reported in the HSE's 'SPADs Report' for September 2002, which is six fewer than in September 2001, and 19 fewer than the average September figures for the last six years (HSE, 2002). However, the understanding of the influence of human factors on SPAD occurrence remains limited. It has historically been limited to driver issues such as competency, shift patterns, and fatigue, and is often based only on subjective opinion. But as Lord Cullen pointed out in his inquiry into the Ladbroke Grove crash of 1999, focusing solely on driver error fails to identify root causes such as poor infrastructure and management failures.

A rational, objective and coherent approach to the investigation of human factors causes of SPADs was therefore needed to create a generic performance-based understanding of human factors, eliminate the need for subjective opinion, and pinpoint the root human factors causes of SPADs. The Human Factors SPAD Hazard Checklist addresses this need.

## Approach

The checklist originates from two main sources of information:

- a cognitive model of the driver task (cognitive task analysis);
- a database of human factors SPAD hazards.

### Development of the Cognitive Model

In order to reduce SPADs, it is important to understand driver performance. Through an examination of exactly what is involved in the driving task and of how drivers organise and process information, it becomes possible to identify where mistakes can be made. Cognitive task analysis (CTA) was the method used to achieve this. Cognitive task analysis allows the train driving task to be analysed from the point of view of the driver, and aspects of the task that place heavy demands on the driver's cognitive resources – their attention, perception and so on – to be identified. By focusing upon the parts of the signal reading task that drivers will find hardest to learn and perform, it becomes possible to see where errors are most likely to be made. The ultimate aim was to use the cognitive task analysis approach either to predict SPAD risk for a given layout or to identify significant human factors within a SPAD event.

The way the driver processes information was modelled in three stages using a combination of established ergonomics techniques:

- task analysis of signal reading – a step by step analysis of the driver task;
- recognise–act cycle modelling – to uncover driver information processing;
- GEMS modelling of human error – to identify the types of errors that drivers could make and their consequences.

### Recognise–Act Cycle Modelling

Information processing can be thought of as a series of repetitions of the *recognise–act cycle*. In each cycle, a driver perceives features (for example, signals) in the surrounding environment, and compares them with rules and procedures stored in memory in order to decide upon the appropriate action (for example, the signal is at single yellow, therefore anticipate that the next signal will be at danger).

By combining knowledge of the driving task from a task analysis with the recognise–act cycle, it is possible to produce a more detailed representation of the signal-reading task in terms of the information processing involved. This is illustrated in Figure 17.1.

### Modelling Human Error

The more skilled we are in performing a task, the more automatic it becomes, but the very nature of this effort-saving cognitive design leaves us prone to making errors. These can occur at any stage of information processing. The GEMS (generic error-modelling system) model developed by Reason (1990) defines three levels of

| *Vigilance* | Signal sighting is a vigilance task that requires sustained levels of attention and alertness |
|---|---|
| *Signal detection* | Surface features of a signal are detected in the environment |
| *Signal recognition* | 1   Signal perception – the signal's form is identified and discriminated from surrounding objects<br>2   Association with line – signal is recognised as appropriate to the driver's route |
| *Signal interpretation* | Signal aspect is read and an appropriate response is chosen |
| *Action* | Driver responds to the signal |

**Figure 17.1   Illustration of driver information processing during the signal-reading task**

error, which have been identified in human performance and error analysis in various industries:

- skill-based: slips and lapses – usually errors of inattention or misplaced attention, when task demands outweigh or outrun cognitive capacity;
- rule-based: mistakes – usually a result of picking an inappropriate rule or procedure, caused by a misjudgment of current circumstances or by making incorrect assumptions based on previous experience;
- knowledge-based: mistakes – due to incomplete/inaccurate understanding of system, or overconfidence.

Our cognitive resources are limited, which explains why skill-based slips and lapses are the most common forms of human error as these occur due to failures in attention and memory. Rule-based mistakes are the second most common form of failure. Because train drivers are highly trained operators performing a very procedural task, knowledge-based mistakes in train driving are rare.

*Developing the SPAD Hazard Database*

Driver performance is influenced by a complex environment that contains a number of 'performance shaping factors' (PSFs). These factors can cause disruption at each stage of information processing, which may cause the types of errors described above. An example is passenger distraction: a passenger could disrupt driver vigilance resulting in the driver failing to detect an upcoming signal. This would be classed as a skill-based error – the driver's attentional capacity was exceeded, resources could not be split between the driving task and the distraction, and a lapse in awareness of lineside signals occurred. Possible factors that could influence driver performance were identified as a result of a literature review (of HMRI and RSSB documents, academic research projects, etc) and a series of site visits to various signals. A total of 52 signals across the seven Network Rail regions were assessed, with SPAD histories ranging from 0 to 17. A combination of cab rides and track walks enabled digital video footage of each signal, its approach and surrounding area to be obtained. Interviews with drivers were also undertaken to ratify our understanding of the driver task.

Three validation criteria were developed to enable consistent qualification and categorisation of human factors SPAD hazards. In order to be included in the final database, a hazard needed to be clearly defined, its effect on human performance and behaviour well understood, and also have evidence (whether real-world or academic) to back it up.

Using these three criteria as a benchmark, 69 PSFs were validated as human factors SPAD hazards. For convenience, these were grouped into factors that better define their effects. These can be related back to the aspect of driver behaviour they impact. The database includes not only infrastructure factors, such as gradients and curved approaches, but also the effects of driver performance, environmental conditions, operational procedures, and train cab design. Some examples are provided in Table 17.1.

**Table 17.1 Examples of hazard factors**

| Stage of driver task | Hazard factor | Example |
|---|---|---|
| *Vigilance* | A  Personal factors | Fatigue |
|  | B  Sources of distraction | Complex track layout |
| *Detection* | C  Visibility factors | Foliage |
| *Recognition* | D  Perceptibility factors | Dark or complex signal background |
|  | E  Association with line | Irregular signal spacing |
| *Interpretation* | F  Reading aspect | Expectation |
|  | G  Interpretation | Flashing yellow aspect sequence |
| *Action* | H  Action/performance failures | Poor train cab ergonomics |

Qualitative and quantitative analyses of the site visit data were carried out. This revealed a significant relationship between the number of hazards present at a signal site and number of SPADs at that signal.

## Outline of the Human Factors SPAD Hazard Checklist

In order to bring the cognitive model and database of hazards together in a practical, usable format, a checklist was developed for use by signal sighting engineers and committees, and SPAD investigators (HEL/RT/02719/RT2, 2002). The checklist follows a simple logic that is rooted in the sequential stages of driver information processing. It is designed to address the human factors SPAD hazards that may be encountered by the driver.

It is assumed that a number of external influences, for example train, infrastructure or signaller failures, are ruled out in any investigation prior to using the checklist. Any SPAD without authorisation is considered to be a violation of the driver rulebook, so the starting point for the checklist is 'Was the violation deliberate?' This allows a distinction to be made between abnormal or criminal behaviour and excusable violations.

Eight outline questions direct the user to the relevant checklist questions. These are summarised in Table 17.2.

Questions within each checklist part relate to the presence or absence of hazards and have been worded to enable simple yes/no answers. This eliminates the need for subjective opinion and promotes a precise, performance-based understanding of SPAD hazards. Help is provided for each checklist question, which explains the argument behind the hazard (that is, the precise description of the hazard and examples of where it could occur), evidence for the hazard, and what it is necessary to do in order to answer the question. Using the help as a guide, the user is required to work his/her way through the checklist, answering yes to any relevant question.

*Determining the Level and Source of SPAD Risk*

Once the checklist has been completed, SPAD risk at a particular signal location can be easily calculated. Each 'yes' answer to a question is assigned one penalty point. These are summed to derive a total for each part of the checklist, and an overall total for the checklist. Very simply, the higher the overall number of resulting penalty points, the higher the SPAD risk at a particular signal. In this way, the signals that require the most urgent attention can be assessed and the investment in SPAD risk reduction measures prioritised. This approach has been designed to complement other risk assessment tools that exist for assisting with decisions about investment and risk reduction.

*Identifying Risk Reduction Measures*

Risk reduction measures have been suggested for the majority of hazards, intended to prevent, control or mitigate the effect of a SPAD. The measures proposed are based on evidence or experience (a risk reduction measure implemented at a multi-SPAD signal that has not since been SPADed, for example). Alternatively, new measures have been suggested based on good ergonomic principles.

The totals for each part of the checklist direct the user to the most appropriate risk reduction measures. A reference book of possible risk reduction measures is provided. Numbering is consistent throughout for ease of use, so, for example, by referring back to the relevant queries in the visibility part of the checklist, the associated prevention, control or mitigation measures can be identified.

**Conclusion**

The Human Factors SPAD Hazard Checklist brings together an understanding of the driver task and human error mechanisms. Based on evidence and subject to a

**Table 17.2 Checklist questions**

| | Checklist part | Outline question | Example checklist question |
|---|---|---|---|
| A | **Personal factors** | Is there evidence to suggest that personal factors are involved? | Had the driver worked successive night shifts in the week prior to the SPAD? |
| B | **Sources of distraction** | Is there evidence to suggest that the driver was inattentive? | Is the signal positioned soon after an OLE neutral section? |
| C | **Visibility factors** | Is there evidence to suggest that the driver could not see the signal? | Is the signal obscured by station structures or furniture? |
| D | **Perceptibility factors** | Is there evidence to suggest that the driver could not perceive it as a signal? | Is the signal viewed against a dark or complex background, for example, buildings, foliage, etc? |
| E | **Association with line** | Is there evidence to suggest that the signal could not be associated with the correct route? | Does the number of lines visible to the driver (including sidings) differ from the number of signals visible? |
| F | **Reading aspect** | Is there evidence to suggest that the driver could not read the signal aspect correctly? | Is the signal normally (that is, >75% of the time) encountered at a proceed aspect? |
| G | **Interpretation factors** | Is there evidence to suggest that the driver could not interpret the signal aspect correctly? | Are flashing aspects used on the approach to the signal? |
| H | **Action/performance failures** | Is there evidence that the driver could not perform the correct action? | Did poor cab ergonomics cause a delay in driver action? |

rigorous hazard qualification procedure, it provides a coherent structure for analysing the human factors causes of SPADs.

The traditionally driver-centric view of human factors has been expanded to cover the infrastructure, train design, environmental and operational factors that are likely to be the underlying causes of SPAD incidents. However, in order that all possible human error causes of SPADs are identified as part of the investigation process, further work is required on the role of signaller error in SPADs.

A major benefit of the checklist is that it allows signals to be assessed both reactively, in support of a SPAD inquiry, and proactively, by signal engineers and sighting committees to evaluate the SPAD risk at a particular location. The scoring method provides a rational and objective basis for allocating resources to address those signal locations that are the most at risk of a SPAD incident.

*Further Validation and Future Development of the Checklist*

As with all new methods, the best way to ensure that the checklist works is to test it in the field. Use of the checklist is being supported with training of signal sighting professionals and SPAD investigators. A period of time will be allowed for the tool to be used and then it will be subject to a review to determine how it has been applied and whether the checklist modified to increase its usability and scope.

A structured three-step process for validating human factors SPAD hazards has been specified. This not only ensures that hazard identification is consistent and open to verification, but also can be adopted as a means of capturing and processing any further SPAD hazards that may arise when the checklist is applied in the field. The information generated as a result of the validation process will also be useful in deciding upon appropriate risk reduction measures.

## References

Reason, J. (1990), *Human Error*, Cambridge: Cambridge University Press.

Health and Safety Executive (2001a), *The Ladbroke Grove Rail Inquiry Report* (The Cullen Report), London: HSE.

Health and Safety Executive (2001b), 'SPADs (Signals Passed At Danger)', Report for September 2001, London: HSE.

Health and Safety Executive (2002), 'SPADs (Signals Passed At Danger)', Report for September 2002, London: HSE.

HEL/RT/02719/RT2 (2002), Human Factors SPAD Hazard Checklist 'User Guide', Report prepared by C. Turner and authorised by W.I. Hamilton for Railtrack plc.

# Chapter 18

# ORAM: A Structured Method for Integrating Human Factors into SPAD Risk Assessment

Paul Hollywell

## Background

This paper describes a method that was commissioned by Network Rail and used on major projects to meet the safety risk assessment requirements associated with trains overrunning signals. These requirements, contained in Railway Group Standards, state that safety risk assessments for designs of new and modified infrastructure need to be undertaken to prevent and mitigate SPADs. The method is called ORAM (Overrun Risk Assessment Method) and uses structured expert judgement within a workshop setting. The workshop involves representatives from various disciplines: signalling, operations, train driving, safety risk assessment and human factors.

This chapter explains how ORAM was developed to support the end users in making structured expert judgements on the human factors issues associated with SPADs. End users' responses to the method are also described.

## Introduction

In recent years there have been a number of high-profile railway accidents involving train drivers passing signals at danger and colliding with another train. Southall on 19 September 1997 and Ladbroke Grove on 5 October 1999 are the most well known accidents involving this type of human error. A train overrunning a signal is commonly known as a Signal Passed at Danger or SPAD.

There are several hundred SPADs on the UK rail network every year. However many of these do not have the potential to cause harm because they are caused by train drivers making minor misjudgements of distance or braking capability, or they occur at low speed during shunting movements. In many cases trains stop within the safety overlap provided at the signal. The overlap is the track's safety margin at the signal that protects against drivers misjudging their braking. The most serious SPADs are those when trains run past the overlap and there is a possibility of a collision with another train.

In 2000 a Railway Group Standard 'Prevention and Mitigation of Overruns – Risk Assessment' (RGS, 2000a) was introduced. Its aim was to ensure that all changes to the rail network are assessed to minimise the risk from SPADs. The Standard defines the requirements for risk assessment of the design and operational use of track and signalling so as to control the risks associated with trains exceeding the limit of their movement authority. Movement authority applies to both line-side signals and in-cab signalling. A Railway Group Guidance Note (RGS, 2000b) was also introduced to support the Standard.

## Background to ORAM

Remodelling work has been taking place as part of the West Coast Route Modernisation (WCRM) Programme to improve capacity and journey times. Conventional Resignalling of the West Coast Main Line (WCML) is also required as part of the WCRM programme. The Conventional Resignalling will provide control of trains up to line speeds of 200km per hour and will employ predominantly four aspect colour light signalling. As part of the design for the new and modified signalled layouts associated with the above works, risk assessments were required to demonstrate that the risks due to accidents such as collision and derailment from SPADs were as low as reasonably practicable (ALARP).

In 2000 there was no generally accepted method that could support the WCRM Project in undertaking SPAD risk assessments. The WCRM Programme comprised eight signalling scheme sections and it was necessary for a total of over 740 signals to be assessed prior to signalling designs being finalised. This was a large task that was required to be completed quickly and efficiently against a challenging timescale.

Mott MacDonald were asked to develop a suitable method that would risk assess signalling options for resignalling work for WCRM. The method needed to comply with the requirements of the new Standard, identifying those signals for which a simple risk assessment is sufficient and those signals that require a more detailed risk assessment. A detailed risk assessment needed to be done in a structured way considering each element of SPAD risk. The method also needed to be easily integrated into the WCRM signalling scheme design process, which comprised a number of design development workshops.

This paper describes the multifaceted nature of SPADs and the method developed by Mott MacDonald for WCRM to meet the relevant risk assessment requirements. The method, based upon structured expert judgement, is called the Overrun Risk Assessment Method (ORAM).

## SPAD Risk Assessment

The nature of SPAD risk assessment is that it is multifaceted and involves knowledge from a range of disciplines: signalling, operations, train driving, risk assessment and human factors. Therefore any method developed to assess the risk from SPADs needs to be (a) systematic and comprehensive so as to provide confidence in the method, and (b) understandable and easy to use so as to support those who will use the method.

SPAD risk comprises a number of elements:

A    frequency of signal being encountered at danger;
B    probability of driver committing a SPAD;
C    probability of overrun protection and mitigation failing to keep train within safety overlap;
D    probability of area of conflict (that is, beyond safety overlap) occupied;
E    consequence of train entering occupied area of conflict.

It is only when all these elements are combined within a risk assessment can a true estimate of SPAD risk be assessed that is, an estimate that does not omit or double count factors. It should be appreciated that many of these elements involve human factors issues.

Figure 18.1 shows the above elements drawn as an event tree, with the sequence of events leading to a serious SPAD presented in a graphical form. Each element will now be discussed in more detail.

*A    Frequency of Signal Being Encountered at Danger*

The frequency with which trains approach a particular signal at danger will affect the risk. Simply, the more often trains approach a signal at danger, the more opportunities there are for drivers to commit a SPAD at that signal. There is, however, an argument that says if drivers are used to seeing a particular signal at danger, the more prepared they are to respond to that signal appropriately. The probability of a driver committing a SPAD due to expectation is accounted for in the next element of SPAD risk.

*B    Probability of Driver Committing a SPAD*

The total number of SPADs on the UK rail network in 2002/03 was 402, a 7.5 per cent improvement over the previous year and a 24 per cent improvement on the six-year average. The total number of serious SPADs in 2002/03 was 144, a 15 per cent improvement over the previous year and a 26 per cent improvement on the six-year average (HMRI website, www.hse.gov.uk/railways/spads.htm). Based on the number of signals on the rail network and the number of train movements per annum, the probability of a SPAD occurring at any one signal can be estimated to be, on average, very low (that is, $10^{-5} - 10^{-4}$). This has led to some people not asking the question 'Why do drivers commit so many SPADs?' but rather to ask 'Why do drivers commit so few SPADs?' It is now generally accepted that a driver's unreliability cannot be reduced much further without engineered measures.

The probability of a driver committing a SPAD needs to differentiate between a driver *regarding the signal* (that is, observing the signal) and a driver's response *stopping the train*. Driver failures in either area can lead to a SPAD. The SPAD risk assessment needs to differentiate between these two types of failure because there are major differences in a driver's behaviour in each case and the resultant difference in train speeds.

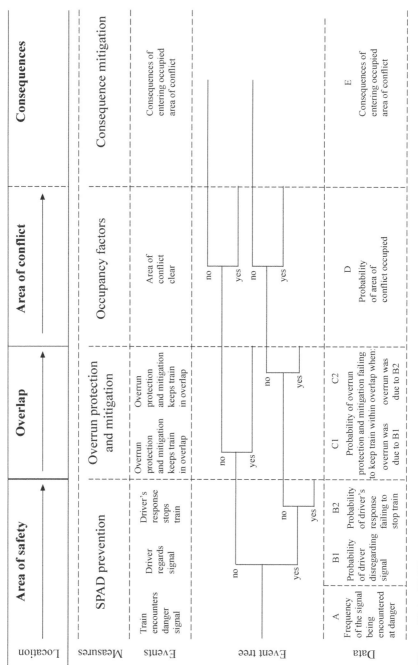

**Figure 18.1   SPAD event tree**

*B1 Probability of driver disregarding a signal*

The causal factors leading to a driver disregarding a signal can be subdivided into those that primarily affect *perception of signal* and those that primarily affect *decision and response selection*.

Engineered and operational aspects of the railway primarily determine the factors that affect a driver's *perception* of a signal. These factors can be summarised using the following keywords and explanatory text:

- *obscured*: overgrown vegetation, overhead line equipment (OLE), bridges, structures, dust or smoke interfering with the driver view of the signal;
- *insufficient time*: viewing time too short for the driver to see the signal properly, possibly due to line curvature, gradient or signal alignment;
- *distractions*: driver distracted by passing through stations, undertaking platform duties, by road traffic, car parks or level crossings in the area;
- *conspicuous*: driver unable to see the signal properly, possibly due to other light sources, sunlight or background lighting in the area;
- *poor contrast*: driver unable to see the signal properly, possibly due to poor contrast against the signal background colour;
- *not expected*: driver does not expect the signal in a particular location, relative to the location of other signals on the same route;
- *environment*: driver cannot see the signal properly, possibly due to cuttings, embankments, bridges, tunnels, buildings or roads.

Engineered aspects of the railway and driver human factors issues primarily determine the factors that affect a driver's *decision and response selection* associated with a perceived signal. These factors can be summarised using the following keywords and explanatory text:

- *anticipation*: driver sees other signals at same location that are visible earlier than this signal. Driver sees other signals visible beyond this signal;
- *delay*: driver could be forgetful due to there being a significant delay between previous signal and this signal, or because of a significant delay between this signal and when action is required to obey it;
- *expectation*: driver could expect the signal to clear on approach to it, as this is normally experienced;
- *confusion*: driver could be confused by other signals or different signal types on adjacent line or same gantry, or by a complex layout. Driver could be confused by other lineside signs or indicators, or by undertaking at the same time other visual and/or aural tasks.

*B2 Probability of driver's response failing to stop a train*

The causal factors leading to a driver failing to stop a train are those factors that primarily affect response execution. The engineered aspects of the railway primarily determine the factors that affect a driver's ability to stop a train at a danger signal. These factors can be summarised using the following keywords and explanatory text:

- *track*: a falling gradient, gradient varying on approach, condition of the railhead, or poor adhesion due to leaves, dust or grease could cause a driver to fail to stop a train at a danger signal;
- *train*: a train with an abnormal braking characteristic or abnormal load could cause a driver to fail to stop a train at a danger signal.

### C Probability of Overrun Protection and Mitigation Failing to Keep Train Within Safety Overlap

The probability of overrun protection and mitigation failing to keep train within the overlap following a SPAD needs to differentiate between two different situations. The first, when a SPAD has occurred because a driver disregarded the signal: the second, when a SPAD has occurred because a driver's response failed to stop the train. The SPAD risk assessment needs to differentiate between these two situations because of the differences in the driver's behaviour and the resultant differences in train speeds.

Engineered aspects of the railway determine whether overrun protection and mitigation keep the train within the overlap following a SPAD. These engineered measures include the following:

- train protection system, such as the Train Protection Warning System (TPWS) or Automatic Train Protection (ATP), ensuring that all trains are fitted with the system;
- train stops;
- SPAD indicators and related measures;
- automatic warning via radio;
- trap points and flank protection;
- additional Automatic Warning Systems (AWS).

The above measures, either by themselves or by alerting the driver quickly enough, will affect the probability of keeping a train within the overlap following a SPAD. It is expected that the network-wide fitment of TPWS will significantly reduce the number of serious SPADs.

### D Probability of Area of Conflict Occupied

The probability of the area of conflict being occupied following a SPAD depends upon a number of inter-related engineered and operational aspects of the railway. These factors include the following:

- resignalled layout;
- number of conflicting moves;
- train speed and train length;
- train timetable.

*E Consequence of Train Entering Occupied Area Of Conflict*

The consequence of a train entering an occupied area of conflict following a SPAD depends upon what is entering the area of conflict and what is occupying the area of conflict. This is primarily based on the types of rolling stock involved and the potential numbers of occupants. A detailed consequence analysis would require consideration of speed of train(s), crashworthiness, type of collision, and whether there is potential for a fire or derailment after the collision.

## ORAM

It can be easily appreciated from the above that a SPAD risk assessment involves knowledge and experience from a range of rail disciplines, including a significant contribution from human factors. A risk assessment method for WCRM needed to satisfy a number of important criteria that are outlined below.

*Method Requirements*

A suitable SPAD risk assessment method is required to have the following characteristics:

- it is systematic, comprehensive, documented and fully auditable;
- it includes contributions from all relevant railway disciplines;
- it is easily understood by all railway disciplines;
- it supports structured expert judgements by all railway disciplines;
- it can be successfully applied providing high uncertainty is not present;
- it applies an appropriate level of risk assessment proportionate to the level of risk;
- it enables signalling design risk assessments for a typical WCRM project site to be achieved within the existing signalling design production process.

*Method Details*

Full details of the SPAD risk assessment method, ORAM (Overrun Risk Assessment Method), are provided in ORAM (2002). The method is undertaken in two main stages.

First, identify all stop signals of sufficiently low risk that do not require the application of any additional risk control measures. These signals do not require further risk assessment and the information is recorded. If a change in design and/or its operational use is proposed following the ORAM assessment, then the signals affected should be reassessed. Second, all stop signals that require the application of additional risk control measures are subject to further risk assessment. A risk assessment workshop is conducted to apply the ORAM method. Each overrun scenario at a particular stop signal will require a separate assessment.

In summary, ORAM is based on the event tree shown in Figure 18.1 and comprises two forms (Figures 18.2 and 18.3). The data for each event in the event tree is given a ranking that is derived using structured expert judgement (SEJ) made by the relevant

| | | SIMPLE RISK ASSESSMENT | | |
|---|---|---|---|---|
| 1 | a | Is the stop signal on plain line with no points or crossings between the stop signal and the next stop signal ahead, meeting all the criteria in clause 4.4(a) of GI/RT7006? | Y ☐ | N ☐ |
| | b | Is the stop signal on a line fitted with a train protection system, meeting all the criteria in clause 4.4(b) of GI/RT7006? | Y ☐ | N ☐ |
| | c | Is the stop signal, preceding caution signal(s), and the associated AWS equipment compliant with the requirements of relevant Railway Group Standards? | Y ☐ | N ☐ |
| | | Are there no significant factors at this signal location (see Table B) that could increase the likelihood of a SPAD? | Y ☐ | N ☐ |
| | | Is there no history of SPADs at this signal location (if existing unmodified signal)? | Y ☐ | N ☐ |
| | | *If the answer to either 1 a OR 1 b is Yes AND all the answers to 1 c are Yes THEN no further assessment is required. Otherwise continue to 2.* | No Further Assessment ☐ | |
| | | **FURTHER RISK ASSESSMENT** | | |
| | | **Characteristics of Signal** | | |
| 2 | | *Which of the following (one or more) describes the signal?* | | |
| | a | Signal protecting a simple diverging move | Y ☐ | N ☐ |
| | b | Signal protecting a conflicting route ahead | Y ☐ | N ☐ |
| | c | Signal at which trains regularly start from rest and protect a conflicting route ahead | Y ☐ | N ☐ |
| | d | Signal protecting junction(s) where there is extended opportunity/time window for conflict | Y ☐ | N ☐ |
| | e | Signal protecting a station platform / level crossing. (Delete as appropriate) | Y ☐ | N ☐ |
| | f | Signal closely following the end of a Permanent Speed Restriction (PSR) | Y ☐ | N ☐ |
| | g | Signal presenting a higher likelihood of SPADs or (if unmodified) has a history of SPADs | Y ☐ | N ☐ |
| | h | None of the above. (Please provide details) | Y ☐ | N ☐ |
| | | **Overrun Scenario Being Considered** | | |
| 3 | | | | |

| 4 | Assessment Summary Table | | | | | | | |
|---|---|---|---|---|---|---|---|---|
| | A | B1 | B2 | C1 | C2 | D | E | |
| | Frequency of the signal being encountered at danger | Probability of driver disregarding signal | Probability of driver's response failing to stop train | Probability of overrun protection and mitigation failing to keep train within overlap (for driver disregarding signal) | Probability of overrun protection and mitigation failing to keep train within overlap (for driver failed response) | Probability of area of conflict occupied | Consequence of entering occupied area of conflict | Qualitative risk ranking (product of five risk rankings) |
| | | | | | | | | |
| | See Table A | See Table B | | See Table C | | See Table D | See Table E | |
| | Total qualitative risk ranking for signal (sum of two risk rankings) | | | | | | | |

**Requirements, Recommendations & Remarks***

* If the workshop decides that a stop signal is not amenable to applying the ORAM, due to the complexity of the layout, an alternative method is to be used. Note that a quantitative risk assessment should always be used in support of the workshop rather than replacing it.

**Assessment Assumptions**

| Name of Attendee: | Representing: | | Telephone Number: |
|---|---|---|---|
| Chairperson: | | | |
| | | | |
| | | | |
| | | | |
| | | | |
| Approval/Acceptance: Recommendations accepted on behalf of Railtrack: | Name: | Signature: | Date: |

| Project and Signal Location: | | Source Record Updated | | | ORAM SHEET No. 1 OF 2 | | | | Railtrack Records Group | RAILTRACK | | |
|---|---|---|---|---|---|---|---|---|---|---|---|---|
| | | Version | Prod. | Check | | | | | West Coast Route Modernisation | Drawing No. | | |
| Scheme Plan and Version Number: | | | | | | | | | Signal Box | Last Full Correlation | Date | |
| | | | | | Produced | | | | Interlocking Area | | Version | |
| Accompanies WCRM ORAM Guidance Document. Ref: RM/CRS/REP/10272, Issue 3, Date 13 May 2002. | | | | | Checked | | | | Overrun Risk Assessment Form | Current Version | | |
| | | | | | Issued | | | | Signal: | | | |

**Figure 18.2 ORAM Sheet 1**

**Table A**

**Rankings for A**

| Frequency Category | Ranking |
|---|---|
| 1/Year | 1 |
| 1/Month | 2 |
| 1/Week | 3 |
| 1/Day | 4 |
| 1/Hour | 5 |

**Table B**

**SPAD Causal Factors**

**Causal factors for 'driver disregarding signal' – perception of signal**

| | | Y | N |
|---|---|---|---|
| Obscured | Vegetation, OLE, bridges, structures, dust, smoke | | |
| Insufficient time | Viewing time too short (line curvature, gradient, signal alignment) | | |
| Distractions | Stations, platform duties, road traffic, car parks, level crossings | | |
| (In)conspicuous | Light levels compared to other light sources, sunlight, background lighting | | |
| Poor contrast | Background colours | | |
| Not expected | Signal not expected in location (relative to other signals on route) | | |
| Environment | Cuttings, embankments, bridges, tunnels, buildings, roads | | |

**Causal factors for 'driver disregarding signal' – decision and response selection**

| | | Y | N |
|---|---|---|---|
| Anticipation | Other signals at same location visible earlier than this signal (reading through) | | |
| | Other signals visible beyond this signal | | |
| Delay | Significant delay between previous signal and this signal | | |
| | Significant delay between this signal and when action required to obey it | | |
| Expectation | Expect signal to clear on approach to signal (consider point of release of signal) | | |
| Confusion | Other signals or signal types – adjacent line, same gantry, complex layout | | |
| | Other signs or tasks – lineside signs, indicators, visual and/or aural tasks | | |

**Causal factors for 'driver's response failing to stop train' – response execution**

| | | Y | N |
|---|---|---|---|
| Track | Gradient – falling, varying on approach | | |
| | Condition – railhead, adhesion (leaf, dust, grease) | | |
| Train | Abnormal braking characteristic | | |
| | Abnormal speed | | |
| | Abnormal load | | |

**SPAD Prevention Measures**

| | | Y | N |
|---|---|---|---|
| a | AWS equipment – standard 183m | | |
| b | RA indicators (if claimed, consider method of operation) | | |
| c | In-cab or platform mounted CCTV display | | |
| d | Alterations to signalling (e.g. banners) | | |
| e | Use of Driver's Reminder Appliance (DRA) | | |
| f | Restrictive interlocking (e.g. additional proving of flank tracks and points/points self normalising/calling points) | | |
| g | Consideration of predominant use of line (lie of points) | | |
| h | Time-tabling to avoid conflicting moves | | |
| i | Countdown markers for approach speeds | | |
| j | Proposed improvement in signal sighting (e.g. repositioning, spacing, sighting distances) | | |
| k | Proposed layout alterations | | |
| l | TPWS (if claimed, ensure all rolling stock is suitably fitted) | | |
| m | Automatic Train Protection (ATP) (if claimed, ensure all rolling stock is suitably fitted) | | |
| n | Others? (Please provide details) | | |

**Rankings for B**

| Probability Category | Ranking |
|---|---|
| Much lower than network average probability | 1 |
| Lower than network average probability | 2 |
| About network average probability | 3 |
| Higher than network average probability | 4 |
| Much higher than network average probability | 5 |

**Table C**

**Overrun Protection and Mitigation**

| | | Y | N |
|---|---|---|---|
| a | SPAD indicators or SPAD magnet(s) and notice boards | | |
| b | Additional AWS | | |
| c | Automatic warning via radio | | |
| d | Train stops | | |
| e | Trap points | | |
| f | Flank protection / diversion onto lower risk route | | |
| g | Point of conflict at great distance from overlap | | |
| h | Proposed layout alterations | | |
| j | TPWS (consider whether just train stop or train stop and speed sensor, if claimed, ensure all rolling stock is suitably fitted) | | |
| | ATP (if claimed, ensure all rolling stock is suitably fitted) | | |
| k | Others? (See Anti-SPAD Toolkit and Railtrack Group Guidance) | | |

**Rankings for C**

| Probability Category | Ranking |
|---|---|
| Very low probability | 1 |
| Low probability | 2 |
| Medium probability | 3 |
| High probability | 4 |
| Very high probability | 5 |

**Table D**

**Rankings for D\***

| Probability Category | Examples (speeds are closing speeds) | Ranking |
|---|---|---|
| Very low probability | Unidirectional with no conflicts | 1 |
| Low probability | Unidirectional with conflicts (converging or crossing) | 2 |
| Medium probability | Bi-directional not in regular use (head-on, all speeds) | 3 |
| High probability | Bi-directional in regular use (head-on, 40 mph or less) | 4 |
| Very high probability | Bi-directional in regular use (more than 40 mph) | 5 |

**Table E**

**Rankings for E\***

| | Occupying area of conflict | | |
|---|---|---|---|
| Entering area of conflict | No train, goods / empty passenger train or authorised person | Partially-occupied passenger train or car, etc. on level crossing | Fully-occupied passenger train |
| Goods/empty passenger train | 1 | 2 | 4 |
| Partially-occupied passenger train | 2 | 3 | 5 |
| Fully-occupied passenger train | 4 | 5 | 6 |

\* Examples shown are for guidance only. Ranking selected should accurately reflect the likelihood/consequence of a collision/derailment occurring in the area of conflict.

Project and Signal
Location:
Scheme Plan and Version Number:
Accompanies WCRM ORAM Guidance Document.
Ref: RM/CRS/REP/10272, Issue 3, Date 13 May 2002.

| Source Record Updated | Version | Prod. | Check |
|---|---|---|---|
| | | | |

**ORAM SHEET No. 2 OF 2**

| | |
|---|---|
| Produced | |
| Checked | |
| Issued | |

| Railtrack Records Group | RAILTRACK |
|---|---|
| West Coast Route Modernisation | Drawing No. |
| Signal Box | Last Full Correlation — Date — Version |
| Interlocking Area | |
| Overrun Risk Assessment Form | Current Version |
| Signal: | |

**Figure 18.3  ORAM Sheet 2**

subject matter experts in the workshop. The ORAM method provides guidance on making these judgements (see ORAM Sheet 2 in Figure 18.3).

The advantage of the workshop environment is that it facilitates a comprehensive and thorough design safety review of the site by all the necessary experts. This is achieved through an examination of all relevant drawings, photographs, other supporting documents, a series of searching questions, and comprehensive selection of mitigation measures. The success of the workshop requires personnel with operational, management, risk, signalling and human factors expertise to be brought together who can provide an accurate view on aspects of the design of the proposed re-signalled layout, led by a competent facilitator.

A qualitative estimate of the overrun risk at a particular signal is made at the risk assessment workshop by selecting ranking for each component of ORAM. The selected rankings for elements A, B, C, D and E are then combined for both types of SPAD to give the workshop the qualitative risk ranking associated with a particular signal. The qualitative risk rankings obtained can be compared across the same signal layout. Thus ORAM is able to give a relative estimate of risk, comparing estimates of risk for signals on the same layout.

Using the total qualitative risk ranking for all the stop signals associated with a particular layout design and/or layout design options enables signal design engineers to:

• obtain a comparative indication of the overall safety of a particular layout design and/or design options;
• compare the relative safety of signals within that layout design and/or design options;
• carry out sensitivity analyses to determine the impact of data uncertainty and/or the effectiveness of different risk reduction measures on the level of risk.

All assumptions made during the risk assessment workshop, whether they are regarded or disregarded, are recorded. Spaces are reserved on ORAM Sheet 1 (Figure 18.2) to record these assumptions and any recommendations resulting from the workshop. Care should be taken with the assumptions made concerning the operational use of a signal in the layout. Future operational changes could alter significantly the overrun risk.

## Conclusion

The ORAM method described in this document meets the criteria for WCRM and the requirements for SPAD risk assessment described in Railway Group Standards. It has also been successfully used on a number of other Network Rail resignalling projects such as Manchester South and East Midlands and has shown to have a number of advantages over other methods of overrun risk assessment. These include:

• a method of differentiating those signals requiring simple risk assessment from those requiring further risk assessment;

- a formal method for applying and recording a qualitative risk assessment that enables a ranking figure to be calculated. this figure can be used for comparison and evaluation purposes;
- a workshop environment where the necessary expertise, documentation and equipment is gathered to enable a thorough risk evaluation to be undertaken;
- a method for evaluating and agreeing appropriate risk mitigation actions to reduce SPAD risk to ALARP. These decisions are formally recorded and agreed on the ORAM form;
- a method that gives an auditable trail to the overrun risk assessment process;
- a method that can be easily updated should any changes to the design and/or its operational use in the future require the overrun risk assessment to be revisited.

Experience in using ORAM has shown it to be a simpler, more efficient and effective method of SPAD risk assessment compared to alternative methods. It successfully integrates human factors into SPAD risk assessment using a simple approach to making structured expert judgements by a group of subject matter experts who do not have a human factors background. Its particular strengths are:

- it is not a 'black-box' approach and workshop members can clearly see the contributions to risk and identify where risk can be reduced;
- it does not overly rely on numbers but emphasises the process for SPAD risk assessment and reduction;
- it uses a series of open-ended questions rather than closed (that is, 'Yes/No') questions. This encourages workshop members to explore all relevant issues associated with SPAD risk assessment, particularly human factors.

**Note**

In April 2003, after a long period of consultation within the rail industry, Network Rail responded to the Railway Group Standard on SPAD risk assessment with a new Company Standard (Network Rail, 2003a). The new standard, supported by documents (Network Rail, 2003b and 2003c), computer-based tools and training courses, is slowly being adopted by the rail industry.

**References**

Network Rail (2003a), 'Prevention and Mitigation of Overruns – Risk Assessment of Signals', Network Rail Company Procedure, RT/E/P/14201, Issue 1, April.
Network Rail (2003b), 'Prevention and Mitigation of Overruns – Risk Assessment of Signals', Network Rail Company Guidance Note, RT/E/G/14202, Issue 1, April.
Network Rail (2003c), 'Prevention and Mitigation of Overruns – Risk Assessment Tool', Network Rail Company Specifications, RT/E/S/14200, Issue 1, April.
Railway Group Standard (RGS) (2000a), 'Prevention and Mitigation of Overruns – Risk Assessment', GI/RT7006, Issue One, December.

Railway Group Standard (RGS) (2000b), 'Prevention and Mitigation of Overruns – Risk Assessment', Railway Group Guidance Note, GI/GN7606, Issue 1, December.

West Coast Route Modernisation Overrun Risk Assessment Method (ORAM) (2002), 'Guidance Document', RM/CRS/REP/10272, Issue 3, 13 May.

# PART 6
# DRIVING – FATIGUE

Chapter 19

# Human Friendly Rosters: Reducing the Risk of Fatigue

Rebecca Ashton and Abigail Fowler

## Background

There is a specific requirement within the 2003/2004 Railway Group Safety Plan for Railway Group members to take steps to minimise the possibility of excessive fatigue amongst their workforce. A poorly designed shift pattern or roster can adversely affect performance and morale, resulting in higher absenteeism and an increased likelihood of errors and accidents.

For over three years, AEA Technology Rail has been actively involved in assisting train operating companies (TOCs) in the generation of optimal driver rosters via the use of our rostering software, IRMA. The package has now been expanded to include the assessment of rosters for compliance with human factors best practice principles in order to minimise adverse fatigue effects. This has been achieved through the incorporation of the Health and Safety Executive's (HSE) Fatigue Index to produce a specific 'fatigue assessment'.

This chapter provides the background to our work and details the way in which the HSE Fatigue Index has been incorporated into our software. Details of the software and examples of its output are provided. The pragmatic approach that we have taken to the assessment of rosters, with respect to human factors issues, is described. Our experiences in using the Fatigue Index, and the key findings that have emerged from our work for TOCs, are discussed in order to provide a snapshot of the current state of the industry. The pros and cons of using the Fatigue Index measurement as a benchmarking tool are also discussed.

Recommendations are made for a 'way forward' for the effective management of driver rosters in order to minimise fatigue and other adverse health and safety effects.

## Introduction: Why Manage Fatigue in the Rail Industry?

Fatigue in humans impairs performance and can lead to reduced attention, poor decision making, delayed reaction time and human errors. All of these types of impaired performance increase the likelihood of an incident or accident occurring. Therefore in a safety critical industry such as the railways, where humans perform key safety roles, it is important to manage fatigue as a source of safety risk.

Much research has been conducted in order to better understand both the causes and consequences of fatigue and the implications for shift work. However, fatigue management remains a difficult area in which to produce definitive best practice guidelines as it is unfortunately not an 'exact science'. Nevertheless, a summary of key fatigue research findings and their applicability to the rail industry is given later in this chapter.

The Railway (Safety Critical Work) Regulations (HSE, 1994) and supporting Approved Code of Practice (AcoP) (HSC, 1996), require Railway Group members, including train operating companies (TOCs), to complete a fatigue risk assessment prior to the implementation of any changes to work patterns. This is intended to minimise the potential for changes to increase fatigue risk amongst safety critical workers (HSE, 1994), which includes train drivers.

The remainder of this chapter concentrates on fatigue management for train drivers. However, the issues raised and the recommendations made could equally be applied to other railway workers.


**What is Fatigue?**

Fatigue causes a loss of efficiency, feelings of weariness, and a disinclination for physical or mental effort (Kroemer and Grandjean, 1997). It adversely affects both our mental and physical performance. Feeling fatigued is not unpleasant if sleep is allowed, but can be distressing if sleep is not an option (for example during shift work). Shift working has been shown to lead to greater fatigue than working fixed hours, particularly if the shift pattern includes night shifts. Much research has been conducted to understand why this should be the case, and how best to manage shift working in order to limit possibly adverse fatigue effects. Some of the key findings from research on fatigue and shift work are briefly described here.

Figure 19.1 summarises the main consequences of fatigue on attention, vigilance, memory and the capacity for effective decision-making. Health can be negatively affected, and reaction time and the likelihood of human error both increase. It has been found that even moderate levels of fatigue produce impaired performance similar to alcohol (Dawson and Reid, 1997).

Figure 19.2 shows the key factors that are known to influence the likelihood of fatigue, although it is not an exhaustive list. While the employer can address 'work influenced factors', the 'other factors' are outside the control of the employer and are dependent on the individual taking a degree of responsibility for their own well-being.

The extent to which *start time of shift* affects fatigue varies throughout the day. For example early shifts are known to cause greater fatigue. Therefore to limit fatigue effects, the general best practice advice is for shifts to start, where possible, after 7 a.m. as it is difficult to fall asleep earlier in the evening to get sufficient sleep before an early shift, and commuting time may compound the early start (Railway Safety, 2003). Research is unclear as to the optimum *length of shift*, however, for the purposes of calculating fatigue risk, as presented in this chapter, any shift over eight hours is regarded as fatiguing. The effect of shift length is exacerbated by the start and end time of the shift; for example, shifts finishing after midnight have been found to cause more fatigue than shifts finishing before midnight (Rogers *et al.*, 1999).

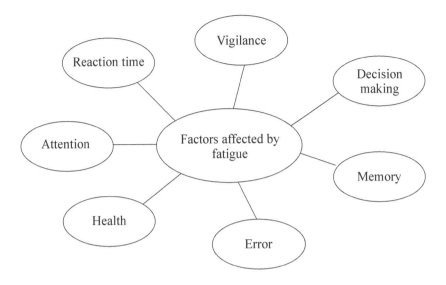

**Figure 19.1  Consequences of fatigue**

| **Work-influenced factors** | |
|---|---|
| Start time of shift | Length of shift |
| Shift pattern | Workload |
| Rest period between shifts | Breaks |
| **Other factors** | |
| Length of sleep prior to shift | Individual factors |

**Figure 19.2  Key causes of fatigue**

Research is unclear as to an optimal *shift pattern*. However, to limit fatigue, the number of consecutive shifts known to be highly fatiguing (for example night shifts) should be limited, as fatigue has a cumulative effect. Also the pattern should incorporate sufficient *rest periods between shifts* for sleep. The length of rest period required depends on the time of day it is taken, the preceding sequence of shifts worked and the next shift to be worked. Staff may have a preference for a particular pattern. However, limiting fatigue may not be the reason for the staff preference. It may be due to the time off or pay they receive, or a particular pattern fits their other commitments. With regard to *workload*, overload can lead to fatigue due to the fatiguing effort (both mental or physical) in performing the task. Conversely, underload can also lead to fatigue if a task provides an insufficient stimulus for the individual to keep his or her attention focused on the task. When calculating fatigue risk in the rail industry, however, it is not possible to include more than an approximate guide level of workload as insufficient information is available on drivers' workload to differentiate the fatiguing effects of different tasks, for example driving different routes.

*Breaks* during a shift help mitigate the effects of fatigue. In the rail industry, train drivers are allowed personal needs breaks (PNBs) during their shift, when the timetable allows, which provide some opportunity for a break from the driving task. However, in fatigue assessment, it is not possible to include a precise factor for the beneficial effects of breaks, as their time and duration cannot be accurately predicted.

The *length of sleep prior to shift* can be affected by a great number of factors. Employers need to ensure the length of rest period between shifts is sufficient, and where possible take into account individual factors for example commuting time. It is then the responsibility of the employee to use the rest period time effectively. However, *individual factors* may result in insufficient sleep, for example commitments outside of work detract from sleeping time, sleep is interrupted (for example when a baby is present at home), or the individual finds it difficult to get to sleep. These individual factors cannot be managed in a roster, however, it may be possible for managers to take these type of factors into consideration when managing individual members of staff and monitor performance or, if possible, change their shifts to reduce the likelihood of fatigue. The employee in return has a responsibility to turn up fit for work and should raise any concerns with their manager or personnel if they have a legitimate reason for not achieving sufficient sleep. Effective fatigue management requires a collaborative approach between employee and employer.

### The Fatigue Index

The Fatigue Index (FI) is a tool for assessing whether intended changes to work patterns will increase the risk of fatigue. The FI concentrates on five key factors all known to have an impact on the level of fatigue, namely:

- shift start time;
- shift duration;
- length of interval between shifts;
- breaks;
- number of consecutive shifts.

The FI is specifically designed for the assessment of rotating shift patterns and is therefore ideal for assessing the rosters of train drivers (N.B. it is not suitable for assessing fixed shift patterns). Following the introduction of the Railway (Safety Critical Work) Regulations (HSE, 1994), the FI was developed by DERA's Centre for Human Sciences, on behalf of the UK Health and Safety Executive (HSE). The developed FI and report (Rogers *et al.*, 1999) drew on research previously conducted on fatigue from a wide range of industries.

The Health and Safety Laboratory (HSL) assisted the HSE in further development of the FI to produce a spreadsheet for its application, which was tested in a range of companies in the rail industry and beyond. Since its issue, the HSE has used the FI to assist the investigation of incidents and accidents. Various fatigue assessment tools, including AEA Technology's IRMA software, have also been developed since then based on the FI.

The FI is a comparison tool that works by attributing a score to a roster for each of the key factors described above. The overall average scores of two shift patterns can then be compared, for example a future shift pattern with the current shift pattern. This comparison provides an initial indication of whether the risk of fatigue is likely to increase if the intended changes of work pattern are implemented. In this way the FI acts as an initial screening tool; if the FI shows a potential increase in fatigue, then further risk assessment should be conducted and measures taken to reduce or mitigate the effects of fatigue.

The FI also provides scores for particular days in a roster so that peaks in fatigue risk can be identified. The FI recommends that scores should not be allowed to exceed 30, and scores above 25 should be carefully considered. These values were calculated following a review of incidents and accidents, which showed that half the incidents occurred when the FI score was greater than 30. The value is provided as an indication only for comparative purposes, and should not be treated as absolute.

If the FI suggests a risk of fatigue, the design of the roster should ideally be modified. However, this may not be possible in practice, if there is a limited availability of drivers for example. However, an understanding of the factors that contribute to fatigue can help employers understand how to manage the risk of fatigue. The identification of drivers at high risk of fatigue can help driver standards managers to decide when to assess drivers. In addition, roster clerks can use the information to support their decision of which driver on a spare shift to assign to which job.

## AEA Technology Rail Fatigue Assessment

IRMA is a flexible, user-friendly computer package designed by roster clerks within the rail industry as a tool to assist in producing base, weekly and daily rosters. It is currently used in ten UK TOCS and other companies within the light rail and bus industries. The system has been developed over the last few years by AEA Technology Rail, incorporating suggestions from users identified through formal user group meetings.

The software is a front end to a database, which stores a variety of information, ranging from sickness records to timetable data. Over the years we have helped organisations to exploit these data in a variety of useful ways. Following customer interest in a fatigue-assessing tool, and through discussions with the HSE, the FI was integrated into the base roster generation area of IRMA.

Base roster data are produced at least every timetable change to correspond to that particular set of services. These rosters include all the work for the timetable. The example in Table 19.1 shows a sample of base roster data as used in IRMA.

The number of rows needed in the base roster is determined by the number of jobs to cover services and therefore the number of drivers required, the local agreements (for example maximum number of hours in a shift, maximum number of work days in a week), and an assessment of how many spare drivers will be needed each day.

**Table 19.1 Sample of type of information used in IRMA to assess fatigue**

| Hours | Sunday On | Sunday Off | Monday On | Monday Off | Tuesday On | Tuesday Off | Wednesday On | Wednesday Off | Thursday On | Thursday Off | Friday On | Friday Off | Saturday On | Saturday Off |
|---|---|---|---|---|---|---|---|---|---|---|---|---|---|---|
| 1 Hugh | 9:29 | 18:53 | RD | RD | RD | RD | 4:56 | 12:01 | 4:56 | 12:01 | SP | SP | 15:02 | 20:57 |
| 2 Pew | SP | SP | 16:35 | 2:45 | SP | SP | RD | RD | RD | RD | 15:02 | 20:57 | 15:02 | 20:57 |
| 3 Barn | 15:02 | 20:57 | 15:02 | 20:57 | 15:02 | 20:57 | 5:27 | 15:53 | 5:27 | 15:53 | RD | RD | RD | RD |
| 4 <vac> | RD | RD | RD | RD | 15:02 | 20:57 | 4:56 | 12:01 | 4:56 | 12:01 | SP | SP | 15:02 | 20:57 |
| 5 Evans | 16:30 | 23:30 | 16:35 | 2:45 | RD | RD | RD | RD | 15:02 | 20:57 | 15:02 | 20:57 | 15:02 | 20:57 |
| 6 Green | RD | RD | RD | RD | 15:02 | 20:57 | 5:27 | 15:53 | RD | RD | RD | RD | 15:00 | 23:51 |
| 7 Pat | 9:29 | 18:53 | SP | SP | RD | RD | 4:56 | 12:01 | 4:56 | 12:01 | SP | SP | SP | SP |
| 8 Smith | RD | RD | RD | RD | 16:25 | 2:45 | SP | SP | 15:02 | 20:57 | 15:02 | 20:57 | RD | RD |
| 9 Jones | 15:02 | 20:57 | 15:02 | 20:57 | RD | RD | RD | RD | SP | SP | 15:41 | 20:41 | 15:00 | 23:51 |

Note: RD = rest day; SP = spare.

The drivers rotate around the roster each week, so that the actual pattern of work for the driver on the first row reads from left to right across the roster. It is the base roster data that is used in IRMA for the fatigue assessment. Clearly this is just a plan, and the introduction of *ad hoc* activities can interfere with the levels of fatigue predicted, and the spare shifts may in reality be at different times.

### Incorporating the Fatigue Index into IRMA

The algorithm from the Excel spreadsheet supplied as the FI tool was extracted and written into a module within the IRMA software. This enables the user to run an automatic fatigue assessment. The software assesses the roster and produces an output and any rows of the roster that score particularly highly are highlighted (see Figure 19.3 for an example output).

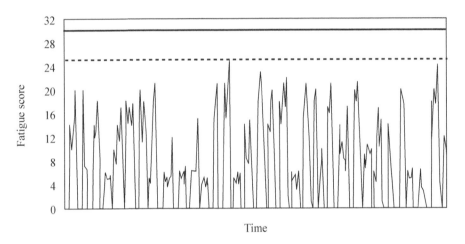

Values above 25 undesirable, above 30 highly desirable

**Figure 19.3  Sample Fatigue Index score report**

Note: One row (row 19) of the roster produced a score of 25.0.

Extensive comparison testing was carried out on the finished tool and on the individual parts of the algorithm, to ensure that the outputs were identical in all circumstances. Many TOC roster databases were used for this testing, along with some of AEA Technology Rail's own test databases. A report of the comparison was produced and supplied to the HSE in order that they were satisfied that the software produced the same output as being the standard HSE FI spreadsheet.

**Assessing Fatigue in Industry**

AEA Technology Rail has used the HSE FI within the IRMA software to conduct a total of 12 fatigue assessments on the rosters of six TOCs and one light rail operator. This has included the analysis of roster data from TOCs who do not use IRMA to produce their rosters. To accomplish this, the TOC's roster data are converted to an IRMA compatible format. A fatigue score is then generated using the software and a report produced explaining the results and providing the TOC with guidance and advice on the actions to be taken to mitigate the fatigue risk.

The graph in Figure 19.3 shows an example graphical output from a fatigue assessment in IRMA. For a ten-row roster there are 70 days of analysis in total (ten rows × seven days in a week). The far left point will be the score for the first Sunday on row 1, the second score is for the Monday of row 1, through to the last point on the far right for the Saturday on row 10 (assuming a Sunday through Saturday roster). The peaks are the risk associated with each day. Along with the graphs, the scores are provided where the fatigue risk exceeds a score of 25 or 30.

Where high scores are identified, AEA Technology Rail look carefully through the base roster to pinpoint the probable reasons for the high score. By changing certain parameters in the roster and reassessing, the particular factors that appear to be causing the high risk of fatigue can be confirmed.

**Findings from Industry**

In our experience the base rosters of TOCS differ in two main ways, namely variations in the Local Agreements for drivers and the amount of night work required. Local Agreements in force in a specific TOC may be tighter than the 'Hidden rules',[1] which cover the length of shifts, rest times between work and number of consecutive days work. We have found that, in general, the fatigue assessments for the companies with tighter rules have less fatiguing rosters. Additionally there are other factors that affect the fatigue scores that are only addressed as part of the local agreement; for example the length of time enforced between shifts when there is a rest day (or even two) in between. Problems of fatigue arise when a driver moves from nights to early shifts with one rest day but have insufficient time on the single rest day to recover.

The most common reasons for a high fatigue score appear to be the presence of: night work, very early shifts, very late shifts, very long shifts, high levels of consecutive work, or a combination of these. There are sometimes more subtle reasons, for example a run of work that starts earlier each day, or involves longer shifts towards the end of the week.

The FI is very unforgiving when presented with night-work. However, this seems to be a reasonable approach, as the majority of workers find this work extremely fatiguing. Nevertheless, night work fatigue may not affect all employees, and in some situations it may be possible for individuals to be put permanently on night shifts. However, this is not usually the situation in the rail industry, as rosters containing night shifts generally also contain day shifts.

There are two factors to consider where night work is concerned:

1   the actual risk of fatigue build up for the employees as they work consecutive night shifts;
2   the placement of rest days around the shifts to ensure that the employee has time to recover fully and not allow the fatigue build that will affect their work later in the week.

Often the night shifts relate only to shed work, and this is in no way comparable to main line driving. Regardless of this, fatigue on these shifts must be properly managed to prevent it affecting the employee on any passenger-driving shifts in the following week. It is therefore essential that the driver be provided with an adequate rest period following a sequence of night shifts. Scenario analysis of the diagrams using the FI can help produce a good estimate of the number of shifts that can be safely strung together, considering the usual lengths and times involved in the particular roster.

Other common areas of contention among almost all the fatigue assessments that have been conducted by AEA Technology Rail include the positioning of long shifts, combination shifts, and use of drivers on spare shifts.

A long shift after several days' work will increase the FI score. Decreasing the shift lengths or putting the longer shifts at the beginning of a string of work can avoid this. If possible the roster should start with the earliest shift and the subsequent days should have equally early or slightly later start times.

Fatigue levels can escalate quickly over consecutive days, particularly if the shifts are very early or continue very late into the night. There are many shifts that are fatiguing because they are long and very early, or long and very late. It may be possible to address this problem at the train planning stage, and if possible this would be the best solution. Alternatively the positioning of these shifts in the base roster needs to be considered carefully, for example not placing them at the end of a run of fatiguing work, nor after a long shift where the rest period is limited.

Finally, if the fatigue score for a driver on a particular day is likely to be high due to fatigue risk build up over previous days, there may be a benefit in allocating that driver to a spare shift. This is because the duties on a spare shift are likely to be less fatiguing than other shifts. However, there is a danger that the work assigned may be equally fatiguing. It is therefore important that roster clerks are aware of the fatigue risk of a driver on a spare shift and understand how to reduce this risk if possible, for example by assigning the driver to a shorter shift or one with more sociable hours.

## Lessons Learnt from Application of the Fatigue Index

The following section provides a summary of findings and lessons learnt from assessing fatigue in base rosters of TOCs using the FI.

*   Overall the feedback from TOCs has been positive. However, in reality, once produced the base roster can rarely be changed due to operational constraints and limits on driver availability. Therefore, recommendations are made in the report to help driver standards managers judge how best to manage the risk of fatigue. For example they may use the information to assist their decision of which driver

they will assess on which day, to coincide with a driver on a shift with a high fatigue risk.

•   In applying the FI to the assessment of rosters it has been found that the first line of a roster will rarely produce a high fatigue score. This is because there are no shifts before the first line for the cumulative effect of fatigue to be assessed. However, in reality, the first line of the roster will be worked directly after the last line of the roster. Therefore, in order to ensure the fatigue risk of the first line can be assessed, we place a second copy of the roster below the first copy and then run the Fatigue analysis. If the first line is likely to have a high fatiguing score, this will be detected at the beginning of the second copy of the roster.

•   The actual hours worked by drivers may vary significantly from the base roster. Reasons for a variation from the base roster include unplanned overtime (for example due to the train running late), and staff taking breaks from working the shifts due to annual leave or time off for safety briefings. Also, drivers may swap shifts with a colleague. It is unknown whether this variation will cause an increase or decrease in fatigue risk. Currently there is no requirement on TOCs to monitor the actual hours of work for the purposes of assessing fatigue, however, the facility exists within IRMA to do this type of assessment once any changes to the base roster have been entered into the data to reflect the actual hours worked.

•   As mentioned previously, the effect workload has on a driver's likelihood of fatigue is not known. Two different routes may have very different workloads, for example long stretches at high speed on an intercity high-speed service, versus a local stopping service. It is not clear whether the differences in task and workload affect the likelihood of fatigue and should therefore be factored into any fatigue assessment. This may not be a factor for future comparison within the FI, however, it may prove of wider use within the rail industry to develop a better understanding of the differing workloads experienced by train drivers.

**The Way Forward**

Currently the requirement for fatigue risk assessment applies only to changes in the working pattern as used in the base roster, not actual hours of work. One key step forward would therefore be the fatigue risk assessment of actual working hours of safety critical staff (including train drivers). The IRMA software and tools for such an assessment are already in place. It would be useful to discover whether the actual working hours and shift patterns of train drivers result in a higher or lower fatigue score than the initial indications provided by an assessment of the base roster.

The issue of fatigue in safety critical work is not limited to the rail industry. Much research has previously been conducted in other safety critical industries such as air traffic control, nuclear, off shore oil and gas, police and military. These are important sources of information and guidance. However, the applicability to rail of findings from other industries varies, as task, workload and work environment differ across industries. There is therefore a need for continued research to produce clear guidance on fatigue specific to the rail industry. Current Rail Safety and Standards Board (RSSB) funded research (Kroemer and Grandjean, 1997) includes a review of fatigue assessment methods and the development of a fatigue monitoring system. In America, the Federal Railroad Administration's (FRA) Office of Research and Development (ORandD) have also begun a Fatigue Research Program (Dawson and Reid, 1997), which may prove a useful source of information. Further suggested areas for research include obtaining a better understanding of the effects of workload on the train driver, and the differences in workload experienced by driving different routes, in relation to the link to fatigue effects.

Currently fatigue assessments are conducted by TOCs to meet the requirements of the Railway (Safety Critical Work) Regulations (HSE, 1994). Reviews of previous incidents and accidents have confirmed fatigue can be a contributory factor and so therefore it remains important to manage the risk of fatigue. However, TOCs are likely to experience greater benefits from applying best practice in their rosters including improved performance, improved morale, reduced sick leave or absenteeism, less incidents, accidents and errors. It would be useful to conduct a comparative study of TOCs before and after they have applied fatigue management best practice to their roster to assess these potential benefits, even though it may be difficult to prove a direct link between managing fatigue and improvements in these areas.

Finally, effective fatigue management requires a collaborative approach between employee and employer. Employers should design shift patterns to minimise the risk of fatigue and employees should be aware of their own responsibility to keep to the roster and use their rest periods between shifts effectively and sensibly.

## Conclusion

Further research is required to provide clearer guidance in the rail industry and an assessment of the actual hours worked by staff in the rail industry (rather than just the base rosters) will provide a clearer picture of the actual rather than predicted levels of fatigue experienced by staff. Providing such information to TOCs will enable them to better manage their staff on rotating roster and minimise the potentially dangerous effects of fatigue.

## Note

1    The 'Hidden rules' were defined following the public inquiry into the Clapham rail crash in 1988, chaired by Anthony Hidden QC, and published on 27 September 1989.

## References

Coplen, M. and Sussman, D. (2003), 'Fatigue and Alertness in the United States Railroad Industry. Part II: Fatigue Research in the Office of Research and Development at the Federal Railroad Administration', as cited in Railway Safety, *Impact of Shiftwork and Fatigue on Safety*, January.

Dawson, D. and Reid K. (1997), 'Fatigue and Alcohol Intoxification have Similar Effects on Performance', *Nature*, 38, 17 July: 235.

HSC (1996), *Railway Safety Critical Work. Approved Code of Practice and Guidance on the Hours of Work of Staff Undertaking Safety Critical Work on the Railways* (L50), Health and Safety Commission, London: HMSO.

HSE (1994), *Railway (Safety Critical Work) Regulations 1994* (L50), Health and Safety Commission, London: HMSO.

HSE (1994), *Guidance on the Definition of Activities Regarded as Safety Critical under the Railways (Safety Critical Work) Regulations*, Health and Safety Commission, London: HMSO.

Kroemer, K.H.E. and Grandjean, E. (1997), *Fitting the Task to the Human: A Textbook of Occupational Ergonomics*, 5th edn, London: Taylor and Francis.

Railway Safety (2003), 'Impact of Shiftwork and Fatigue on Safety', January.

Rogers, A.S., Spencer, M.B. and Stone, B.M. (1999), 'Validation and Development of a Method for Assessing the Risks Arising from Mental Fatigue', HSE report by DERA, Ref. 254/1999.

Chapter 20

# The HSE Revised Fatigue Index in the Rail Industry: From Application to Understanding

Rob Cotterill and Helen Jones

## Introduction

The relationship between impaired human performance and fatigue is now well established. Therefore it is important to ensure that employees who undertake safety critical work are not exposed to high risk of fatigue through the shift patterns they work. DNV conducted a study for a national rail freight company to investigate whether the proposed changes to working conditions would increase the risk of fatigue and hence impact on the safety of their operations. This chapter describes how an integrated risk assessment approach was taken which employed the HSE's Revised Fatigue Index to assess the proposed changes to shift patterns and the SWIFT hazard identification technique to determine additional risk factors that could impair driver performance. The approach highlighted areas for further investigation and improvements to the driver management systems.

Fatigue has been implicated in many accidents, such as Selby, Southall, and Clapham Junction. This is one of the reasons why shiftwork and fatigue are amongst the HSE's top ten list of human factors issues.

One of the recommendations that came out of the HSC's inquiry into the Southall crash was to collect data and research and to conduct work streams on SPAD mitigation, human factors, and fatigue studies which will seek to develop a better understanding in these areas. As part of this, a project was undertaken to identify the relative importance of factors that influence driver inattention and fatigue, which has highlighted the importance of shiftwork planning.

### Fatigue

Fatigue management needs to be considered when implementing a change to working practices. Typical factors to be considered include the following:

* limits on duty lengths;
* biocompatible shift scheduling;
* working environment design;
* risk assessment of critical tasks;

- changes to work patterns;
- planning of work;
- shiftworker education;
- cultural changes.

It is generally accepted that fatigue produces a variety of performance decrements all of which are directly relevant to safety within the train driving task. These include:

- variability in performance;
- slowed reaction time;
- increased number of work-related errors;
- increased tendency to repeat behaviours;
- increased false reporting,
- increased memory errors;
- decreased vigilance; and
- reduced motivation.

*Fatigue in the Railway Industry*

The issue of fatigue is of particular concern in the rail industry, where long journeys are common, and require constant vigilance by a single driver. A momentary lapse in concentration can have potentially disastrous results. In the rail freight sector, journeys are longer and frequently occur at night. As shall be made clear, these factors lead to an increase in the level of risk from fatigue.

The effects of shift patterns and overtime are now better understood and issues like driving without attention mode (DWAM) (May and Gale, 1998), the 2–4 hour shift phenomenon (HSE, 2000a) and working on extended night shifts have been studied in some depth (Miles, 2000). Shift patterns for safety-critical staff therefore need to be designed to avoid fatigue.

**Background to the Study**

Train operating companies are required by the railways inspectorate to provide evidence that changes to working practices do not result in any increase in the levels of risk over existing working arrangements. It is therefore necessary to perform a more holistic assessment of risks resulting from any proposed changes and how these are to be managed. The benefits of any changes must be considered in parallel with any potential impacts on the safety of train company operations. Changes in work practices may affect risk from sources like boredom, training, workload, communications, organisational culture and fatigue.

English, Welsh and Scottish Railways (EWS) are the UK's leading rail freight operator. One thousand five hundred rail freight services are operated every day by EWS, powered by over 650 locomotives. When EWS were looking to introduce changes to the working practices, they were aware of the need to assess whether the introduction of the new practices would potentially have a detrimental impact on safety. If this were to be the case, necessary changes, and appropriate mitigation or management strategies would also need to be identified to ensure that existing high levels of safety were not degraded.

Due to increased market pressures in the years following privatisation, changes have been made to drivers' working arrangements that have been wide ranging and have challenged certain traditionally accepted limits and parameters. These changes have been deemed necessary to allow EWS to operate at an optimal level within the current competitive railway environment. In order to ensure that these changes were not detrimental to safety, EWS commissioned a number of safety studies to assess the potential impact of the various changes (for example DNV, 1996). In addition to these studies, a wide base of knowledge has been developed from human factors research in areas such as shiftwork, boredom, distraction, fatigue, time of day effects, etc. (HSC, 2002). One of the outcomes of this knowledge base has been the development of the RARE (RAilway REsearch) database (HMRI, 1997).

*Aims*

The aims of this study were to provide EWS with the following:

- recommendations derived from literature, where such recommendations were considered to be applicable;
- recommendations derived from best practice risk management, where such recommendations were considered to be applicable; and
- the development of a set of additional controls and monitoring procedures to act as a precursor in the event that the changes were having a detrimental effect.

This chapter discusses the approach adopted, the main tool used, and how the results were made robust by their interpretation and use of supporting material.

**Approach**

The proposed changes to working conditions were studied using an integrated toolkit of methods, illustrated in Figure 20.1.

The four components of the assessment were:

- literature review;
- HAZID (hazard identification);
- Revised Fatigue Index;
- review of driver monitoring systems.

This integrated approach was required to assess how the proposed changes would impact on all aspects of driver operations and thus identify all the potential risks associated with these changes. The review of the driver monitoring systems aimed to identify whether the current safeguards would be adequate to control the risks associated with the proposed operations or whether additional measures would be required.

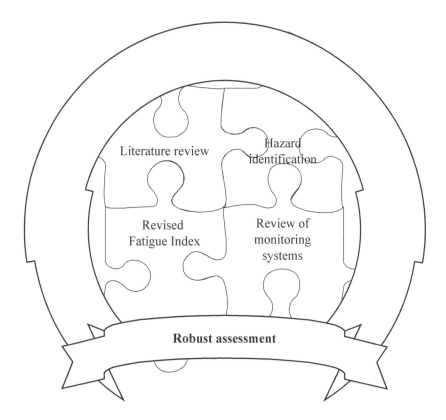

**Figure 20.1   Robust assessment**

*Literature Review*

The literature search was specifically targeted at information relevant to:

- fatigue;
- boredom;
- drugs and alcohol policy;
- lodging issues; and
- remotely supervised booking on.

Some key elements of best practice as suggested by the literature include:

- the number of consecutive night shifts should be restricted to two or three, to avoid cumulative fatigue effects;
- a sufficient rest period should be allowed, following night shifts, of two full night's sleep;
- early starts should similarly be restricted to two or three in succession;

- rest periods between shifts should be at least 14 hours;
- issues affecting driver attitude and morale should also be considered.

*Revised Fatigue Index*

The Revised Fatigue Index (HSE, 2000b) was developed by DERA on behalf of the HSE to identify when shift patterns would cause excessive levels of fatigue and indicate how patterns might be adjusted to ensure fatigue is minimised. It was originally produced to support Railway (Safety Critical Work) Regulations 1994 and assesses daily and cumulative fatigue risks associated with rotating shift patterns.

The tool is backed by wide-ranging research studies and investigations, which are detailed in the research report, and it is based on a system of rating scales, which have been validated by comparison with data from rail, air traffic control, military and other industries. It is composed of five factors:

- F1: the time of day a shift starts;
- F2: the duration of the shift;
- F3: the amount of rest gained before the shift started;
- F4: breaks taken during the shift;
- F5: cumulative fatigue effects.

The predicted fatigue caused by factors 1–4 is measured and the cumulative level of fatigue (factor 5) for each shift is calculated by summing these factors.[1] The main causes of high fatigue levels can be summarised as follows:

- early and night shifts (those starting before 05:00 or after 18:00);
- long shift duration (in excess of nine hours, when starting before 06:00 or after 16:00);
- short rest periods between shifts (less than 13 hours);
- long periods without breaks during a shift (above four hours);
- successive late, early or night shifts;
- rotating shifts without including rest days.

Using the Revised Fatigue Index, fatigue levels can be identified for each worked shift period, and can be compared between shift patterns to see how proposed changes affect the levels of fatigue on a shift. The causes of high fatigue can also be identified and steps taken to avoid shifts with high levels.

While there is no 'benchmark' figure above which fatigue affects safety, a shift pattern with several peaks of fatigue is indicative of potential problems in the strategy for setting shiftwork rosters. A Fatigue Index score of 35 can be considered to be high (HSE, 2000c) but attention should be placed on a whole shift pattern, in ensuring fatiguing shifts are not clustered together, and are followed by periods of recovery, rather than ensuring that individual shifts have low fatigue scores. In other words, a single shift with a very high fatigue score could be less serious than a week of shifts that have a moderately high fatigue score.

*Hazard Identification*

SWIFT is the Structured **What-IF T**echnique for hazard identification. It has been adopted by Network Rail Major Projects Department, in their risk management system, as a means of hazard identification for safety risk assessment. SWIFT is a particularly good technique for considering human and organisational factors that may affect safety. It is a brainstorming approach, supported by a prepared checklist of issues that may be relevant to the study. The intention of the SWIFT exercise was to identify all factors that may adversely influence driver performance and identify the potential consequences, current safeguards and any recommendations for additional safeguards.

*Review of Control and Monitoring Systems*

It is essential that an organisation has in place a robust and effective monitoring system that is able to pick up any increase in risk levels prior to adverse consequences being realised and in a manner that allows effective management intervention to redress the increase in risk. Monitoring systems might include:

- traincrew monitoring and journals;
- driver fitness (booking on and off);
- incident/accident review and recommendations arising;
- CIRAS;
- daily incident log;
- weekly accident reports;
- safety performance indicators;
- regional safety reports;
- period end safety report;
- safety tours and briefings;
- human resources systems;
- event recorders;
- safety critical assessments;
- employee performance profiles.

**Results**

For each of the proposed changes to working practices, the implications were identified from the literature, the SWIFT and the RFI analysis. The effectiveness of the current safeguards was then assessed against these implications for safety, and proposals were drawn up to strengthen the control and monitoring systems.

The SWIFT exercises were conducted with a representative cross section of personnel, including drivers, safety representatives, roster planners, traction officers and management representatives. Key issues identified were related to abnormal operations (that is, late running or cancelled services) and the monitoring of actual hours worked by drivers.

The review of the control and monitoring systems identified where the existing systems would need to be revised to match the proposed changes to working practices, and where additional safeguards might be put in place. The recommendations from the SWIFT exercise were used to guide this.

The implication of the Revised Fatigue Index for each proposed change was identified. This included an analysis of how that change would impact on the fatigue levels of drivers, whether it was beneficial, detrimental or had no impact.

In addition to the analysis of each individual change, an analysis was made of sample shift patterns from several depots, both under existing and the proposed working practices. Using this analysis it was possible to show how the proposed shift pattern had an overall lower Fatigue Index score and therefore had a lower level of risk from fatigue.

Figure 20.2 shows a graphical representation of the results of an RFI analysis of a three month link diagram for one train depot.

The topmost line in the figure shows the overall Fatigue Index score, which exceeds a value of 35 on the occasions where several successive night shifts are worked. The main cause of the high fatigue level, is the late starting time and duration of the shift, and the cumulated effect of working several successive night shifts.

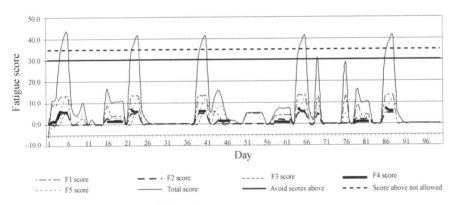

**Figure 20.2  Results of an RFI analysis**

**Discussion**

*Uses for the RFI*

The Fatigue Index assists in performing a fatigue risk assessment, however, it is an initial screening tool to identify where more detailed assessment may be necessary. As such, it can be used to compare different proposed shift patterns, compare old and new shift patterns, ensure that individual shifts are not excessively fatiguing (ironing out the peaks) and it can help identify what actions will reduce fatigue scores.

The guidance score of 35 is an indicator of a high risk level from fatigue. Where shift patterns can be altered, this can be used an indicator, and amendments made to shift length, start time, breaks, rest periods and days worked so that the risk from fatigue is reduced. In the railway industry, it may not be possible to easily change shift patterns due to operational requirements. This is particularly notable in the freight sector where many of the journeys are carried out at night, or early in the morning. In this case, mitigating factors can be sought or made in the amount and length of rest breaks taken or time not spent driving. Also, the RFI can be used to guide the allocation of spare turns and overtime to drivers, so that the amount of fatigue experienced is kept to a minimum.

*Limitations of the RFI*

When using the RFI, it is necessary to be wary of certain factors that will affect the calculated fatigue levels. These include informal shift swapping amongst drivers where this is not monitored by the company, drivers with second jobs or who work excessive amounts of overtime, and on-call arrangements where the actual hours worked may not reflect the shift patterns modelled in the RFI. All these factors affect the likely levels of fatigue, so where these circumstances exist, account should be made of them in the fatigue risk model.

The RFI does not account for individual differences between people, that is, those who may prefer early shifts over late ones, or those who prefer late shifts over early ones. Family and personal circumstances are also not measured, and it is assumed that a driver is able to get a reasonable amount of quality rest between shifts. Fixed shift systems where night shifts are permanent rather than temporary, and split shift systems are also not modelled by the RFI. It should also be remembered that fatigue can have causes other than shift length and times.

The RFI is not a physiological measure and it is not a predictor of performance on a task. It is a tool for assessing the risk from shiftwork related fatigue, predicted from planned shift patterns.

**Robust Fatigue Management**

The RFI has limitations in its application, but is nonetheless a very useful tool for predicting factors in a shift pattern that are likely to lead to high levels of driver fatigue. By combining the application of the RFI with a hazard identification methodology and a review of control systems, an integrated approach can be applied.

For each element in a change to working practices, the implications need to be considered on not only the Revised Fatigue Index levels, but also the hazard identification investigation, and on how the monitoring systems would work to control any risks. Since the fatigue tool is not necessarily fully applicable in this case, the predicted levels of fatigue are not sufficient to demonstrate that the risks are being controlled. Here it is vital to ensure that there are systems in place to detect high levels of fatigue should they arise.

**Conclusion**

In identifying where the risks are in a shift system, the RFI is a useful tool for seeing where potential problems might occur. However, it is not the whole story: many aspects of fatigue need to be properly managed and a holistic approach is necessary.

The combination of a hazard identification study, shiftwork fatigue review, review of control and monitoring systems as well as identifying the most up-to-date research on the causes of fatigue, attempted to satisfy this requirement. It provided a robust method for assessing the risks from changes to working conditions and using the Revised Fatigue Index provided a clear way of quantifying the risks and indicating the reasons why particular shift patterns were more prone to fatigue.

The result of this investigation was that we were able to demonstrate to the satisfaction of HMRI that the proposed changes in working practices did not add to the risk involved in train driving.

## Acknowledgement

The authors would like to acknowledge the efforts of Peter Maguire in facilitating the research in this field.

## Note

1    The method for the calculation of each factor is explained in detail in HSE (1999).

## References

DNV (1996) 'Safety Evaluation of Changes in Working Conditions', Job No. 621101.019, Revision 4, July.

HMRI (1997), *RARE (Railways Research) Database*, Ref. 27014001, version 2.0, March.

HSC (2002), T*he Southall Rail Accident Inquiry Report: Summary of Progress at February 2002*, London: Health and Safety Commission.

HSE (1999), *Validation and Development of a Method for Assessing the Risks Arising from Mental Fatigue*, Contract Research Report 254/1999, London: HSE Books.

HSE (2000a), 'Studies into the 2–4 Hour Shift Phenomenon', *Contract Research Report 299/2000*, London: HSE Books.

HSE (2000b), 'Validation and Development of a Method for Assessing the Risks Arising from Mental Fatigue', *Contract Research Report 254/1999*, London: HSE Books.

HSE (200oc), Briefing Note – Dr Deborah Lucas, prepared on 27 June.

May, J.L. and Gale, A.G. (1998), 'How Did I Get There? Driving Without Attention Mode', in M. Hanson (ed.), *Contemporary Ergonomics*, London: Taylor and Francis: 456–60.

Miles, R. (2000), 'Developments in the Understanding of Working on Extended Nights Offshore', Society of Petroleum Engineers.

# PART 7
# SIGNALLING AND CONTROL
# FACILITIES

Chapter 21

# Meeting Human Factors Challenges During the Design of the New Integrated Electronic Control Centre (IECC) in York

Mike Stearn, Martin Hazell, John Robinson and Steve McLeod

## Introduction

By the mid 1990s, Railtrack (now Network Rail) was actively investigating the benefits of a national railway control strategy based on a small number of network management centres (NMCs) providing integrated signalling, strategic planning and infrastructure management control across the UK rail network. In support of this control philosophy, Network Rail's London North East (LNE) Region initiated a project to develop a new control centre in York which, for the first time in the UK, would control signalling, regional control, very short term planning (VSTP), timetable planning (ARS input) and electrical supply from a single, open plan operating floor. Strategic planning for the new control centre began in August 1999 and the vision of a fully integrated control environment was finally realised in December 2001 when electrical control joined signalling and regional control functions on the operations floor of the new York IECC.

The main aims for the York IECC project were to:

- integrate the disparate control functions within a visually and operationally seamless operations floor;
- optimise operational efficiency and reduce operational risk by improving working environment, communication and working culture;
- expand York signalling control by integrating routes covered, at the time, by Leeds Power signal box/Leeds North West, Stourton and Church Fenton signal boxes (all of which would subsequently be closed);
- create a focus for LNE regional control and set the standard for regional controls across the UK rail network.

At the start of the project in 1999, York signal box was located to the north of York station and accommodated two IECC Signallers and a duty signalling manager (DSM). Leeds Power signal box was located in the centre of Leeds and accommodated a large NX panel, five on-duty signallers and a variety of support staff. Church Fenton signal box was located between York and Leeds and accommodated a switch panel and one on-duty signaller. Regional and VSTP controls were located in two separate offices within LNE's headquarters building in the centre of York and accommodated up to 16 on-duty Controllers. Timetable planning (ARS input) was located in Newcastle and accommodated four on-duty staff. Electrical Control was split between two electrical control rooms (ECRs) in Hornsey and Doncaster and accommodated four on-duty staff. Doncaster ECR operated a mimic panel control system and Hornsey ECR operated a supervisory control and data acquisition (SCADA) control system.

Human factors support was brought on-board at the beginning of the project. From the very first project meetings it was apparent that there were a number of human factors goals that needed to be achieved to meet the main project aims:

- integrate functions, roles and responsibilities;
- address user concerns and gaining user buy-in;
- create an acceptable open plan operations floor;
- integrate new technology and working practices;
- create a high quality working environment.

*Integrating Functions, Roles and Responsibilities*

The York IECC project aimed to bring together experienced operating staff who were used to working within function specific controls located remotely from one other. After speaking at the beginning of the project with nearly all of the staff likely to be involved in the new control, over 80 per cent expressed concerns over the benefits of integrating roles within a single operating floor. These concerns ranged from personal worries (for example about relocating or retraining) to specific concerns about safety (for example staff not going through the correct recorded communications procedures if they were located close enough to be able to walk over and speak directly with a colleague) and efficiency (for example increased noise created by other functions masking communications). Staff also expressed concerns about integration of roles leading to job losses. It was the responsibility of the human factors specialists to understand these concerns, to represent user views within the project and to address integration issues within the physical design and layout of the operations floor, job descriptions and operating procedures specified for the control.

*Addressing User Concerns and Gaining User Buy-in*

The type of user concerns expressed at the beginning of the York IECC project meant that it was essential to establish a formal process for collecting user views, encouraging user involvement and communicating issues and decisions relevant to users. Experience in control projects has shown that user buy-in to decisions taken during design and development is an important factor in determining the success of a final scheme. The best way to gain user buy-in is to involve users at all stages of the project. This should mean that staff who eventually work on the operations floor have ownership of its specification, awareness of its compromises and responsibility for its success. Human factors specialists were responsible for establishing user groups, user communication processes and user involvement work programmes.

*Creating an Acceptable Open Plan Operations Floor*

One of the biggest user concerns (and human factors challenges) at the beginning of the project centred on the creation of an open plan operating floor. Users expressed concerns about possible visual and noise distractions, improper communications from staff talking directly with each other rather than on recorded telephones and interference in work processes (especially during busy or perturbed operation) by staff being overseen by other staff and visitors. The majority user view at the beginning of the project was that each of the main functions (signalling, regional control and electrical control) should be separated by floor to ceiling partitions. This did not, however, match the regional management vision for a 'visually and operationally seamless operations floor. The human factors specialists were responsible for investigating human and operational implications of the open plan concept and specifying workstation designs, room layout and environment design to best support user and management views.

*Integrating New Technology and Working Practices*

From the beginning, the York IECC project faced enormous operational and human factors challenges not only from the need to integrate people and roles into a single environment but, at the same time, from the need to radically change the way in which staff worked. The business case identified efficiency and asset benefits from new technology introduction and standardisation. Signallers from Leeds, Stourton and Church Fenton were not only asked to move to York but also asked to change from panel- to VDU-based IECC signalling operation. Electrical control operators from Doncaster were asked to relocate to York and change from mimic- to VDU-based SCADA operation. The pressures on users from a change in location, working arrangement and working process needed to be carefully managed and supported. The human factors specialists were responsible for bridging the gap between meeting technical and user requirements. This meant ensuring that upgrades in technology were matched by equivalent upgrades in user support (that is, comprehensive training, correct workstations, supportive environment and compliant working practices).

*Creating a High Quality Working Environment*

The human factors specialists were responsible for specifications, concept designs, user testing validation, detailed designs, technical specifications, manufacture drawings, manufacture tender documentation and manufacture/installation support for workstations, furniture, storage, finishes and whole room environment. It was important to the project that human factors specifications could be quickly and accurately transformed into practical designs that met user needs, health and safety guidance and human factors standards. To this end, the project contracted a human factors specialist with in-house industrial design and engineering skills and with experience of managing entire control centre design projects. As the human factors specialists were involved throughout the project lifecycle, and in all aspects of the physical and social work environment, human factors was often seen as the 'glue' that brought together the many diverse disciplines and skills within he project team (that is, civils, power, project management, engineering, telecommunications, HR, IT etc). Human factors work programmes such as user workshops, user group meetings, workstation and whole room user trials (Figure 21.1), operational scenario testing and CAD walkthrough simulations (Figure 21.2) often became the focus of project team meetings and the most visual indication of project progress.

**Figure 21.1   Signalling workstation user trial**

**Figure 21.2   Operations floor CAD walkthrough**

**Establishing the Vision**

Human factors took a lead in defining the operational, functional and aesthetic vision for the York IECC. Human factors specialists ran two workshops at the start of the project which brought together stakeholders representing 'those providing a service from the operations floor' and 'those receiving a service from the operations floor'. Stakeholders included train operating companies (TOCs), freight operating companies (FOCs), infrastructure maintenance contractors (IMCs), architects, building contract managers, systems and equipment suppliers, user representatives, project delivery managers, control centre managers and regional managers. The aim of the workshops was to allow open and frank discussion on the needs of each of the main stakeholder groups. These included:

- short, medium and long term operating concepts;
- definition of roles, responsibilities and numbers of staff working on the operations floor;
- business objectives;
- definition of individual stakeholder objectives;
- emergency and visitor handling procedures;
- visual aesthetic and image;
- definition of stakeholder concerns and project risks.

The workshops proved useful in a number of respects. Firstly, they involved stakeholders who were not to be located on the operations floor but who were to receive a service from the operations floor. These stakeholders provided valuable insight into ways by which business objectives (such as providing a better service) could be met by, for example, arrangement of working groups on the operations floor, integration of communications and IT systems, creation of meeting areas and use of an overview display facility (ODF). Secondly, the workshops brought together individuals who had been working with each other, sometimes for many years, but who had never actually met face-to-face. This lead to lively and sometimes confrontational exchanges but started the process of defining user needs and establishing a user group which would provide invaluable input throughout the project. Finally, the workshops brought together regional managers with 'overall vision' and users with 'practical requirements'. Again, there was lively and far reaching debate covering areas as diverse as noise control, relocation packages, supervision, training needs and future proofing. The benefit of the workshop approach was that representatives from each camp had an opportunity to hear the views, wishes and concerns of their colleagues. Although many issues remained unresolved at the end of the workshops, at least the issues had been aired openly, recorded and identified.

The human factors specialists were responsible for facilitating the workshops and getting agreement on key issues that could move the project into its next phase.

### Defining the Operational Concept

The operational concept defines exactly how the operations floor will operate to deliver the business objectives. The human factors specialists were responsible for producing the operational concept by undertaking a number of standard work packages that included:

- audit of sites due to migrate onto the operating floor;
- discussion with staff at each site;
- discussion with union and health and safety representatives;
- discussion with regional and headquarters HR managers;
- task analysis of each role and responsibility designated for the operations floor;
- communication and link analysis (predicted);
- review of existing and future (predicted) technology, IT and telecommunications;
- review of standards, guidance and best practice;
- integration of vision workshop requirements.

From this work, an operational concept document was produced that specified the number of operators to be accommodated on the operations floor and defined job roles and responsibilities for each operator. Operational links were established to identify best location of each role relative to each other (Figure 21.3).

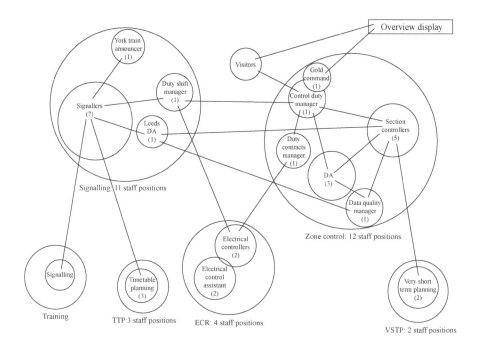

**Figure 21.3  Operational links diagram**

Lines of management and supervisor responsibility were defined. A technical specification for equipment required to support each role was produced. On and off workstation storage requirements were defined for each role. Workstation layout drawings (Figure 21.4) and peripheral furniture layout drawings were produced for each role.

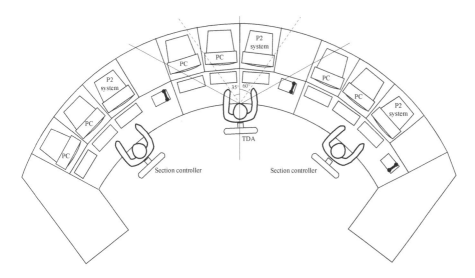

**Figure 21.4   Human factors workstation layout drawing**

An environmental specification was produced that covered the physical and aesthetic environment as well as access, visitor handling and DDA compliance. Requirements for operations floor facilities (that is, training room, ODF, drinks and administration areas) and control suite facilities (that is, locker rooms, mess rooms) were defined. Finally, all aspects of the operational concept were brought together to produce a number of operations floor layouts (Figure 21.5) for discussion with stakeholders.

The operational specification became a pivotal document in the York IECC programme. Although it developed throughout the project lifecycle, the first issue document (produced three months into the project) formed the foundation for all human factors specifications. Work undertaken to produce the Operational Concept identified a number of key risks that required more detailed human factors input. For the York IECC project these were;

* operator workload;
* user support;
* ODF;
* noise and distraction.

**Minimising Risk**

*Operator Workload*

Concerns were expressed about the level of operator workload (especially signaller and ECR workload) that may result from integration of functions and changes in

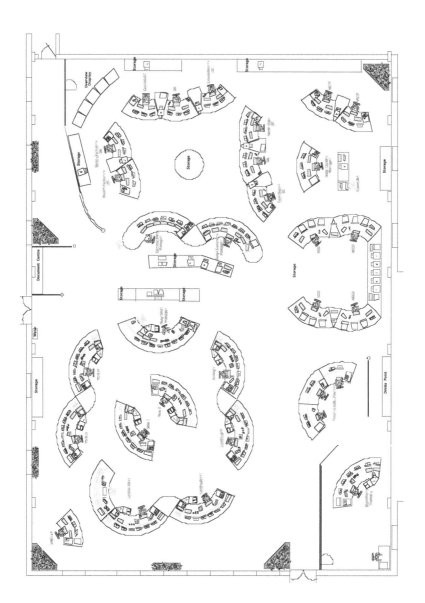

**Figure 21.5  Initial operation floor layout**

operating technology in the York IECC. The project required early and predictive assessment of signaller and ECR operator workload to define the effects of new control areas and phased technology introduction on the number of workstations and on-duty operators required on the operations floor. The human factors specialists were responsible for developing a method for assessing signaller and ECR operator workload to answer specific questions concerning manning levels and workstation/ technology duplication. A review of workload studies undertaken in UK railway controls (pre-1999) revealed no standardised method being used or prescribed for use. A workload method was, therefore, developed specifically for this project. The method incorporated direct observation and activity recoding combined with measurement of existing and predicted operational demand. The method allowed the production of predictive timeline workload charts (Figure 21.6) for different technology and manning level scenarios.

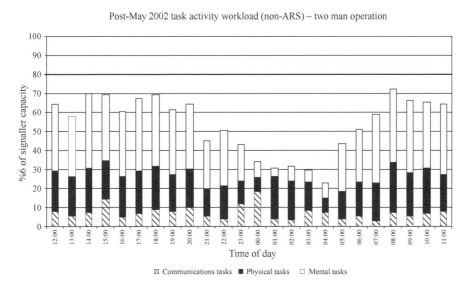

Post-May 2002 task activity workload (non-ARS) – two man operation

☒ Communications tasks  ■ Physical tasks  ☐ Mental tasks

**Figure 21.6  Example of a predictive timeline workload chart**

Workload was defined in terms of mental, physical and communications activity required by an operator, or operators, to undertake each operating scenario. A nominal threshold of 80 per cent time capacity was defined above which a workload risk was identified. Predictive workload charts were produced for defined signaller roles and all ECR roles. The workload assessments had a number of consequences for the project:

• an additional signalling workplace and staff roster was introduced from day one to the station area. Workload issues were predicted to be caused mainly by the number of possessions and manual running required during the infrastructure upgrades around Leeds station. To this end the additional workstation was specified to be temporary so that it could be removed when the infrastructure stabilised;

- the importance of reliable technology provided by IT and control systems was highlighted as a major factor in reducing operator workload. The workload assessments showed, for example, that when automatic route setting (ARS) falls out or when ARS is phased onto a signalling control area, as was the case in the York IECC, the level of predicted operator workload significantly increases. Tight specifications were put on ARS integrity and a fast track process for ARS introduction was implemented by the project. Likewise, workload assessment showed that the network management system (NMS) within the new ECR SCADA was predicted to reduce the workload required to plan isolations. However, the NMS was not being introduced until a year after electrical controllers started operating from the York IECC operations floor. Higher manning levels were specified from day one as a result.

The workload assessments gave project managers confidence in the specification of manning levels, numbers of workstations and consequently size of the operations floor. The workload method developed for this project has subsequently been used as an element within Network Rail's standardised signaller workload assessment method.

*User Support*

A user group was established to provide input to the design, assessment, sign-off and decision-making process. The user group comprised representatives from the various roles and functions to be located on the operations floor. The user group met on a regular basis throughout the project lifecycle and took part in user trials to specify workstations/furniture designs and operations floor layout (Figure 21.7).

The involvement of the user group in user trials was found to be extremely important in gaining user buy-in. Users threw themselves whole heartedly into participation in the trials and provided guidance and suggestions that could not have been established without their involvement. In addition to the user group, three subject matter experts (SMEs) were seconded from operational duties to join the project team and work alongside the human factors specialists. These SMEs were current signallers and regional controllers who, in addition to representing the views of users in all project meetings, provided technical expertise to all human factors work streams. The contribution made by the SMEs was invaluable not only in contributing technical expertise but in providing a link to operators, increasing operational credibility of the human factors team and sorting out the many logistical problems that arise in such a project.

*Overview Display Facility (ODF)*

The human factors specialists were responsible for undertaking a feasibility study and subsequently developing a technical specification for the ODF. It is fair to say that the requirement for an ODF came from the regional managers vision rather

**Figure 21.7  Operations floor user mock-up trial**

than from the user requirements, however the feasibility study showed there to be a number of potential operational benefits:

- improved information to regional control;
- improved information during major incident management;
- improved information for proactive decision making.

Human factors specialists and SMEs researched available technologies and provided technical and user specifications into the tender process.

SMEs produced requirements for information to be displayed and for operator control of ODF information. The ODF is installed and operating on the York IECC operations floor. There has yet to be a formal evaluation of the benefits or effectiveness of the ODF.

*Noise and Distraction*

A concern expressed by users during the vision workshops was that operational integrity and accountability could be disrupted by the integration of roles within an open plan operating floor. The major concerns centred on:

- auditory distraction from noise generated by people and equipment;
- visual distraction from operators and visitors walking around the operations floor.

The human factors specialists were responsible for specifying finishes, screening, operating procedures, annunciators, a room layout and an environment specification to minimise noise disruption. Acoustic experts were consulted on materials for wall, ceiling, window, floor and work surface finishes. Human factors specialists worked closely with building architects and equipment suppliers to consider acoustic properties at all times. Studies were undertaken to predict acoustic performance and 'live' acoustic measurements were subsequently taken to check that the populated operations floor met acoustic requirements set out in the operational concept document. The open plan operations floor was found to meet all acoustic requirements. Operators questioned after having worked on the operations floor for about six months expressed satisfaction, in the main, with the acoustic performance. Apart from some minor alarm noise distraction, users have not experienced any significant noise distraction.

The main solutions for reducing visual distraction within the open plan operations floor were defined as:

* managing visitors entering and moving around the area;
* implementing work practices and adopting an operating culture that respects the safety critical nature of the operating floor and the individual privacy of operators.

A visitor holding and reception area was specified at the entrance to the operating floor (Figure 21.2). This area was screened to allow visitors a clear view of all operating areas without having to enter the main operations floor. A procedure for visitor handling was defined to ensure that all visitors were booked onto the operating floor and reported directly to the operator in overall charge of the operations floor, who was located closest to the visitor holding area. Walkways were defined to minimise the need for visitors to walk directly behind operators at their workstations.

## Conclusion

Feedback from visitors to, and users of, the York IECC operations floor have to date been positive. The end product is the result of teamwork between human factors specialists, SMEs, involved users and a forward thinking and dynamic project team. Not all aspects of the human factors input went smoothly and much was learnt for future project, especially in the areas of environment control and infrastructure flexibility. However the York IECC does, we hope, provide a satisfying working environment that sets the standard for future Network Rail controls.

Chapter 22

# The Application of Ergonomics to Standards Development for VDU-based Signalling Control Systems

Anthony Slamen and Nick Coleman

## Introduction

Across the Network Rail infrastructure a number of alterative approaches to the development of VDU-based signalling control systems (VSCS) are being adopted by different companies in order to meet the demands of improved railway safety and performance. The systems incorporate different signaller – system interaction design models, interface functionalities, alarms systems and schematic layout conventions.

While Railway Group and Network Rail Company standards largely address the technical and functional requirements to ensure compatibility across the network and between its systems, they have historically left important HMI safety critical elements (for example alarms, aspects of visual coding) open to designer interpretation. Further, VSCS designers complain that where interface requirements have been specified they are occasionally restrictive – being too prescriptive and based on older technologies – and in some cases, actually promote poor interface design solutions.

This chapter introduces the work that the Network Rail Ergonomics Group has undertaken in order to ensure that *ergonomics requirements* and guidelines are properly incorporated into revised and new Network Rail *standards* for VSCS design. It outlines the issues and processes involved, the adopted approach, and provides specific examples concerning the development of requirements for signalling schematic maps and alarms systems.

VSCS are being widely adopted by Network Rail as an alternative to more traditional signalling display and control technologies. They are generally cheaper to install and maintain, require less space than 'NX' (entry-exit control) panels, and can integrate computerised automatic signalling systems, to enable signallers to operate larger sections of railway.

This change in technology has led to fundamental changes in the nature of the signalling task, the signalling environment and the design of the signallers' workstation. These changes mean that new challenges have been set for both signallers and system designers. The aim of this chapter is to highlight a number of ergonomics issues associated with VSCS design. The study provides examples of guidance that can be applied to VSCS in order to support the development of future signalling systems which better assist signallers in maintaining the safety and performance of the railways.

## Background

The introduction of new technologies and systems with regard to VSCS should enable system designers to provide more support to the signaller, through the development of new on screen information and search facilities, integration of secondary systems (such as voice communications) and more appropriate divisions of labour between human and machine and co-workers. The introduction of VDU-based signalling systems in the UK during the 1980s has left a legacy of signalling symbology and display precedents that are based on the limitations of these older display screen technologies. Any future signalling display systems based on existing standards would fail to capitalise on the benefits of modern display screen capabilities.

Carey (1992) identified a number of ergonomics areas, which would need to be considered when designing a VDU-based signalling system. These include:

• provision of automatic signalling systems;
• interaction dialogue design;
• graphical display design;
• alarms and alerts handling.

These are key to the development of effective working environments within which signallers can complete their job safely and efficiently. In order to ensure that future systems can achieve this it is necessary that standards and guidance address these issues in a manner which is not prescriptive but forward thinking and grounded in application of ergonomics 'best practice'.

There is a need to investigate what a signaller needs to know and see when operating VSCS. Clearly any changes to operating practices (such as those brought about through the introduction of automatic signalling systems) will necessitate a change in the information requirements needs of the users. As signalling centres begin to cover larger sections of railway the need for accurate information about the actual track layout, gradients and signal positioning will increase.

## Approach

In order to understand the signallers' information requirements it was necessary to assess what information was currently provided to the signaller both within current VDU-based systems and traditional NX panels. Observing how signallers make use of the current systems available to them affords a better understanding of what information is missing or is available but is not presented correctly or used to its full potential. It was also essential to gather user views of those operating systems in order to form future thinking about user needs; signalling personnel are best placed to inform interface design requirements due to their daily interaction with a given system.

With these points in mind a number of visits were undertaken to signal boxes (VDU- and panel-based) across the Network Rail infrastructure. Visits involved detailed observation of signalling tasks and structured interviews with signallers, trainers, signalling managers and SMEs in order to elicit views about design and set-up of VSCS. Current Network Rail standards were reviewed in order to understand the limitations and current thinking surrounding the design of VSCS.

## Findings and Exemplar Guidance

The major findings of the work can be categorised under the following headings:

* automatic signalling systems;
* contextual awareness;
* information display requirements;
* reduction of visual clutter;
* alerts and alarms.

The following section will discuss each of these issues in turn providing examples of specific ergonomic guidance.

### *Automatic Signalling Systems*

Automatic Signalling Systems (ASS) represent an important step forward in helping to maintain and improve the performance of the network. However, they require careful design to ensure that the allocation of functions between the technology and the human operator is appropriate.

The most common problem with automated systems (as demonstrated across a wide range of technologies and industries) is that they exclude the human operator from system operations and decisions, relegating their role to one of monitoring. This can mean that when a problem occurs, and the operator is required to take control and the initiative they:

* have been so far abstracted from operations up until that point that they are unable to make appropriate decisions within the time available; and
* are unable to operate the system controls as effectively and efficiently as they should because they have become de-skilled after extended periods of non-use.

It is clear that more research is required in this field in order to investigate the most appropriate allocation of functions between the signallers and ASS, however, the following guidelines are examples of ergonomic 'best practice' in this area:

*ASS should be designed to keep signallers actively engaged with the signalling task and not result in periods of low workload*

Even the most diligent signallers under the tightest management regime are unlikely to be able to maintain full awareness of signalling operations if they are not actively involved with the signalling task and decision making. ASS that do not actively involve signallers in normal signalling operations need to support signallers during degraded operations or incident management and perturbations without modifying the way that they interact with the system. A common signaller complaint is that ASS work well during normal operations ('*when we don't need it*') but not at all during degraded operations ('*when we do*').

*The sections of track under ASS control should be immediately obvious at the user interface*

Figure 22.1(a) shows an example of the current method of displaying and controlling ASS sub-areas, each sub-area being described by codes that the signaller must associate with sections of the railway infrastructure held in memory. The load placed on the signallers to do this means that they often report frequent errors in activating and de-activating areas of ASS control. Figure 22.1(b) shows an improved solution

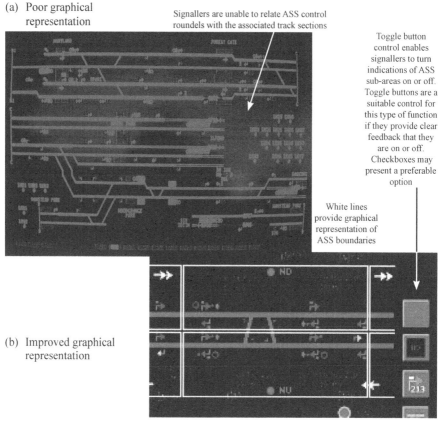

(a) Poor graphical representation

Signallers are unable to relate ASS control roundels with the associated track sections

Toggle button control enables signallers to turn indications of ASS sub-areas on or off. Toggle buttons are a suitable control for this type of function if they provide clear feedback that they are on or off. Checkboxes may present a preferable option

White lines provide graphical representation of ASS boundaries

(b) Improved graphical representation

**Figure 22.1   Demarcation of ASS areas**

where ASS controls are more closely associated with their areas of control; and signallers are provided with the option of turning visual indications of ASS sub-area boundaries on or off.

*Contextual Awareness and System Integration*

Retaining a broad contextual awareness of railway operations is of fundamental importance to signallers for managing workload and maintaining safe and efficient operations. Classic VSCS provided this information typically through the train describer (TD) map, but signallers have indicated that this solution is not ideal.

Where possible, signallers would select a TD map that displays traffic that is about to enter, and traffic that has just left, their area of control. However, signallers complain that only one TD map can be selected, despite there often being multiple entrance or exit points to other areas. Additionally, TD maps rarely align themselves adequately in order to provide a clear visual extension of their control area. Many of the signallers interviewed reported rarely, if ever, using the functionality.

When trains are running to time the current information providing methods are manageable, but when they do not (for example during peak periods) even the most experienced signaller can become confused by what train is arriving and where it is supposed to be going. It is clear that signallers have a requirement to gain access to information regarding whether trains are running according to plan for example whether they are early or late, whether they are in the correct sequence.

The following example guidance was developed in order to ensure that contextual awareness is supported by the VSCS:

1   signallers should be kept sufficiently aware of the movements of trains outside of their area so that they are able to maintain optimum train running performance and safety within their own area of control;
2   at any time, signallers should be able to see the next two or three trains about to enter their area of control from any associated entrance or exit point.

While signallers recognise that responsibility for running trains on time and setting priorities lies ultimately with Control and TOCs, they consistently report that they would be able to make better decisions, communicate more effectively with Control and drivers, and help maintain improved railway services (especially during perturbations or degraded operations, and when managing freight) if they had more immediate access to this information.

Figure 22.2 provides an example of a VSCS display solution that could support the signallers' need to maintain a contextual awareness of both their own and (perhaps more importantly) neighbouring control areas.

*System Integration*

Occasionally incidents have occurred on the network as a direct result of signallers misinterpreting the location and identities of drivers when giving instructions. This is not surprising given the number of communications signallers are required to manage during major perturbations and degraded operations (see Figure 22.3). Analyses and

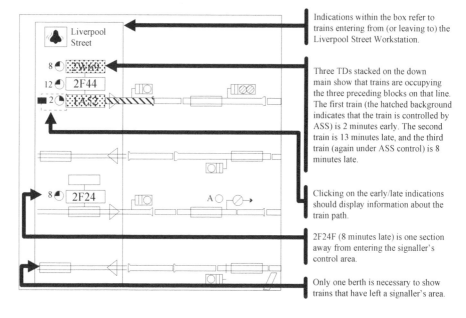

**Figure 22.2   Example of integrated system data within a VDU signalling
system**

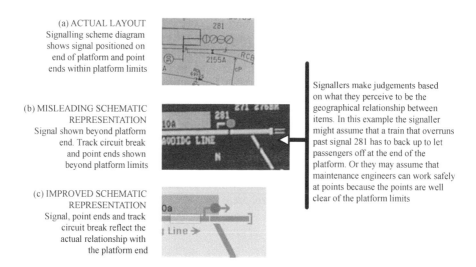

**Figure 22.3   Example of improved positioning of signals and junctions with
respect to platforms**

observation of signaller activities show that that they constantly refer to the schematic displays during communications. Therefore some form of visual feedback indicating the location of the driver that a signaller is talking to would provide an important and useful indication to the signaller. This integrated system approach would reduce the number of systems the signaller would need to interact with whilst also providing an important step in mitigating against the risks of miscommunication.

*Information Display Requirements*

VSCS user interfaces must be designed to optimise signallers' perceptual and cognitive load. The user interfaces and associated interaction design models need to take proper account of the physiological, perceptual and cognitive capabilities and limitations of signallers, and ensure the safety and efficiency of signalling operations. Signallers and signalling experts have raised a number of ways in which VSCS user interfaces could better provide the information they require. Improvements to the information provided at the VSCS user interface will help optimise signaller workload, minimise response times and decrease the risks of error by reducing memory demands.

Despite the fact that signalling display schematics do not provide an accurate representation of track layout (see Figure 22.3), signallers make assumptions about the locations of items based on their relative positioning compared to other schematic indications. Schematic indications that do not adhere to the principles of relative positioning can lead to signallers misinterpreting railway events or status; miscommunications between signallers and drivers/trackside staff; and/or inappropriate signalling decisions.

The following example guidance was developed in order to ensure that information contained within the VSCS is displayed in the most appropriate way to aid the user:

1    signals should be positioned on the correct side of the track on the schematic to reflect their actual position on the side of the track outside;
2    signals and junctions should be positioned on the schematic to reflect their actual location with respect to platforms/junctions and other signals. See Figure 22.3 and Figure 22.4;
3    boundary indications (for example interlocking boundaries, Group replacement boundaries, ASS boundaries) should provide an unambiguous display of the signals and control elements they contain (See Figure 22.1).

Signallers also indicated that they would like the ability to search for information that may not always need to be on display. For example drivers and trackside workers frequently attempt to communicate locations to signallers in terms of the trackside information they see or use. Since signallers are not privy to the same information, they typically have to either run through a series of question and answers (increasing the time and risk of miscommunications) to identify the actual location, look up detailed signalling maps, make an educated guess, or apply safety restrictions to an area of track far larger than required. A quick search function that highlights the corresponding parts of the signaller's display would be more efficient and ensure more effective and swift communication could be undertaken.

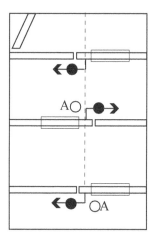

**Figure 22.4  Subtle gantry indications help show relationships between signals as well as providing information about the infrastructure**

*Reduction of Visual Clutter*

Signalling displays are perceived by signallers to vary in complexity and clutter depending on:

*   their experience;
*   the relevance of the information to the tasks that have to be performed; and
*   the way that the information is designed and presented on screen.

To minimise the perceived density and clutter of visual displays it is essential that only the information that operators require be displayed and presented in the most appropriate manner for effective and efficient interpretation.

The density of information displayed on VSCS schematic displays is frequently raised as a concern. Often information requirement needs are not raised in case the inclusion of that information adds to already cluttered screens. However, high information density and cluttered display screens are not a necessary evil of VSCS. Modern systems, designed in accordance with contemporary principles of good human-computer interaction (HCI) design can provide a greater depth of information on displays that are subsequently perceived as less complex and cluttered.

The following example guidance provides information about how the layout of a VSCS could be designed in a manner that would reduce the perceived visual clutter:

Track schematics, by their very nature do not accurately reflect the real world. Their purpose is to display only the information required in the simplest form. Schematics have arisen as the basis for signalling displays primarily because of the complexities associated with displaying geographically accurate map displays. For schematics to be easy to interpret and devoid of clutter they should:

- show only the information required;
- follow simple and transparent display rules that users are able to consistently apply when interpreting the information display.

The following rules should be considered when developing track schematics for VSCS:

1    lines of track that run across screens should be perfectly aligned. The alignment of track across screens is essential for the optimisation of visual tracking performance. Misaligned tracks increase perceptual load and can lead to read across errors when signallers scan across displays. Figure 22.5 shows the degree of misalignment present in many VSCS displays. Figure 22.6 shows the same section of track laid out across four displays with no misalignment of track, which is immediately more intuitive;

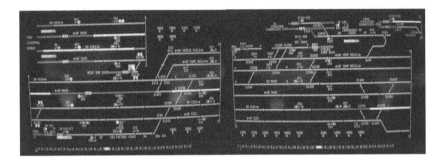

**Figure 22.5    A view of the track alignment across set of detailed displays for an IECC VSCS**

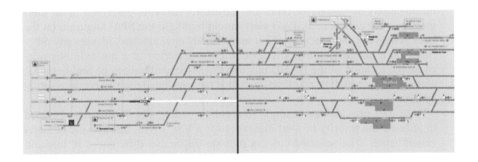

**Figure 22.6    Shows how the detailed displays in Figure 22.5 can be redesigned using the guidance provided above**

2    running lines should be kept straight. The simplicity of track schematics is
     largely achieved by eliminating the geographical data associated with shape
     (for example bends) and distance. However, classic VSCS reintroduce the
     complexities associated with line shape for the sole purpose of fitting more
     information on the display. Excessive, irrelevant and unnecessary kinks and
     changes in track direction only serve to complicate the display while adding no
     discernable benefit to the signaller;
3    rules determining the gaps between running lines should be consistently applied
     across the display. Random variation in the gaps between running lines adds to
     visual complexity and should be avoided. However, consistent differences in
     the distance between sections of track can be employed to provide useful visual
     clues (for example pairs of running lines, sets of sidings) that improve display
     clarity;
4    schematics should clearly show the geographical relationship between different
     sections of track so that the association between lines is immediately obvious.

*Alarms and Alerts*

Alerts management and design is a critical aspect of signalling control system
development. It has a direct impact on workload and the ability to respond effectively
in an emergency. The way that all indications and alerts are presented must be
consistent with the user interface design model and interaction processes associated
with each specific VSCS. A detailed analysis of the alerts generated and displayed to
the signaller has allowed a set of requirements to be generated which should ensure
that alerts and alarms are effective and allow the signaller to take any appropriate
action as required.

   The following example guidance has been developed with regard to the visual
display of alerts and alarms within a VSCS environment:

1    where an urgent priority alert refers to a location specific event, the location
     of the event shall be highlighted on the track schematic to optimise signallers'
     response time.
        In emergency situations signallers should not be relied upon to remember
     the locations of signals or other infrastructure elements described in an alert
     message. For example, a SPAD alert should highlight the SPADing train on the
     track schematic, its direction (if not immediately obvious) and, ideally, any other
     trains on a potential collision course with it; detection of a hot axle box may
     cause the location of the HABD unit and the associated train to be highlighted;
2    alerts messages shall be displayed such that the associated priority is immediately
     apparent.
        Table 22.1 provides an example of a colour coding system that may be used
     to indicate the priority of alert messages for a VSCS with three levels of alerts
     classification. The selected colours provide distinctive differences in hue and
     brightness;
3    alert message texts should be worded so that their meaning is immediately
     apparent, and should include all the core information that the signaller will need
     to respond effectively and efficiently to the alert.

**Table 22.1 Example colour coding system that may be used to indicate the priority of alert message texts**

| Alert priority | Colour | RGB values | | |
|---|---|---|---|---|
| | | Red | Green | Blue |
| Urgent | ■ | 220 | 0 | 0 |
| High | | 255 | 221 | 77 |
| Normal | | 185 | 252 | 252 |

Signallers should not have to filter redundant information or interpret meaning from alert messages. For example, the alert: 'WARNING: HABD [line name] ACTIVATED' could be more effectively presented so that redundant words (that is, WARNING and ACTIVATED) are eliminated. Signallers see the most important information first (that is, the type of alert) rather than the word WARNING (which would already be apparent). The alert message would be better presented as: 'HABD on [train] at [line name]'.

## Conclusion

The development of ergonomic signalling display guidance has enabled Network Rail to detail some of the key user needs for VSCS design. It is important to ensure that future system design is not constrained by legacy systems and prescriptive standards, but is instead managed and influenced, where appropriate, through the application of ergonomic 'best practice'.

It is necessary to ensure that display designers are aware of all the different states, types and conditions associated with primary VSCS elements (for example track,

**Table 22.2 Segment of signal states and indications matrix**

| Possible signal indications | Colour aspect (x2) | Direction/ association with track | TPWS fitted or not | TPWS failed | Reminder fitted (for regulation) | Inhibit fitted | Possession/ track block from signal |
|---|---|---|---|---|---|---|---|
| Colour aspect (x2) | ■ | ✓ | ✓ | ✓ | ✓ | ✓ | ✓ |
| Direction/association with track | | ■ | ✓ | ✓ | ✓ | ✓ | ✓ |
| TPWS fitted or not | | | ■ | ✓ | ✓ | ✓ | ✓ |
| TPWS failed | | | | ■ | ✓ | ✓ | ✓ |
| Reminder fitted (for regulation) | | | | | ■ | x | x |
| Inhibit fitted | | | | | | ■ | ✓ |
| Possession/track block from signal | | | | | | | ■ |

signals, trains and points) that can exist. To this end a number of matrices have been developed (see Table 22.2) that provide a framework from which designers can determine the types and indications that may be appropriate depending upon the capabilities and interaction design models of each VSCS.

As technology advances so too does the need for future analysis of the user needs. With the advent of moving block signalling systems and the efficiencies they are expected to bring, the signallers' role and information needs will change again. For example it is likely that geographically accurate representations of the rail network will be required. It is important that when this change occurs user needs are at the forefront of system development.

**References**

Carey M.S (1992), ;A Position Paper on Human Factors Approaches for the Design of VDU Interfaces to Computer-based Railway Signalling Systems', Research Report No. 46/1992, HSE.

# Chapter 23

# Using Visual Layering Methods to Design Rail Network Control Room Displays

Darren Van Laar, Ian Andrew and Mathew Cox

## Background

Good colour coding of complex control room displays can be used to improve the time taken to complete tasks, to reduce errors, increase learnability, decrease workload, and to raise subjective ratings of the displays. However, poor colour coding can lead to a reduction in these measures.

The 'visual layering' (VL) method for developing and applying good colour codes for use in complex displays, based on principles of colour science and perception, has been used for years in the nuclear industry but until recently has not been applied to the rail industry.

This chapter outlines the general human factors considerations involved in the process of designing a modern colour coded rail network display using the VL method and presents a possible design solution and evaluation. This chapter will be of interest to any human factors engineer or designer who is involved with the design of complex control room displays for the rail or related industries.

## Introduction

Network Rail's West Coast Mainline is Europe's busiest mixed traffic railway, with over 2,000 passenger and freight trains running every day. Network managers within the new West Coast Rail Traffic Control Centre (WCRTCC) at Saltley, Birmingham, will eventually control all train signalling and operations between London Euston and Crewe. Within the control rooms, computer presented Rail Network Displays (RNDs) are the only visual interaction available between the Network Managers and the railway infrastructure. The RNDs are used in a safety-related environment and are in operation 24 hours a day, seven days a week. It is therefore essential that the configuration of the displays is designed to optimise the interaction between the operator and the displayed information. Rail network diagrams have been in use for some years and conventions for the colour coding of the displays on VDUs have been laid down. However, a recent project undertaken by CCD for Network Rail provided an opportunity to investigate these recommendations to optimise the usability and performance of the displays.

At the simplest level, allocating colours to elements of an information display has long been known to provide a quick way of making the display more attractive and acceptable to users (Tullis, 1981). In its most highly developed form, the process of colour coding can be used as a sophisticated methodology for structuring and designing the whole user interface and the tasks to be supported by it (Van Laar *et al.*, 1996, 2002).

Good colour coding can be used in complex displays to improve the time taken to complete tasks, to reduce errors, increase learnability, reduce workload and to raise subjective ratings of visual comfort, usability and pleasantness (Van Laar and Deshe, 2002). However, poor colour coding can impair these measures. Previous work in cartography has long advocated the use of conceptual layers as a means to organise task relevant data elements on a map or display (Wood, 1968). Conceptual layers contain display elements that have been grouped on the basis of task importance or relevance. The most common method of communicating information within and between conceptual layers is through the use of colour coding; for example by convention most maps use blue to represent areas of water, and green for wooded areas.

It has already been shown in cartography that visual layering effects may be generated through colour codes by following principles of colour science and colour perception (Wood, 1968; Robinson *et al.*, 1995). In addition to being useful for generating layering effects, colour appears to be amongst the most flexible and effective of all of the coding dimensions available to the designer of computer displays, especially when the task to be supported is concerned with searching or locating information (Christ, 1984; Silverstein, 1987).

Van Laar (2001) used colour coding to communicate conceptual layers through a hierarchy of visual effects in nuclear power station control room displays. A set of colour palettes were produced which, when combined in the recommended way, produced the effect of information appearing on perceptually different layers of the display. The main principle of this method was that more important objects and areas appeared 'higher' and more conspicuous to the user. Importance was defined as the extent to which the displayed object supported the operator's task, the frequency of its usage, and its relations with the efficient, effective and safe operation of the system operated through the display.

Researchers at the University of Portsmouth have been working for many years on predicting suitable schemes for developing and applying colour codes for use in complex displays such as control rooms (Van Laar *et al.*, 1996; Van Laar and Deshe, 2002). Specially written software has been developed to calibrate computer monitors and then to produce an effective palette of colours. Such a palette of colours can then be used to code the display using a number of principles based on the use of visual layering techniques and a simple task analysis.

**Colour Coding Analysis**

Given that the RND prototype programme is an extremely complex safety critical system, it was not possible in the early piloting work that this study reports on to test the colour coding system with an RND representing the whole rail network. The process of applying colour coding to the type of RND to be employed at the WCRTCC therefore began with the development of a single typical screen display from a prototype RND screen showing the gross spatial layout of the display. To achieve this, a particularly graphically complex area of the West Coast Main line near Euston was selected as containing most classes of screen element which needed to be coded. All colour and spatial codes were copied from this screen and graphically represented within a computer package (MS PowerPoint) where the different elements could be easily manipulated and colour coded. In order to ensure that all elements of the RND would be assessed through the use of this 'Test' display, a number of other aspects were introduced artificially into the track layout. This included extra track circuit categories, signal states, a level crossing and a station/platform. Two bridges on the original screen were removed. Evaluative feedback received from the human factors and operations teams on this proposed non-coloured screen display provided a basis for amending and improving a number of the originally proposed spatial codes, and a final revised screen arrangement, prior to colour coding was agreed (see Figure 23.1).

## Conceptual Layer Task Analysis

Experienced signallers and members of the Network Management Centre (NMC) human factors team were contacted and were asked to rate each screen element on its importance or relevance to the general task likely to be carried out with the RND. The process of allocating importance levels to the screen elements was useful to the human factors and the operations team as it gave a context in which to more fully explore the task with the operators and to investigate the general and meta level relationships between elements displayed on the screen. For example, initially six levels of importance were available to be allocated to the screen elements, but after negotiations and discussion it was decided that only three levels of rated importance were required. These ratings enabled screen elements to be grouped into layers on the basis of their task importance. The three main conceptual layers generated through this process were:

high importance:        items that had dynamic status, and that also move, that is, trains;
medium importance:   items that had dynamic status, but did not move, for example signals, points;
low importance:        items that did not have dynamic status and also did not move, for example bridges, platforms, direction arrows.

Once provided with the gross spatial coding display and the general levels of importance, the first pass colour coding could begin.

## Creating Visual Levels with Colour

The general aim of the colour coding method is to emphasise the most relevant information, and de-emphasise the remaining information according to levels of importance established for the operator's task, whilst maintaining the overall

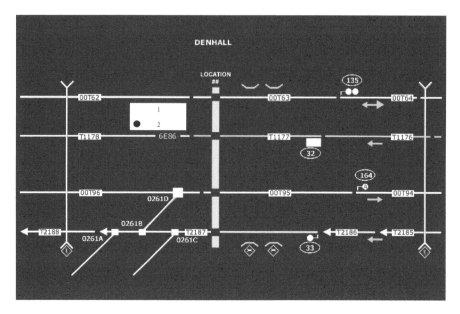

**Figure 23.1  Test RND screen, showing gross spatial coding**

**Key**

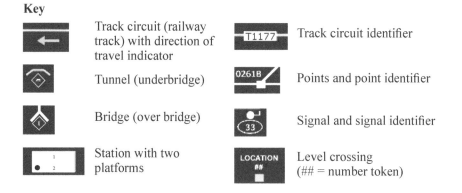

| | | | |
|---|---|---|---|
| | Track circuit (railway track) with direction of travel indicator | | Track circuit identifier |
| | Tunnel (underbridge) | | Points and point identifier |
| | Bridge (over bridge) | | Signal and signal identifier |
| | Station with two platforms | | Level crossing (## = number token) |

conceptual and spatial integrity of the display.

Colour coding can be used in many ways to provide structure to the elements of a display. For example, screen elements possessing a similar hue (for example light and dark green) will be seen as related, elements of dissimilar hue (red, green) will be seen as unrelated. Screen elements with high levels of colour saturation or vividness will appear more conspicuous and therefore important, and elements coded with extremes of light or dark will appear more emphasised relative to elements coded with middle degrees of lightness. Nelson (2000) studied the relationship between spatial coding and colour dimensions and provides evidence for the relative use of colour in defining separate and separable coding dimensions.

Based on theories of colour perception and principles of graphic design, a methodology for structuring complex information displays for control rooms had previously been established (Van Laar, 2000). The methodology is supported by a design tool that enables the designer to select up to eight hues, to define four saturation lightness combinations which give rise to different perceptual levels within the display, and then generate examples of the 32 (8 × 4) colours, along with their relative perceptual difference and colour blindness index between the colours. Once an acceptable palette of colours is produced, the designer can allocate colours to screen elements. Two other palettes are also generated by the design tool, one where highly saturated colours can be specified, the other where the selected saturation lightness combinations can be forced onto specific hues (for example the exact blue hue used for a particular track circuit status) in order to place them within specific perceptual levels.

The colour characteristics of the monitors to be used in the actual control rooms were measured. These were used to develop the colour codes to a known level. These data are necessary to replicate the colour schemes used in this study on different VDUs.

**Allocating Colour to Screen Elements**

The process of allocating colours to screen elements requires a measure of the task importance of each screen element, a definition of the conceptual levels within the display, and a palette of colour codes with which to design relationships between the screen elements. Given this information is available, the next stage of the methodology is to decide which colour from the palettes to allocate to which screen element. Our experience of using this methodology in the past suggested that the first elements to be colour coded are those which have conventional colours attached to them. For example, track circuits coded red indicate by convention 'Track occupied' status, and yellow signal graphics and identifiers indicate the signal aspects as they appear in the infrastructure.

The next stage in the methodology is to go through the hierarchy of task importance allocating colours required to generate the visual layering effect. In most tasks, the most important screen element on the RND is the dynamic position and status of each train, and so this element requires a high level colour – in this case a bright green.

It was also necessary that the track layout and track circuit status should be easily identified at all times. Some restrictions were placed on colour coding in order to comply with conventional colour schemes that are currently in operation. Many of the conventional colour codes will be selected from the highly saturated palette of colours (as they will be of high importance), or from the specific hues palette (if they are to be seen as lying within a conceptual/perceptual layer).

Signals and signal identifiers (ID) were the next most important level of information. With the exception of level crossings, which had to be very easily identified, all other elements including points (and their IDs), track circuit IDs, bridges and tunnels were of a lower level of importance.

Table 23.1 provides a list of screen elements ordered by levels of task importance, and containing conventional colour codes which need to be incorporated into the final design solution.

**Table 23.1  Layers of importance of screen elements and conventional colour codes**

---

**Layer 1: Dynamic status, moving**

| Screen elements | Level of importance |
|---|---|
| Train headcodes | Medium to high |

**Layer 2: Dynamic status, static**

| Screen elements | Level of importance |
|---|---|
| Track circuit status | Red: high, yellow: medium to high |
| | Blue: medium to high, green: high |
| | White: medium to high, grey: medium |
| Signal graphics and identifiers | Red: very high, yellow: medium to high |
| | Green: medium to high, grey: low |
| Points graphics | Medium |
| Reminders | High |

**Layer 3: Non-dynamic status, static**

| Screen elements | Level of importance |
|---|---|
| Direction arrows | Medium to high |
| Level crossings | Medium to high |
| Track circuit identifiers | Medium, and de-emphasised |
| Signal identifiers | More important than point IDs, but may still be relatively de-emphasised. (After comments by operators that signal IDs might be confused with point IDs, an oval shape was set around signal IDs to differentiate and to emphasise them in relation to the point IDs.) |
| Points Identifiers | Low to medium, and de-emphasised |
| Location Names | Medium to high |
| Platforms | Low to medium |
| Bridges (over) | Low to medium |
| Bridges (under) | Medium |
| Signal legs | Low |

---

**The Colour Coding Schemes**

Two colour coding schemes were produced – one optimised for 0 per cent white background (negative polarity) and the other with a 100 per cent white background (positive polarity) display. The hues of the display elements were kept as similar as possible between the displays, only the lightness was manipulated to provide good contrast with the backgrounds. Negative polarity displays have been used in network control rooms for many years, positive polarity displays, however, are the standard set up for most general computer applications such as word processors or spreadsheets.

Previous research comparing different polarity screens has been inconclusive. Snyder *et al.* (1990) found significant advantages in visual task performance for positive polarity screens over negative polarity screens. Both accuracy and times were improved when completing visual search and reading tasks. However, Coren, Ward and Enns (1994) found the opposite effect for reading and search times. Young and Miller (1991) found that for extended screen viewing distances (1520mm), a black background provided much better visual discrimination when more luminous symbols were used. From a human factors perspective it is impossible to conclude whether negative or positive polarity displays enhance performance for a particular application and context. Given the small differences found between positive and negative polarity screens, the Canadian Centre for Occupational Health and Safety recommend that user preference should be the determining factor when setting display polarity (CCOHS, 2002).

The full RGB (red, green, blue) values to create the colours used in the screen studies appear in Table 23.2.

**Evaluation**

Thirteen experienced signallers viewed the colour coded displays under two lighting conditions within a control room simulator. The displays were allocated a score which comprised four ratings given by each participant. In accordance with the advice from the CCOHS, subjective ratings were measured in the experiment for later data analysis. This consisted of four visual analogue rating scales anchored by opposing statements: not at all comfortable/very comfortable, not at all sharp/very sharp, not at all usable/very usable, not at all pleasing/very pleasing. Each of these scales generated scores from 0 to 100, with the ratings on all four producing a maximum score of 400 for the displays for each participant.

*Equipment*

An NEC 20.1″ MultiSync® LCD 2010X, monitor running at a resolution of 1280 × 1024 was used in the study – the same type of monitor as is planned to use within the WCRTCC. Maximum white was a D65 colour white at 180 cd/m².

*Procedure*

The participant was seated at a workstation simulating those designed for the WCRTCC. Instructions were read to the participant from an information sheet concerning the nature of the study and the participant's involvement. The participant was asked to sit comfortably with their eyes approximately level with the edge of the desk (approx. 700mm from the screen). The participant was shown the displays in a random order. For each screen presented the participant was required to fill out a question sheet. The questions asked the participants to rate the comfort, usability, sharpness and pleasingness of the screen with consideration to the background colour. There was also space for them to make any extra comments which they felt were relevant. This was repeated with two levels of illuminance (high: 320 lux and low: 127 lux).

**Table 23.2  RGB screen elements values for negative and positive polarity**
**displays**

| Screen element | Negative polarity RGB Value | Positive polarity RGB Value |
|---|---|---|
| **Screen background** | | |
| % max. white | 0 | 100 |
| Background | 0,0,0 | 255,255,255 |
| | | |
| **Signal** | | |
| Red | 255,0,0 | 255,0,0 |
| Yellow | 245,230,42 | 245,230,42 |
| Green | 0,184,7 | 0,184,7 |
| Grey | 192,192,192 | 145,145,145 |
| | | |
| **Track circuit** | | |
| White | 205,205,205 | 0,0,0 |
| Red | 255,0,0 | 255,0,0 |
| Green | 0,255,0 | 0,255,0 |
| Blue | 41,151,248 | 41,151,248 |
| Yellow | 245,230,42 | 255,255,0 |
| Grey | 93,93,93 | 206,206,206 |
| | | |
| **Identifiers** | | |
| Points and signal text | 255,255,255 | 0,0,0 |
| Points ID frame | 75,75,75 | 222,222,222 |
| Track circuit text on background | 255,255,255 on 14,0,51 | 48,48,48 on 244,242,253 |
| | | |
| **Track features** | | |
| Level crossing | 214,138,0 semi-transparent | 214,138,0 semi-transparent |
| Level crossing text | 255,201,105 | 255,201,105 |
| Platform outline and bridge under | 214,138,0 | 214,138,0 |
| Fill for station platform and bridge over and icons | 75,75,75 | 222,222,222 |
| Platform outline | 214,138,0 | 255,135,42 |
| Reminder | 120,248,249 | 120,248,249 |
| Location and train head code | 0,255,0 | 0,255,0 |
| Train head code background | 0,0,0 | 0,0,0 |
| Direction arrow | 106,106,106 | |

## Results

Each screen received an overall score which comprised the sum of the four ratings in the questionnaire. Figure 23.2 shows the mean score for the two background types and two luminance levels.

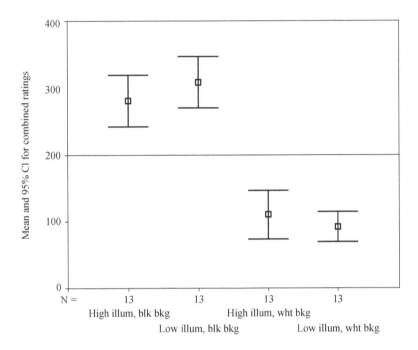

**Figure 23.2** **Average rating of the four combined evaluation scales by level of illumination and background (blk = black, wht = white)**

The displays rated above 200 are rated positively, those below 200 negatively by operators.

A repeated measures ANOVA found no difference between luminance levels $(F(1,12)=0.24, p>.05)$, but screens with the black (negative polarity) backgrounds were significantly better rated than those with the white backgrounds (positive polarity) $(F(1,12) = 72.3, p < 0.001)$. Figure 23.2 also illustrates the significant interaction found between luminance and background $(F(1,12) = 7.6, p = 0.02)$, which suggests that the black background was preferred over the white background in the low illumination conditions, but that the opposite was true for the high illumination condition.

**Discussion**

This chapter has described a method of colour coding Rail Network diagrams using a methodology based on producing visual layers. The method was used to produce two versions of a Test RND, one with a black background, one with a white background. These were evaluated, and subjective ratings of the two colour coding solutions were collected. These ratings covered the fundamental factors which should be considered when designing a good user interface. Therefore, screens receiving higher scores can be considered to provide a better interface than those with lower scores.

The negative polarity display (black background) was rated very highly by operators as being comfortable, sharp, usable and pleasing. The average rating of the negative polarity displays was 294.4 out of 400, whereas the average rating score for the positive polarity display was 100.7. There was no significant difference between levels of illumination of the screens. Informal comments by operators suggested that the white background provided too much contrast, with details being lost through glare.

The results appear to indicate that colour coding can provide highly acceptable displays to users, or displays with very low user acceptability depending on the type of colour coding employed. That the negative polarity displays were so highly preferred over the positive polarity displays goes counter to much of the previous literature. This may possibly be explained by the previous experience of operators in using these displays, or possibly by the fact that an artificial environment of the control room simulator was used.

However, much of the criticism of negative polarity screens in the past has stemmed from interference by screen reflections. However, with the more modern LCD screens, like the ones to be used in the Network Management Centre, reflections are no longer a problem. These conclusions are also supported by the findings of Young and Miller (1991) who found advantages of black backgrounds particularly at extended viewing distances (1520mm). The Network managers will be required to monitor multiple screens and will, therefore, often be viewing the RNDs from similar extended distances.

**Recommendations**

1   A colour coding methodology based on visual layering effects can provide control room displays which have high user acceptability.
2   Railway network displays should employ light foreground colours on a dark background.
3   Future work should investigate non-operator personnel and operators working within real control room environments.
4   Further evaluation trials should be carried out using the recommended RND over a prolonged period to confirm that there are no negative effects from this extended exposure.

# References

CCOHS (2002), *Computer Monitors and Display Colours*, Canadian Centre for Occupational Health and Safety, http://www.ccohs.ca/oshanswers/ergonomics/office/vdtcolor.html.

Christ, R.E (1984), 'Research for Evaluating Visual Display Codes: An Emphasis on Color Coding', in R.S. Easterby and H. Zwaga (eds), *Information Design*, Chichester: Wiley.

Coren, S., Ward, L. and Enns, J. (1994), *Sensation and Perception*, New York: Brooks/Cole.

Nelson, E.S. (2000), 'Designing Effective Bivariate Symbols: The Influence of Perceptual Grouping Processes', *Cartographic and Geographic Information Science*, 27 (4): 261–78.

Robinson, A.H., Morrison, J.L., Muehrcke, P.C., Kimerling, A.J. and Guptill, S.C. (1995), *Elements of Cartography*, 6th edn, New York: Wiley.

Silverstein, L.D. (1987), 'Human Factors for Color Display Systems: Concepts, Methods and Research', in H.J Durrett (ed.), *Color and the Computer*, New York: Academic Press: 27–62.

Snyder, H.L., Decker, J., Lloyd, C. and Dye, C. (1990), 'Effect of Image Polarity on VDT Task Performance', *Proceedings of Human Factors Society 34th Annual Meeting*, Santa Monica, CA: Human Factors Society: 1447–51.

Tullis, T.S. (1981), 'An Evaluation of Alphanumeric, Graphic and Colour Information Displays', *Human Factors*, 23 (5): 541–50.

Van Laar, D.L., Flavell, R., Umbers, I. and Smalley, J. (1996), 'A Methodology for Producing Maximally Discriminable, Nameable Colours in Control Room Displays', in C. Dickinson,, I. Murray and D. Carden (eds), *John Dalton's Colour Vision Legacy*: London: Taylor and Francis: 673–6.

Van Laar, D.L. (2001), 'Psychological and Cartographic Principles for the Production of Visual Layering Effects in Computer Displays', *Displays*, 22 (4): 125–35.

Van Laar, D.L., Deshe, O. (2002), 'Evaluation of a Visual Layering Methodology for Colour Coding Control Room Displays', *Applied Ergonomics*, 34: 123–50.

Wood, M. (1968), 'Visual Perception Map Design', *Cartographic Journal*, 5: 54–64.

Young, H.H. and Miller, J.T. (1991) 'Visual Discrimination on Colour VDTs at Two Viewing Distances', *Behaviour and Information Technology*, 10 (3).

# Chapter 24

# Baseline Ergonomics Assessment of Signalling Control Facilities

Mike Stearn, Theresa Clarke and John Robinson

## Introduction

In 2002, Network Rail's Ergonomics Group commissioned CCD Design and Ergonomics Limited to develop a simple, standardised and holistic methodology for assessing ergonomics performance in signal control facilities. This has become known as the Baseline Ergonomics Assessment programme. To date, the programme has been used to assess 292 signal box panels/IECC workstations in 135 signal control facilities across the UK rail network. Drivers for the programme were: i) Network Rail's response to Recommendation 43 of Part 1 of the Cullen Report; ii) the need for comparative ergonomics performance data between signal boxes; iii) the need for a standardised ergonomics assessment method usable both by professional ergonomists and by Network Rail staff; iv) the need for a coordinated signal box assessment and upgrade strategy within Network Rail. A review was undertaken of over 150 ergonomics signal box studies conducted in the UK between 1997 and 2002. A number of interesting issues emerged to support the need for a new method of assessment: i) studies showed inconsistencies between assessment methods; ii) studies tended to be reactive rather than proactive; iii) studies focussed on specific ergonomics problems rather than on overall ergonomics performance; and iv) studies focused on individual signal boxes rather than on the 'larger picture'. This chapter outlines the process by which Network Rail's Baseline Ergonomics Assessment programme was developed, piloted, implemented and reviewed for compliance. It discusses the measures developed to assess and compare ergonomics performance in signal boxes. It reviews the problems encountered in developing a non intrusive assessment method and in assessing signal boxes over a wide geographic area. Finally it presents deliverables from the Baseline Ergonomics Assessment programme, discusses how these deliverables were communicated to Network Rail's operating Regions and developed into action plans for signal box upgrade.

Network Rail operates over 1,000 signalling control facilities across the UK rail network. Building accommodation, signalling workstations and methods of signaller operation support a wide variety of control systems ranging from older 'mechanical lever frames' to 'electronic entry-exit panels' (NX panels – see Figure 24.1) and on to the more recent VDU-based Integrated Electronic Control Centre (IECC – see Figure 24.2) and Network Management Centre (NMC) systems. A review of ergonomics studies undertaken in UK signalling facilities between 1997 and 2002

**Figure 24.1 NX panel signal control facility**

**Figure 24.2 IECC workstation signal control facility**

showed inconsistencies in the methods adopted to assess ergonomics performance in such facilities. This was due, in the main, to the large number of different ergonomics consultancies or in-house specialists undertaking the studies and the nature of the study briefs which tended to focus on assessing a particular ergonomics issue or problem rather than overall ergonomics performance. Although useful in addressing site specific problems, the ergonomics data set produced from the studies reviewed did not allow confident comparison of ergonomics performance between control facilities or identification of holistic ergonomics issues prevalent across the UK network. Network Rail's Ergonomics Group identified the need for comparative ergonomics performance data to proactively prioritise actions and investment in existing and future signal control facilities. To this end, CCD was commissioned to develop and implement a simple, standardised and holistic methodology for assessing ergonomics performance in signal control facilities – the Baseline Ergonomics Assessment programme.

## Baseline Ergonomics Assessment Programme Method

The Stage 1 Baseline Ergonomics Assessment programme brief was to develop a method to assess Integrated Electronic Control Centres (IECCs) and higher grade panel signal control facilities across all seven Network Rail operating Regions. One hundred and thirty five signal control facilities were identified for Stage 1 Baseline assessment. Many of the larger signal control facilities accommodate more than one operating panel/workstation. As ergonomics issues such as operational demand, equipment interfaces, workstation design, hardware and manning levels differ significantly between panels/workstations, it was decided to assess each operating panel/workstation separately under Baseline. Within the 135 signal control facilities, 292 individual panels/workstations were identified for Stage 1 Baseline assessment.

A set of 30 ergonomics variables were identified as influencing ergonomics performance within signal control facilities. These variables were developed from a review of published ergonomics studies, standards and best practice guidance and in consultation with Network Rail signalling managers, Ergonomics Group and risk assessment managers. The variables were allocated into five ergonomics performance categories as shown in Table 24.1.

One hundred and fifty two questions were developed to assess the ergonomics variables, and ultimately the ergonomics performance of each panel/workstation (see Table 24.1). Where possible, questions were objective (i.e., answers were collected by a physical measurement or by direct observation) however, it was not possible to collect all of the required information objectively and so some information was collected subjectively from on-duty signallers and signal box managers. Approximately 70 per cent of the Stage 1 Baseline information collected is from objective questions and 30 per cent from subjective questions.

Because so many of the questions required direct measurement or observation, it was decided that the questions should be asked and recorded by an on-site Baseline assessor. The 150 questions were developed into a Baseline checklist and piloted by four CCD Baseline assessors in 18 signal control facilities in East Anglia and Southern regions. The checklist and assessment procedure were reviewed with Network Rail and refined before undertaking the main Stage 1 assessments.

**Table 24.1  Ergonomics variables assessed under the Baseline Ergonomics Assessment programme**

| Ergonomics performance category | Ergonomics variables | No. of questions in Baseline |
|---|---|---|
| Panel/workstation | Design | 6 |
| | Display and control interfaces | 13 |
| | Alarms | 7 |
| | Level crossings | 3 |
| | Secondary task activity | 5 |
| Environment | Lighting | 6 |
| | Glare | 4 |
| | Noise | 4 |
| | Thermal | 8 |
| | Vibration | 2 |
| People and organisation | Signaller experience | 4 |
| | Training | 2 |
| | Roster and manning | 4 |
| | Handover and planning | 7 |
| | Supervision | 5 |
| | Work breaks | 2 |
| Welfare | Comfort | 2 |
| | Furniture and posture | 15 |
| | Space | 5 |
| | Access | 4 |
| | Facilities | 4 |
| Operational demand | Control complexity index | 3 |
| | Train numbers and speed | 2 |
| | Stations | 5 |
| | Level crossings | 4 |
| | Yards and depots | 2 |
| | Paperwork | 6 |
| | Communications | 6 |
| | Alarms | 6 |
| | Interventions | 6 |

The pilot established a procedure and method for Stage 1 Baseline Ergonomics Assessment which is outlined in Table 24.2.

The pilot established a simple scoring system for each of the questions covered in the Baseline checklist. Answers to each question were scored on a scale of 1 to 3, where 1 equated to ergonomically effective and 3 equated to ergonomically ineffective as judged against standards and best practice requirements. The scoring system was developed with input from Network Rail and CCD risk assessment specialists. All questions were scored equally that is, no weighting was applied. An Excel analysis model was developed to take answers from each question and automatically assign a score. Scores were then combined (again automatically within the analysis model) to

**Table 24.2　　　Baseline Ergonomics Assessment programme method**

| When | What | How |
|---|---|---|
| Prior to assessment | Arrange assessment with signal box manager | Telephone |
| | Provide communication to signallers to explain aims if the assessment | E-mail |
| On-site assessment | Interview with signal box manager, health and safety representative, union and staff representatives | Record key issues on checklist form A |
| | Take photoset for each panel/ workstation and whole operations floor | Digital camera |
| | Take measurements of operations floor, panel/workstation dimensions and access spaces | Record on checklist form B |
| | Work through 152 questions on checklist with on-duty signaller | Record answers, key issues and comments on form C |
| Off-site post assessment | Input findings from on-site assessment into analysis model | Input into Excel analysis model |

give an overall score for each variable and for each ergonomics performance category. The analysis model also allowed key issues and comments, recorded during the on-site assessment to be captured and linked to the relevant ergonomics performance category.

After the pilot and development of a final methodology, four CCD ergonomists/ designers were trained to undertake Baseline assessments. CCD assessors were experienced in undertaking signal control facility ergonomics projects but had no professional railway experience. In addition, CCD produced a one day training course to train Network Rail staff to undertake Baseline assessments. In total, twelve Network Rail personnel (covering Southern Region, Great Western Region and Headquarters staff) were trained in Baseline assessment.

Stage 1 Baseline assessments were undertaken between March and October 2002. 90 per cent of assessments were undertaken by CCD Assessors and 10 per cent by Network Rail Assessors.

**Analysis and Deliverables**

The following deliverables are produced from the Baseline Ergonomics Assessment programme.

1　　Baseline ergonomics report (for each panel/workstation).
2　　Regional report.

3    Regional action plans.
4    National report.
5    National photoset database.
6    Regional support, training programme and user guide.

*Baseline Ergonomics Reports*

The Baseline Ergonomics Report is automatically produced from the analysis model for each panel/workstation. The Baseline Report provides scores for each of the ergonomics performance categories below:

i)    panel/workstation performance (scale 1–3)
ii)   environment performance (scale 1–3)
iii)  people and organisation performance (scale 1–3)
iv)   welfare performance (scale 1–3)
v)    operational demand (scale 1–3)
vi)   ergonomics effectiveness (scale 4–12) calculated as:
      $\sum$ [panel performance score + environment performance score + people and organisation performance score + welfare performance score];
vii)  Baseline rating (scale 4–36) calculated as:
      ergonomics effectiveness score $\times$ operational demand score.

The Baseline Ergonomics Report also provides a summary of key issues and a summary of comments and observations for the panel/workstation assessed.

*Regional Report*

Baseline Ergonomics Report scores are combined into 'league tables' for each Network Rail operating Region (East Anglia, London North East (LNE), Midland, Southern, Great Western, North West, Scotland). 'League tables' order panel/workstations assessed within an operating Region by worst to best score in each ergonomics performance category. 'League tables' are presented, along with an analysis commentary, to Network Rail regional managers in regional reports. The 'league tables' allow the regions to identify their worst performing signal control facilities, compare ergonomics performance between signal control facilities and assess and prioritise key ergonomics issues across the region.

*Regional Action Plans*

Network Rail's Ergonomics Group and regional managers develop together an action plan for each panel/workstation assessed. The action plan takes the key ergonomics issues from the Baseline Report and provides a formal structure for development, tracking, costing and implementation actions to address poor ergonomics performance. A blank action plan is presented as Figure 24.3.

   Network Rail's Ergonomics Group and regional managers discuss a timescale for action planning and implementation either at a regional or national level.

| BASELINE REGIONAL SIGNAL BOX ACTION PLAN | | | | | | | |
|---|---|---|---|---|---|---|---|
| Reference: | **AP** | | Date of Assessment: | | | | |
| Signal Box: | | Region: | | Signal Box Baseline Ranking: **/135** | | | |
| **ACTION PLAN:** | | | Ergonomics Effectiveness Ranking: 292 | Panel Operational Demand Ranking: /292 | | | |
| **Key Issues** | *Suggested Action* | *Action Method* | *Agreed* | *Regional/ National Role* | *Priority:* | *Cost prediction (£)* | |
| **Welfare Performance** | | | | | | | |
| **Environment Performance** | | | | | | | |
| **Panel Performance** | | | | | | | |
| **People and Organisation** | | | | | | | |
| **Operational Demand** | | | | | | | |
| | | | | | | | |

**Figure 24.3   Blank Regional Baseline Ergonomics Assessment programme action plan**

*National Report*

Data from all Baseline assessment checklists (that is, for all operating regions) are loaded into an Excel spreadsheet to provide a national analysis of ergonomics performance within signal control facilities. Results are presented to Network Rail in a national report. Ergonomics performance scores for each panel/workstation assessed are presented in national 'league tables'. Key ergonomics issues common to all operating regions are combined to provide a priority list of key national issues. These key national issues are those considered for central, rather than regional funding and action implementation. Scatter charts are presented for each ergonomics performance category to show regional performance plotted against national performance for each panel/workstation assessed. An example is shown in Figure 24.4.

Each triangle in Figure 24.4 represents an individual panel/workstation which can be identified and labelled on the chart. The further a triangle appears to the right of

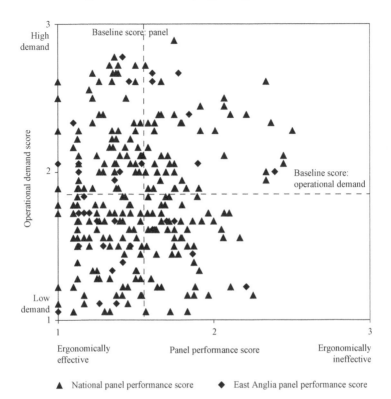

**Figure 24.4   Analysis of East Anglia panel performance against national scores**

the scatter chart, the worse its ergonomics performance. The higher a triangle appears on the scatter chart, the higher its operational demand. The dotted lines on each scatter chart represent the average score for all panels/workstations for the ergonomics performance measures on the x and y chart axes. The panels/workstations at most ergonomics risk are those in the top right quadrant of any chart (i.e., those with worst ergonomics performance and highest operational demand). The aim of any regional action plan is to implement actions that move the panel/workstation triangle nearer to the bottom left corner of the chart (point 1/1). Benefits of investment in upgrade actions can be assessed by re-Baselining a panel/workstation after action implementation to see if its location on the scatter chart has moved towards point 1/1.

*National Photoset Database*

During Baseline assessment, a standard set of digital photographs are taken in each signal control facility. These cover exterior and interior of the building, access routes, panel/workstation, furniture, environment, signaller posture, control and display interfaces and facilities. An Access database has been developed to hold the Baseline photosets. The main menu page is shown in Figure 24.5.

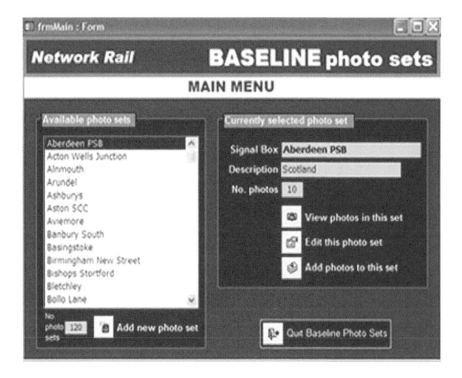

**Figure 24.5   Baseline photoset database**

The database allows users to select individual signal control facilities and view photosets, either by labelled photograph or as an animated walkthrough slideshow. The aim is to make Baseline photoset database accessible within Network Rail via the Intranet.

*Regional Support, Training Programme and User Guide*

Action plans produced from the Baseline Ergonomics Assessment programme have financial, resource and planning implications for the operating regions. Network Rail's Ergonomics Group provides a resource to support the regions in understanding and implementing Baseline recommendations. Regular liaison meetings are held with regional managers throughout the action plan development and implementation phases. Financial and logistical support is provided within a coordinated national plan for signal control facility upgrade. Training courses are provided to the regions to explain Baseline objectives and process as well as to train regional baseline assessors to undertake follow-up Baseline assessments. Finally, a Baseline Users' Guide has been developed to document all aspects of the Baseline programme.

**Future Developments**

The Stage 1 Baseline programme covered the larger IECC and panel signal control facilities. It is planned that Stage 2 will provide further coverage of panel facilities not included in Stage 1 and provide assessment of larger lever frame facilities. This will allow interesting comparison of ergonomics issues between different facility types, ages and operation methods. A Baseline assessment programme will be offered to the regions to allow particular facilities of concern to be assessed centrally. Baseline training will be rolled out to all regions. Hopefully this will allow smaller facilities to be assessed locally and included within the regional and national Baseline dataset. One of the original Baseline aims was that the programme should be transferable to other types of control facilities. It is planned that the Baseline programme will be developed and piloted in electrical control rooms (ECRs) and Regional Controls. It is important that the Baseline programme interfaces with other assessment programmes within Network Rail. Links have already been established with Risk Assessment and Panel Engineering Lifecycle programmes and further links will be developed with staff welfare and building/asset management programmes.

Chapter 25

# Photo-biological Lighting: An Avenue for Maintaining Alertness

Martin Reid and Martin Barnes

## Introduction

The rail industry relies on shift workers for control of safety critical systems. The performance of these workers is critical for the economy and safety of both staff and the public. Unfortunately shift work can have both health effects and operational safety implications. A review of industrial accidents in Australia concluded that work related fatalities are more than twice as likely to occur at night (Williamson and Feyer, 1995) and sleep deficit has also been implicated in many major catastrophes (Mitler *et al.*, 1988).

As an example consider railway control centres and signal boxes. Signal boxes in the UK vary in age from new to over 100 years old. The signallers can usually control the internal environment in these boxes to some extent. Generally the environment can be characterised as follows:

*   warm – signal boxes are often very warm, ancillary heaters are often used to boost temperature;
*   dark at night – most signal boxes are very dark at night. in some cases this is to reduce glare on screens but generally it is to give an environment that operators perceive as more comfortable because of poor lighting design;
*   quiet – as the night wears on, traffic reduces along with the need for communication this means that signal boxes become very quiet.

These three factors are likely to increase the natural tendency to sleep between around 0000 and 0500. It would not be surprising, in these conditions, if signallers took a nap during quiet parts of the night shift. Napping is not uncommon, between 10 and 20 per cent of night workers report falling asleep at work (Akerstedt, 1995). In a Swedish survey of 1000 train drivers, 11 per cent admitted to 'dozing off' on most night trips (Akerstedt *et al.*, 1982) and an additional 59 per cent had experienced this at least once. In some circumstances controlled napping may be a reasonable strategy for managing fatigue. In one study the effect of allowing pilots to nap was examined and it was found that compared to pilots who did not nap, those that did were better able to perform at the end of the flight (Rosekind, 1994).

In order to minimise the negative effects of shift work, a number of strategies have been employed. These fall into the following three categories:

- optimisation of shift patterns;
- controlled or managed napping;
- improvements in the working environment.

Optimised shift patterns have been shown to be effective and are described in Monk (1992). The use of napping may be considered but clearly needs to be controlled and carefully considered. It is also likely to be unfeasible in some operational environments. Environmental improvements can be made through better heating and ventilation but the major avenue for action is through lighting. The remainder of this chapter considers the effect that lighting can have on maintaining alertness at night.

### The Effect of Light on Circadian Rhythms

The lighting of safety critical areas must be adequate to support the work undertaken, this can generally be taken to mean the provision of sufficient light to adequately illuminate work areas and provide a comfortable environment. However, light can have biological effects over and beyond illumination and these can be used to improve both performance and well being.

Many animals and in particular mammals (including humans) follow a daily cycle. The human circadian cycle has been measured to be a little over 24 hours (24.18 hours (Czeisler *et al.*, 1999)). This cycle is exhibited through a number of measurable functions, for example body temperature and hormone levels. While this is a self regulated mechanism it requires an external stimulus to re-set it. The re-set is provided by exposure to light (Hatonen, 2000). Shift workers continue to be exposed to daylight and so their body clock continues to be 're-set' to the local time. It has been shown that permanent night workers usually do not lose this orientation of their internal clock (Roden *et al.*, 1993). Their sleep/wake pattern is permanently out of sync with their work/rest cycle. Day sleep, after a night shift is usually shorter than night sleep by two to four hours (Akerstedt, 1995). Loss of sleep is associated with performance impairment and a loss of alertness (Carskadon and Roth, 1991, quoted in Rosekind *et al.*, 1997).

One key hormone that varies through the 24-hour cycle is melatonin. Melatonin depresses body temperature, facilitates sleep onset and if administered as a pill can advance or delay the circadian rhythm. Melatonin varies through the day and affects both perceived alertness and sleep. Irrespective of whether a person is awake in dim light or even asleep, melatonin is usually secreted between 2100h and 1000h, with peak levels occurring between 0200 and 0600 (Hatonen, 2000).

Light of the correct intensity and spectrum can suppress the production of melatonin and shift the phase of its production. The mechanism by which production of melatonin is suppressed is not fully understood. However, it is likely that the primary route involves the retina,[1] SCN (suprachiasmatic nucleus) and the pineal gland which secretes melatonin.

Light not only suppresses melatonin level, it can be used to phase shift the melatonin rhythm. A light that suppresses melatonin production and shifts its phase would allow shift workers to adapt to night working. This would make them more alert whilst at work and their phase shifted melatonin level would rise when they were resting during the day, improving the quality and length of sleep. The qualities of light required to achieve these effects are described below.

**Photo-biological Lighting**

There are a number of aspects of lighting that influence its effect on the circadian rhythm including quantity, spectrum, timing and duration. These factors, which are described below, are not isolated; rather, they act together in combination.

*Light Quantity*

Light quantity is measured in lux and is generally measured at the working plane but can also be measured in the plane of the eye. Outdoor illuminance can vary significantly through the day, depending on the time and location, ranging from 2000 to 100,000 lux. Even on cloudy days it is generally much brighter outside than inside where lighting rarely exceeds 500 lux (Rea *et al.*, 2002). The effect of light on circadian rhythms depends on brightness but also on spectrum (see below). This means that for a given illuminance, the effect on circadian rhythms will differ depending upon the spectrum of the light delivered. Until recently it was believed that very bright light was needed to affect the circadian rhythms. However, it has been shown that light of the order of 200 lux (Monk and Embrey, 1981) has some effect on the production of melatonin and 1000 lux provides near maximum suppression. It should be noted that this effect was found using cool white fluorescents with a colour temperature of 4100K (see below for an explanation of colour temperature), which is much higher that that generally used in offices (2700K). Much higher light intensity would be required with lower colour temperature fluorescents see Figure 25.1.

It is interesting to compare the light level required for effective visual performance with those required for melatonin regulation. Figure 25.2 (Rea *et al.*, 2002) illustrates the illumination levels required at the eye for visual performance (of a high contrast task) and for suppression of melatonin. If the sole purpose of lighting were for illumination, then there would be no need to exceed 100–200 lux for the majority of tasks. However, if melatonin suppression is required illuminance should be at least 1000 lux.

*Spectrum*

All light sources have a range of intensity over their spectrum. The effect of these different spectra is to give the appearance of a different colour. Humans perceive shorter wavelengths as blue whilst longer wavelengths are seen as reds and yellows.

Daylight intensity is relatively uniform over the visible spectrum, however artificial sources generally vary. Fluorescents for example, tend to be subject to a number of peaks and incandescent bulbs give more light in the red end of the spectrum. Manufacturers can manipulate the output of fluorescent tubes to give the appearance of different colours. In order that these are easily comparable they are referred to as a colour temperature. This is the temperature at which a black body's radiation best matches the fluorescent output.

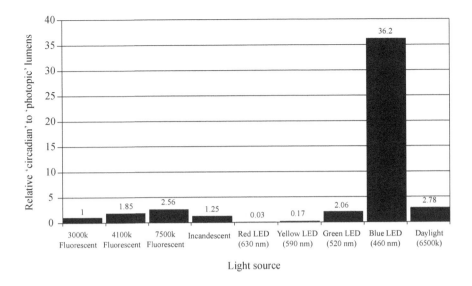

**Figure 25.1   The relative effect of various light sources on the circadian system**

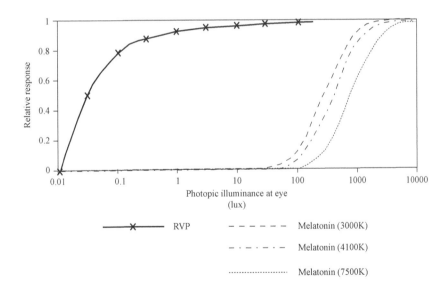

**Figure 25.2   Relative visual performance (RVP) for high contrast reading and relative melatonin suppression as a function of illuminance at the eye**

*Source*: Rea *et al.*, 2002.

It has been shown that the effect on the circadian system varies considerably depending on the spectrum of the light (Rea *et al.*, 2002). Data from Rea *et al.*, 2002 is shown in Figure 25.2. This illustrates the relative effect of various light sources on the circadian rhythm.

It is clear that blue light has by far the greatest effect on circadian rhythms. This means that a far lower intensity of blue light could be used to have the same effect as, say, a fluorescent light.

*Timing*

Depending on the timing of exposure, light can either advance or delay the phase of the circadian cycle. Light exposure in the early night will delay the timing of the clock while light exposure in the late subjective night and early morning will advance the timing of the clock (Czeisler *et al.*, 1989). For night shift workers this suggests that light should be applied in the early part of the night[2] to move the circadian phase back so that they are ready for sleep the following day.

*Duration*

The effect of light in suppressing melatonin appears to be dose dependent, the brighter and the longer the exposure to the light then the greater the effect. Even short exposure to bright light can have an effect on melatonin suppression. The effects of duration are illustrated in the table below. This summarises the effects of exposure to 1000 lux for 20, 30 and 40 minutes on melatonin excretion, where melatonin suppression is expressed as a percentage of the level which would have been produced without the light (Rea, 2002).

**Table 25.1  Light dose and melatonin reduction**

| Time exposed to 1000 lux | 20 minutes | 30 minutes | 40 minutes |
|---|---|---|---|
| Reduction in melatonin produced | 30% | 45% | 55% |

*Circadian Light Exposure in Action*

The principles outlined above have been used in a number of contexts both in laboratory tests and in industry. Various different lighting systems are described within the literature including a number specifically developed for the working environment (for example, CLS lighting (Figueiro *et al.*, 2002) and SIVRA). Whilst such systems vary in their technical detail they have the common aim to shift the phase of circadian rhythms in order to promote alertness whilst at work while at the same time leaving staff better able to sleep during subsequent rest periods.

Reported studies of this type of lighting fall into two groups: those that concentrate on the biological effect of light and those that look at the psychological effects. A good example of the former is the study by Boivin and James, 2002. Nurses working regular night shifts were exposed to bright light during their night shift and compared to a control group of nurses working the same shifts but without the light exposure.

The circadian rhythm phase of the nurses subject to the light was delayed by 9.5 ± 1.06 hours (core body temperature) and 11.3 ± 1.13 hours (melatonin). These were significantly longer than the delay for the control group (4 ± 2 hours and 5 ± 2 respectively). Any effects on the nurses' well being or performance is not reported.

Examples of studies which look at the psychological effects of light include a study of an office environment (NRC headquarters) (Baker *et al.*, 1994) and laboratory experiments of the SIVRA system based on control room tasks (Boyce *et al.*, 1997). The study of the NRC headquarters evaluated the effects of using a 'circadian lighting system'(CLS). The CLS varied light intensity over the shift between 800 and 5500 lux (the colour temperature of the tubes used is not stated). Comparing measures before the light system was fitted they report:

* less subjective effort required performing duties between 0100 and 1300;
* lower sleepiness;
* higher and more even alertness;
* lower reaction time (over a number of test types);
* longer and more refreshing sleep;
* better adaptation to night work.

**Figure 25.3   iGuzzini SIVRA**

iGuzzini, the Italian lighting manufacturer, have developed the SIVRA lighting system specifically to provide a high quality lighting system that is both comfortable and has a Photo-biological effect (Figure 25.3). The SIVRA system has a number of fluorescent lamps of differing colour temperature. These are controlled to give a dynamic light programmed around the shift pattern, time of year and latitude of the installation. This is generally programmed to be bright in the early night and to reduce during the night. This has the effect of shifting the circadian phase back, increasing workers alertness at work but allowing them to rest after work. This was demonstrated in an evaluation of SIVRA based on typical control room tasks in a laboratory. In this study 20 subjects carried out a series of tests over 12 nights in four lighting conditions as follows:

1   low constant light (250 lux) – typical of many control rooms;
2   high constant light (2800 lux);
3   dynamic light gradually increasing from 200 to 2800 lux;
4   dynamic light gradual decreasing from 2800 to 200 lux.

The results of the experiments can be summarised as follows:

*   the high and increasing light levels improved night shift workers performance on complex cognitive tasks relative to the low and decreasing lighting conditions;
*   perceived arousal was greater in the high, increasing and decreasing lighting conditions than the low condition;
*   body temperature dipped as expected under the low lighting condition but did not under the other conditions;
*   the change in performance of the complex cognitive tasks with different lighting conditions is consistent with exposure to bright light changing the subject's wakefulness.

These and other studies clearly show that the use of appropriate light can have measurable biological and psychological effects, and that this can be used to help minimise the negative impact of night work. Whilst there remains some debate as to the type of tasks where benefits are greatest (for example a further laboratory trial of SIVRA showed improvement in the performance of cognitively complex tasks but not for simple tasks) there is mounting evidence that the technology can be used to improve morale, alertness and job satisfaction. This is reflected in the greater recognition and application across a number of industries, in particular the rail industry where it has recently been fitted in the new Signalling Control Centre at Rugby and at the Stoke Signalling Centre.

## Design Issues

The design of any lighting system in safety critical areas needs careful consideration. The design of a Photo-biological lighting system needs to provide a high quality visual environment that supports the operations as well as providing a biological effect. These two requirements, while not contradictory, mean that specialist design skills are needed if both requirements are to be met. The key areas to consider are summarised here.

*   Shift pattern – the lighting control programme should be unique to each application so that it corresponds to the specific shift pattern. This should include consideration of 'hand over' periods where both 'starting' and 'finishing' shifts overlap. The aim of the programme is to give each shift a 'normal day' starting at their shift starting time.

- Specific industry safety requirements – certain industries such as nuclear power generation will consider that the lighting of a control room is safety critical. This could influence how the fittings are wired that is, the power switch element of the control panel could be by-passed therefore should any problem with the control panel occur then the light output would default to 100 per cent output.
- Operator location – where will the operator be positioned for the majority of the working shift? If the operator spends most of their shift in a very local area such as around a desk/control console, the system could be used as a 'task light', only lighting a localised area. If not, a general Photo-biological lighting scheme will need to be designed for the room as a whole.
- System technology and display types – these could include CRT, TFT, plasma, LCD, large format displays (front and rear projection), hardwired illuminated alarm panels, mosaic mimic panels etc. Due to the high lumen output from Photo-biological lighting systems care needs to be taken in their positioning.

  The fittings have a partially polarising diffuser, which greatly reduces problems associated with direct glare but does not eliminate reflections. This problem is most acute in environments where there are a number of 'negative screen' (predominately black background) operating systems, which are common in DCS and SCADA systems.
- Surface finishes – both colour and reflectance values, especially on the desk/ control console need to be carefully selected. Light coloured, highly reflective surfaces could lead to both operator retina and general fatigue.

**The Future**

The potential benefits that photo-biological lighting appear to offer in counteracting some of the effects of night work may be considerable. These would apply not only to enhancing safety, performance and general well being but also to more flexible use of buildings (in particular the more effective use of windowless rooms).

To fully understand and capitalise on these potential benefits, more investigation is needed. It is clear that melatonin excretion can be delayed by application of light. However, further work is required on the relationship between melatonin level, lighting and performance so that the anticipated benefits are fully understood and fully documented.

Armed with such evidence it may be possible to offer a significant improvement to the practical control of many of the problems which frequently effect night workers.

Photo-biological lighting has recently been applied on the West Coat Main Line Signalling Centres at Rugby and Stoke. The technology could also be applied to other areas in the rail industry such as:

- control rooms;
- train cabs;
- rest areas.

## Notes

1   Other mechanisms have been suggested such as skin exposure to light (Campbell and Murphy, 1998). However, this remains unproven with other workers unable to repeat these results (Lockley *et al.*, 1998).
2   When bright light is applied before the minimum core body temperature is reached the circadian clock is phase delayed (Rea *et al.*, 2002). The core body temperature starts to drop from around 2200 with a minimum at around 0400 (Monk and Embrey, 1981).

## References

Akerstedt, T. (1995), 'Work Hours, Sleepiness and the Underlying Mechanisms', *Journal of Sleep Research*, 4 (supplement 2): 15–22.

Akerstedt, T., Torsvall, L. and Froberg, J. (1982), 'A Sleep Questionnaire Study of Sleep/Wake Disturbances and Irregular Hours', *Sleep Research*, 12: 358.

Baker, T., Morisseau, D., Murphy, N., Buckley, K and Persenky, J. (1994), 'Use of Circadian Lighting System to Improve Night Shift Alertness and Performance of NRC Headquarters Operations Officers', in *Proceedings* of the US Nuclear Regulatory Commission Twenty-Second Water Reactor Safety Information Meeting, Bethesda, MD, 24–26 October, Vol. 1, Publication No. NUREG/CP-0140; Vol. 11995, Washington, DC: US Nuclear Regulatory Commission.

Boivin, D. and James, F. (2002), 'Circadian Adaptation to Night-shift Work by Judicious Light and Darkness Exposure', *Journal of Biological Rhythms*, 17 (6), December.

Boyce, P.R., Beckstead, N.H., Eklund, R., Strobel, R. and Rea, M. (1997), 'Lighting the Graveyard Shift: The Influence of a Daylight-simulating Skylight on the Task Performance and Mood of Night-shift Workers', *Lighting Research Technology*, 29 (3).Monk, T. (1992), *Making Shiftwork Tolerable*, London: Taylor and Francis.

Campbell, S and Murphy, P. (1998), 'Extraocular Circadian Photo transduction in Humans', *Science*, 279, 16 January.

Carskadon, M.A. and Roth, T. (1991), 'Sleep Restriction', in T.H. Monk (ed.), *Sleep, Sleepiness and Performance*, Chichester: Wiley: 155–67.

Czeisler, C.A., Duffy, J.F., Shanahan, T.L., Brown, E.N., Mitchell, J.F., Rimmer, D.W., Ronda, J.M., Silva, E.J., Allan, J.S., Emens, J.S., Dijk, D.-J. and Kronauer, R.E. (1999), 'Stability, Precision, and Near-24-hour Period of the Human Circadian Pacemaker', *Science*, 284: 2177–81.

Czeisler, C.A., Kronerauer, R.E., Allan, J.S., Duffy, J.F., Jewitt, M.E., Brown, E.N. and Ronda, J.M. (1989), 'Bright Light Induction of Strong (Type 0) Resetting of the Circadian Pacemaker', *Science*, 244: 1328–33.

Figueiro, M., Eggleston, G. and Rea, M. (2002), 'Effects of Light Exposure on Behaviour of Altzheimer's Patients – A Pilot Study', *Light and Health*, EPRI/ILO 5th International Lighting Research Symposium, Palo Alto, CA.

Hatonen, T. (2000), 'The Impact of Light on the Secretion of Melatonin in Humans', dissertation, University of Helsinki.

Lockley, S., Skene, D., Thapan, K., English, J., Ribeiro, D., Haimov, I., Hampton, S., Middleton, B., von Shantz, M. and Arendt, J. (1998), 'Extraocular Light Exposure Does Not Suppress Plasma Melatonin in Humans', *Journal of Clinical Endocrinology and Metabolism*, 83 (9).

Mitler, M.M, Carskadon, M.A, Czeisler, C.A, Dement, W.C, Dinges, D.F and Graeber, R.C. (1988), 'Catastrophes, Sleep, and public policy: consensus report', *Sleep*, 11: 100–109.

Monk, T.H. and Embrey, D.E. (1981), 'A Field Study of Circadian Rhythms in Actual and Interpolated Task Performance', in A. Reinburg, N. Vieux and P. Andlauer (eds), *Night and Shift Work: Biological and Social Aspects*, Oxford: Pergamon Press: 473–80.

Rea, M.S., Figuerio, M.G. and Mullough, J.D. (2002), 'Circadian Photobiology: An Emerging Framework for Lighting Practice and Research', *Lighting Research and Technology*, 34 (3): 177–90.

Rea, M. (2002), 'Light Much More Than Vision', *Light and Health*, EPRI/ILO 5th International Lighting Research Symposium, Palo Alto, CA.

Roden, M., Koller, M., Vierhapper, H. and Waldhaus, E.R. (1993), 'The Circadian Melatonin and Cortisol Section Pattern in Permanent Night Shift Workers', *Regulator and Comparative Physiology*, 265.

Rosekind, M. (1994), 'Fatigue in Operational Settings: Examples from the Aviation Environment', *Human Factors*, 36 (2): 327–38.

Rosekind, M., Neri, D. and Dinges, D. (1997), 'From Laboratory to Flightdeck: Promoting Operational Alertness', in *Fatigue and Duty Limitations – An International Review,* The Royal Aeronautical Society: 7.1–7.14.

Williamson, A. and Feyer, A. (1995), 'Causes of Accidents and the Time of Day', *Work and Stress*, 9 (2/3): 158–64.

Zeitzer, J., Dijk, D., Kronauer, R., Brown, E., and Czeisler, C. (2000), 'Sensitivity of the Human Circadian Pacemaker to Nocturnal Light: Melatonin Phase Resetting and Suppression', *Journal of Physiology*, 526 (3): 695–702.

# PART 8
# PLANNING AND CONTROL

Chapter 26

# Analysis and Support of Planning in the Dutch Railroad Company

René Jorna, Wout van Wezel, Derk Jan Kiewiet and Thomas de Boer

**Introduction**

One of the outstanding issues in planning research and practice is the relation between the cognitive aspects of the task of human planners, the organisational aspects of planning, and computer support of planning. In this contribution, we discuss two projects at the Dutch Railroad Company where we try to use the synergy of a multidisciplinary approach. In the first project, cognition is related to the representation of the planning domain and reasoning steps. In the second project, cognition is related to task support.

During the last ten years two basic starting points have guided our research into planning. In the first place we believe that planning is so complex and so much involved with organisational and cognitive issues that, despite the power of many modeling and algorithmic techniques, a mathematical/formal approach to planning alone is not sufficient (van Wezel and Jorna, 2001; van Wezel and Jorna, 1999). One has to look at the way planners as cognitive systems within organisations solve planning problems. This implies a cognitive task analysis as well as an analysis of organisational settings. The second starting point is that, despite the fact that much variety exists in planning situations, scheduling domains and human cognition, the planning task itself is a generic task (Breuker and van de Velde, 1994; van Wezel and Jorna, 1999; Schreiber, *et al.*, 2000). This means that in planning as a task – in which a configuration is created or an assignment is realised – generic components can be distinguished. At the problem solving level a planner working with lorries and freight loads is doing similar things as a ward nurse assigning shifts and staff. This is not to say that domains are not important, but it means that domain abstraction from a cognitive or organisational point of view is possible. This perspective facilitates reuse of conceptual modules as well as planning support components.

During the last three years we conducted several empirical research projects at the Dutch Railroad Company (DRC – in Dutch: *de Nederlandse Spoorwegen NS*). We are studying the rolling staff planning, that is to say the assignment of engine drivers and train guards to trains, and the shunting planning at railroad yards in terms of task execution supported by software. In the next section, we will discuss the theoretical background of our studies before outlining the rest of the chapter.

## Planning From a (Cognitive) Task Perspective

Our perspective on planning is task oriented. A task is defined as a sequence of actions in order to reach (various) goals taking into account (various) constraints (Waern, 1990). A cognitive task implies that the acting entity has cognition. The relevance of the human planner is not disputed in many discussions about planning (Zweben and Fox, 1994). What is often missing, however, is the fact that the real life study of the way planners and schedulers execute the planning task requires a more detailed *cognitive* task analysis approach (Schraagen *et al.*, 2001).

Cognitive task analysis of planning contains the study of human (intelligent) activities (tasks) (Miller, Galanter and Pribram, 1960; Hoc, 1988; van Wezel and Jorna, 2001). Planning and cognition are closely connected, which can be seen in two closely related topics. The first deals with the question of whether or not planning is a kind of problem solving (Newell and Simon, 1972; Das, Kar and Parrila, 1996). Because planning as well as problem solving means searching for routes, that is, sequences of actions which lead to a solution or a goal state, the explicit distinction between planning and problem solving disappears (Newell and Simon, 1972). The second is about the question whether human planners work hierarchically or opportunistically (Newell and Simon, 1972; Hayes-Roth and Hayes-Roth, 1979). In contrast to the classical view of planning as a hierarchically ordered decision process, Hayes-Roth and Hayes-Roth (1972) found that humans do not work this way. Rather, they plan opportunistically with the help of some kind of mental blackboard where pieces of information, relevant cues, and possible sub-goals are stored.

The discussion about the cognitive aspects of planning is mixed up with the issues related to planning support, the interaction between planner and algorithms and the reuse of conceptual modules and support gradation. It is quite common that task models of users and (software) designers often do not match.

In this article, the organisation in which the studies are conducted is the Dutch Railroad Company (DRC). We will discuss this organisation and its planning domains in the next section, followed by issues related to what planners in practice in organisations discussed in the context of our empirical research. The issues related to software support are also discussed followed by conclusions.

## Planning in the Dutch Railroad Company

In 1995 the Dutch Railway Company (DRC) was liberated from direct government influence in response to European regulations. The DRC is still owned by the Dutch government, but behaves like an independent company. Within the boundaries of serving the general task of providing public transportation, the DRC strives towards the maximisation of profit. The main office of the DRC is located in Utrecht. The DRC daily transports one million passengers. Transportation takes place with the help of 2,700 railroad carriages, which run approximately 5000 train services per day. The trains run between 384 stations in The Netherlands. The DRC itself consists of several independent business units, like DRC-stations, and DRC-real estate, of which DRC-Passengers (DRC-P; in Dutch: *NS-Reizigers*) is the most important one. This business unit is responsible for the transportation of all passengers. DRC-P has three departments: Production, Marketing, and Staff and Organisation. Within Production, logistics is responsible for all planning and scheduling of DRC.

Within DRC four kinds of planning distinctions can be found: timetables and other plans; rolling stock and rolling stock staff; partitioning in local planning and central planning (of stock and staff); and year plans (long term) and day plans (short term), again of stock and staff. Overall approximately 300 planners are continuously involved in making plans and schedules.

The overall planning process is as follows. The first step is making a timetable. This timetable has all departure and arrival times, stations to stop at and kinds of trains that run between the stations, such as intercity trains, local trains, etc. The timetables are largely copies of earlier years. Once every five years, major changes take place within the timetable. The timetable is the basis for the material plan, indicating which kind of trains and how many carriages are used for transportation.

We conducted our research in two different planning domains within DRC-P. The first planning domain concerns the rolling stock staff planning at central level (main office in Utrecht) for Day and Year Plan. We empirically studied the domain representations and reasoning steps of planners. In more detail the rolling staff planning concerns the planning of all train tasks into shifts for engine drivers and railway guards.

The second planning domain concerns the planning of rolling stock at a local level (station Zwolle) for Day Plan and Year Plan. This is a design-oriented research on software support for shunting planning at a train yard. In this shunting problem the issue was to design a software support system such that a good balance was realised between the intelligence of the planners in task execution and sophisticated algorithms to create optimal occupation and shunting time on the train yard.

### The Rolling Staff Planning in the DRC: Cognition at Work

Basically the rolling staff planning follows the rolling stock planning that is realised in two ways. In parallel, but interactively, Central and Local and also Day and Year Plan make plans for the rolling stock. After that the rolling staff planning starts, in the same parallel but interactive way, between Central and Local Plan and also between Day and Year Plan. In this staff planning we are only talking about staff in an abstract sense. No real train guards and engine drivers are scheduled but tasks are allocated to shifts. Taking into account certain constraints, tasks, such as being a guard on a train from A to B and from there to C, are lined up and calculated as not exceeding shift length. This planning is the input for the 'real' staff planning involving individual guards and engine drivers. Such (concrete) staff scheduling occurs at 23 major stations in The Netherlands. Within every process phase feedback loops take place. None of the plans is built up from scratch. For the most part, existing plans or schedules form the basis for new plans and schedules. Although it is not considered as such, most planning within the DRC-P is not planning, but re-planning. This does not, however, make planning easier; quite the contrary.

*Questions*

The starting question was as follows: 'What are the differences and similarities between Day Plan and Year Plan in the view they have of planning and in the way they make the concrete plans?' The first part of the question refers to the view of the planners of the domain and the conceptual structures, objects, and relations that the planners see between the various parts of what they are working on daily. In cognitive terms we are talking about various representations the planners have of the planning domain. The second part of the question relates to the reasoning steps, the patterns of thinking when they solve planning problems. This second part also implies looking at the outcomes, the results of the problem solving activity.

*Planning Units*

Year Plan (YP) makes plans until eight weeks before execution. Day Plan (DP) makes the plans for rolling stock staff from about eight weeks until 36 hours before execution. DP consist of two subunits: Projects and Extra transport. The work at DP is very much comparable to Year Plan (YP), with the exception that YP uses a dedicated software program called CREWS, whereas the program for DP is VPT, an older and less sophisticated program. The DP unit consists of 17 planners and the YP unit of 12 planners .

*Material and Kinds of Data*

For the sub-question related to the domain view we used a pile of cards with 27 object words and five relation words. They were developed by studying handbooks, talking to planning experts and to the planners in the units. Examples of these words are: guard, train journey, team, train, passenger task, sunday, local task, supervisor or station. Relation words are: applicable to, leads to, consists of, forms. To get an understanding of the domain representations of planners we asked them to put the cards into a graph.

For the sub-question concerning the reasoning patterns of planners we used the Common-KADS framework (Breuker and van de Velde, 1994; Schreiber *et al.*, 2000). The basic step in KADS is an inference. An inference is a mental activity that changes an input into an output. The complete set of inferences we used consisted of: counting; concluding; classifying; evaluating; making hypotheses; identifying; introducing; dissolving; selecting; sorting; joining; comparing; erasing. The 13 inferences represent possible reasoning steps in solving a planning or scheduling problem.

The structure of the complete empirical research resulted in six kinds of data. Concerning the domain representations of the planners we collected a) the three piles of cards of the individual planners, b) the individual graphs, and we determined c) the aggregated graphs of the various units. Concerning the reasoning steps we have d) the solutions for each of the planning problems, e) the time to find a solution, and f) the scoring and classification of the inferences for each of the planners per planning problem.

*Method and Procedure*

With regard to the representations of the domain, each planner received a pile of 27 cards. They were asked to categorise the cards and then to make a graph of the cards and take as many cards as they thought were relevant. Twenty-five planners participated. A photograph was taken of every graph. For examples see Figure 26.1 where two graphs are depicted.

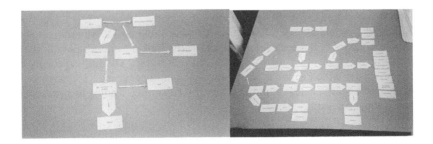

**Figure 26.1   Two examples of graphs (combinations of concepts)**

For the plan execution we developed four representative planning orders. Thirty-five planners, including staff and senior planners, solved the problems by thinking aloud. The data consisted of 140 solutions, 140 time latencies (solution times) and 140 verbal protocols.

*Results*

Large differences were found in the domain representations, not only between units but also within units. Eighteen planners used one sub-graph, six planners two sub-graphs and one planner three sub-graphs. The number of used objects varied between eight to 27 of the 27 possible objects. The units YP and DP varied very much in the objects used. Only six objects were mentioned by at least 75 per cent of the planners in both YP and DP: rolling material composition, shift, train, star, train guard, trajectory. For seven objects it was clear that YP and DP strongly differed, that is to say more than 35 per cent difference. The objects are: *transportation task, local task, train related task, staff plan, station, staff and shift roster*. Large differences existed also within the units. The conclusion clearly is that planners from YP and DP do not have similar representations of the domain they are working in.

Because we had photographs of the domain representations of all the planners, we looked at the graphs not only in terms of the objects used, but also in terms of the connections between the objects. For example, the data showed that 28 per cent of the planners directly connected the objects material composition and train, whereas none of the planners directly combined trajectory and train guard. In this way we made a similarity matrix for all the object combinations. We used INDSCAL to calculate the points in the space for all planners and we also made a distinction between the planners from YP and the planners from DP. The differences for YP and DP consist in the weighting of the axes. For DP the horizontal axis loads only for 0.07, whereas the vertical axis loads for 0.89. In contrast, for YP the horizontal axis loads for 0.96 and the vertical axis for 0.05. For DP the vertical axis is much more important in placing the objects in space, whereas for YP this is the horizontal axis.

Three sets of results were obtained from the protocols developed during the think aloud exercise: the chosen solution, the time to solve it, and the frequencies and sequences of the inferences.

Both the between-unit and within-unit variation of the solutions found is large. The planners have no agreement about what is the best solution for any of the order problems, even if they are in the same unit.

DP solved the problems much faster than YP (see Table 26.1). We interpreted this outcome as follows. YP are more oriented towards looking into all possibilities of the problem; they are more involved in 'deep reasoning'. DP and especially DP-Projects are faster which is due to their experience in improvising. 'Deep reasoning' to find the best solution is less important, but 'fast reasoning' leading to an acceptable solution is most relevant.

**Table 26.1 The solution times for all planners and all orders**

| Year Plan | Day Plan | Projects | Extra transport | Average |
|---|---|---|---|---|
| 15 min. 46s | 12 min. 17s | 10 min. 40s | 15 min. 34s | 13 min. 34s |

In terms of the inferences, the think aloud protocols resulted in 140 verbal statements (four orders for 35 planners) of about one A4 page each. In these protocols the inferences used by the planners were scored. The following inferences are more often used: dissolving, joining, concluding, evaluating and making hypotheses. Erasing, introducing, selecting and classifying are infrequently used. Furthermore, we see that counting, concluding, and evaluating are more often used by YP. We also see that YP make more explicit reasoning steps than DP.

*Conclusion and Discussion*

The domain results showed very well that planners within and between units working in the same domain, with similar constraints and goal functions, did not have the same representations. Interestingly enough the MDS-analysis also showed that the differences between the units had systematic characteristics.

The outcomes of the solutions and the reasoning steps of the order problems showed that planners do not choose the same solution. This means that they do not interpret and prioritise the constraints and goal functions in the same way or, perhaps worse, they were not aware of the constraints and goal functions to be used.

The 'internal' DRC-view on planning is that planners have the same domain representations and reasoning structures. That is not true. The question is whether this variation is positive or negative in its implications. If planners have to cooperate more, which is expected to happen in the future, this is negative. If the planners have to become multi-employable it is also negative. Concerning training and practice our results show that a training programme of a couple of months should be implemented. Finally, we believe that a gap exists between planners' reasoning practices and software support. However, parts of the domain representations and certain inference combinations and inference sequences can also be implemented into software support. How this can be accomplished will be discussed next.

## Shunt Scheduling: Bottom-up Planning Support for Task and Cognition

In a number of research projects, we analysed the planning of train shunting at the Zwolle station. In one of these projects, we extensively analysed the task performance of one of the planners in order to develop task oriented algorithmic planning support. The aim of the shunting project at the DRC was to include algorithms in the planning support, but to let the human planner be in control. The optimal balance between human control versus automatic plan generation is not clear from the literature. Advocates of analytical models argue that humans do a poor job at planning. Advocates of a more human centered perspective, however, state that analytical models cannot deal with uncertainty and instability in the real world (McKay *et al.*, 1988). Specifically, the latter state that the lack of applications of scheduling systems in practice is due to the black box nature of such systems (McKay *et al.*, 1989; Sanderson, 1989). In the mixed initiative approach, the focus is on improvement of the solution by establishing a coalition between the computer and the user. Several authors report the use of the mixed initiative approach for planning support, for example, McKay *et al.* (1995), Sundin (1994), Dorn (1993), Prietula *et al.* (1994), Lui (1993), and Smith and Becker (1997). In the DRC Shunting project, we try to circumvent the need for a priori rules that determine what tasks are done by the computer and what tasks are done by the human planner. In the same way as in the rolling staff project, we apply a bottom-up approach: analyse the planning task, and decide per sub-task what kind of support is needed.

In applying the mixed initiative approach for scheduling support, our line of reasoning is the following (van Wezel, 2001; van Wezel and Barten, 2002). Task analyses show that the overall problem is (nearly) decomposable and solved in a number of steps. In each step, a sub-problem is handled by a sub-task. We can decide per sub-task whether a planner needs to be able to interfere with the search process or not, what kind of algorithm is needed, and create mixed-initiative planning support in which both the computer and the human planner have a role in making decisions. In the remainder of this section, we describe the preliminary results of the application of this approach for the DRC shunting problem.

Trains that arrive at a station do not necessarily leave in the same configuration. During train shunting, trains are split in their individual carriages and these carriages are combined again in trains. For example, the train from Groningen to Zwolle consists of two carriages and the train from Leeuwarden to Zwolle consists of one carriage. In Zwolle the trains are connected, after which they leave as one train to Den Haag. During the night the passenger trains stay at the station. The 'storage' capacity, however, is limited. Additionally, all trains must be washed during the night at a track that contains the washing equipment. Shunting planning is part of the planning of rolling stock. A shunting plan must be made for each station in The Netherlands with shunting tracks. The plans are made locally at the stations, where multiple planners make parts of the shunting plan. So, throughout The Netherlands, dozens of planners are involved in shunting planning.

The task of the planner is to plan the movements of the trains and carriages and to decide on which track trains stay during the night. To plan the movements, the planner must also assign engine drivers, train shunters (the ones who connect and disconnect carriages), and routes of the trains in the station. In the research project, we looked at the local 'day-planner'. His task is to adjust already created plans because events happen, for example, a track that needs maintenance, or an extra train that needs to be shunted. We have extensively analysed several sub-tasks. Currently, the planning tasks are performed manually. Some computer programs are used to collect information, but the plan itself is made on paper before the outcome is put in the computer. The analyses show that there is a large number of relatively small basic sub-tasks that recur in multiple more complex tasks, for example, routing a train, scheduling a train driver, finding a free track, etc. In our approach, we create algorithms to support these small sub-tasks rather than trying to create a black-box algorithm that finds a whole plan without the planner's collaboration. The essence of the approach can be summarised as follows: we make the move from complex algorithms in simple human task structures to simple algorithms in complex human task structures. Currently, a prototype is implemented using an extensible architecture for scheduling support systems (van Wezel et al., 1996; van Wezel, 2001). The prototype provides a rich set of graphical views and manipulation possibilities (Figure 26.2) with a blackboard and real-time constraint checking.

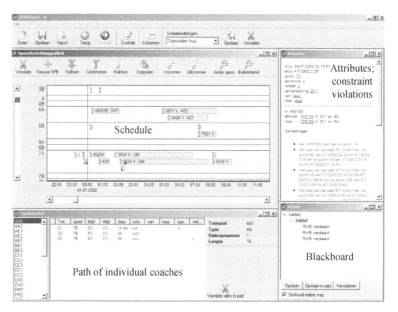

**Figure 26.2  Graphical user interface for shunt scheduling**

In addition to the GUI, constraint checker, and blackboard, several algorithms are available in the prototype, for example: low level algorithms such as finding a free track for a certain interval and determining the direction to drive a train; medium level algorithms such as rescheduling all trains that are on tracks that need maintenance (this latter algorithm uses the lower level algorithms mentioned above); a high level (black box) integer programming algorithm to schedule some or all trains from arrival to departure.

The bottom-up approach of algorithmic design provides an extension to the mixed initiative approaches found in the literature. By describing the creation of the plan as a hierarchical problem-solving process, we place the discussion about human versus computer at the level where it belongs: an algorithm should not be created for a planning problem, but for a planner's sub-task. Algorithms should be able to communicate in terms of the dimensions that the planner uses to define the problem space, and the operators that the planner uses to go from state to state. Current research focuses on an extensive experiment with the prototype to compare manual planning with the different levels of plan generation support.

## Conclusion

In this chapter we discussed two projects at the Dutch Railroad Company (DRC) where we applied a multidisciplinary approach. The projects differ in goal, methodology, scope, and application, but they share several things: seeking improvements in functioning for the DRC by combining planning approaches, taking the planner seriously, looking at the planner's division of the task in sub-tasks, and approaching the planner as a cognitive actor.

In the first project, the goal was to analyse the differences and similarities of approximately 30 rolling staff planners at two different organisational units. Contrary to what was expected, the planners do not seem to share domain representations and reasoning structures. The results can be used to design supporting software, determine training needs, and can help in redesigning the organisation of the planning departments.

The goal of the second project was to create task support, including algorithms, for shunting planners. A combination of cognitive science (for the task analysis), computer science (reuse of system components), and operations research (for algorithms) resulted in a prototype planning system. Because of the bottom-up approach that was taken, algorithms can be tailored to the task performance of the planners.

The two projects are interesting and promising in themselves and will therefore be continued, for example by investigating the consequences of the variety in domain views and problem solving behavior for planning support, and extensive experiments with various levels of planning support. Our overall concern, however, is to generalisation our experiences into a comprehensive planning framework. How are cognitive aspects of the planning task related to the way in which the planning task is divided within an organisation? How can we get a symbiosis between a planner and a scheduling system? How can advanced planning systems be used to redesign the organisation of the planning? Where do algorithms make a contribution? The answers can only be found by crossing the boundaries of the traditional research fields in planning. We hope we have given good examples, and that the DRC will continue to provide us with a rich empirical domain.

## References

Breuker, J. and W. van de Velde (eds) (1994), *CommonKADS Library for Expertise Modeling: Reusable Problem Solving Components*, Amsterdam: IOS Press.

Das, J.P., Karr, B.C. and Parrila, R.K. (1996), *Cognitive Planning*, New Delhi: Sage.

Dorn, J. (1993), 'Task-oriented Design of Scheduling Applications', in J. Dorn and K.A. Froeschl (eds), *Scheduling of Production Processes*, Chichester: Ellis Horwood.

Hayes-Roth, B. and Hayes-Roth, F. (1979), 'A Cognitive Model of Planning', *Cognitive Science*, 3: 275–310.

Hoc, J.-M. (1988), *Cognitive Psychology of Planning*, Academic Press: San Diego.

Lui, B. (1993), 'Problem Acquisition in Scheduling Domains', *Expert Systems with Applications*, 6: 257–65.

McKay, K.N., Safayeni, F.R. and Buzacott, J.A. (1988), 'Job-Shop Scheduling Theory: What Is Relevant?', *Interfaces*, 18 (4): 84–90.

McKay, K.N., Safayeni, F.R. and Buzacott, J.A. (1989), 'The Scheduler's Knowledge of Uncertainty: The Missing Link', in J. Browne (ed.), *Knowledge Based Production Management Systems*, North-Holland: Elsevier Science Publishers.

McKay, K.N., Safayeni, F.R. and Buzacott, J.A. (1995), '"Common Sense" Realities of Planning and Scheduling in Printed Circuit Board Production', *International Journal of Production Research*, 33 (6): 1587–603.

Miller, G.A., Galanter, E. and Pribram, K.J. (1960), *Plans and the Structure of Behavior*, New York: Holt, Rinehart and Winston.

Newell, A. and Simon, H.A. (1972), *Human Problem Solving*, Englewood Cliffs, NJ: Prentice-Hall.

Prietula, M.J., Hsu, W. and Ow, P.S. (1994), 'MacMerl: Mixed-initiative Scheduling with Coincident Problem Spaces', in M. Zweben and M.S. Fox (eds), *Intelligent Scheduling*, San Francisco: Morgan Kaufman.

Sanderson, P.M. (1989), 'The Human Planning and Scheduling Role in Advanced Manufacturing Systems: An Emerging Human Factors Domain', *Human Factors*, 31 (6): 635–66.

Schank, R. and Abelson, R. (1977), *Scripts, Plans, Goals and Understanding*, Hillside, NJ: Lawrence Erlbaum Associates.

Schraagen, J.M., Chipman, S.F. and Shalin, V.L. (eds) (2000), *Cognitive Task Analysis*, Mahwah, NJ: Lawrence Erlbaum Associates.

Schreiber, G., Akkermans, H., Anjewierden, A., Hoog, R. de, Shadbolt, N., van de Velde, W. and Wielinga, B. (2000), *Knowledge Engineering and Management: The Common-KADS Methodology*, Cambridge, MA: The MIT Press.

Smith, S.F. and Becker, M. (1997), 'An Ontology for Constructing Scheduling Systems', in *Working Notes of 1997 AAAI Symposium on Ontological Engineering*, Stanford, CA: AAAI Press.

Sundin, U. (1994), 'Assignment and Scheduling', in J. Breuker and W. van de Velde (eds), *CommonKADS Library for Expertise Modeling: Reusable Problem Solving Components*, Amsterdam: IOS Press.

van Wezel, W. (2001), *Tasks, Hierarchies, and Flexibility. Planning in Food Processing Industries*, Capelle a/d IJssel: Labyrint Publication.

van Wezel, W. and Barten, B. (2002), 'Hierarchical Mixed-Initiative Planning Support', in T. Grant and C. Witteveen (eds), *Plansig 2002. Proceedings of the 21st Workshop of the UK Planning and Scheduling Special Interest Group,* Delft: Delft University of Technology.

van Wezel, W. and Jorna, R.J. (1999), 'The SEC-system: Reuse Support for Scheduling System Development', *Decision Support Systems,* 26 (1), July.

van Wezel, W. and Jorna, R.J. (2001), 'Paradoxes in Planning', *Engineering Applications of Artificial Intelligence,* 14 (3): 269–86.

van Wezel, W., Jorna, R.J. and Mietus, D. (1996), 'Scheduling in a Generic Perspective', *International Journal of Expert Systems: Research and Applications,* 3 (9): 357–81.

Waern, Y. (1989), *Cognitive Aspects of Computer Supported Tasks*, Wiley and Sons: Chichester.

Zweben, M. and Fox, M.S. (1994), *Intelligent Scheduling*, San Francisco: Morgan Kaufman.

Chapter 27

# Future Train Traffic Control: Control by Re-planning

Arvid Kauppi, Johan Wikström, Peter Hellström, Bengt Sandblad
and Arne W. Andersson

## Introduction

The train traffic control system of tomorrow must be able to handle more frequent traffic, higher speeds and several different companies operating on the infrastructure. Performing train traffic control today involves working on a technical level and solving problems as they occur. By shifting the control paradigm to a high-level control strategy many of today's problems may be avoided. Operators will be able to gain a more accurate understanding as to the dynamic development of the traffic process, thereby being able to prevent disturbances more efficiently. The main goal for an operator will be to ensure that there always is a valid plan for the train traffic. This plan is executed, normally by an automated system or, if so required, manually by the human train traffic controller. In order to keep the human in-the-loop and to avoid automation surprises, automatic functions should be unambiguous and predictable to the operator even when conditions are changing. Automatic functions must therefore not autonomously change train routes or the order in which trains are allowed to run. Some potential benefits from this high-level control strategy are the reduction of delays, improved punctuality and better utilisation of the infrastructure.

Modern research on human-computer interaction in complex and dynamic systems provides a framework for how to design an interface implementing these ideas. Important aspects concern workload, situation awareness and automated cognitive processes, limitations in human memory capacity, cognitive work environment problems, human error performance and dynamic decision processes. Throughout the research reported here a user-centred approach has been applied. New proposed interfaces are designed to integrate all decision relevant information into one unified interface and to support a continuous awareness of the dynamic development of the traffic process. Prototypes of new train traffic control interfaces have been implemented and tested in a laboratory environment. Initial tests made with train dispatchers show promising results. Based on the experiences and results from this new control strategy the Swedish National Railway Administration has initiated a feasibility study for the near future implementation of a full-scale operative demonstrator system.

The train traffic system of Sweden is geographically divided into eight control areas. Each area is controlled by a traffic control centre. At the traffic control centre, information about the traffic process status is presented in track diagrams on large distant panels and/or on several regular computer screens. Operators, called train dispatchers, monitor the train movements and control train routes by automatic or manual remote blocking. Track usage is controlled either by ordering automatic functions or by manually executing interlocking routes for each station. Today's control systems are often designed to support the operator's possibilities to react, and to solve disturbances and conflicts when they occur. In order to meet increasing future demands, new principles and technical solutions are required for an efficient train traffic control. Operators should be able to follow the dynamic development of the traffic system over time and prevent disturbances. We call this control strategy *control by re-planning*. This can be compared to the more passive strategy of today, *control by exception* (Sandblad, Andersson, Frej and Gideon 1997; Andersson, Sandblad and Nilsson 1998). One main objective during this research has been to shift the control paradigm from low-level technical control tasks into higher-level traffic re-planning tasks. Re-planning tasks must be supported by efficient user interfaces that allow the train traffic controller to be continuously updated and able to evaluate future traffic conflicts so that these can be taken care of in time. Improving train traffic control could then be a very cost efficient way to improve utilisation of existing and future infrastructure.

## Control Strategies

*Today's Train Traffic Control – Control by Exception*

The results from this project are based on a very detailed description and analysis of how train traffic is controlled today, the work of the dispatchers and the strategies they use for decisions and control actions. Today's control systems are often designed to support the operator's possibilities to react to disturbances and to solve problems and conflicts when they occur (control by exception). It is not unusual that one workstation consists of several separated information systems, keyboards etc. A paper-based time-distance graph can be used as a tool for planning and/or documentation. Train dispatchers can perform traffic re-planning by drawing time table lines in the graph. The paper graph is then used by the dispatcher to remember what needs to be done and at which time. Since the new traffic plan is not automatically introduced into the system there is a great risk that the many different automatic functions in the system will work against the new plan. Automatic support systems are not predictable enough to the dispatchers, because of their internal complexity. Automation can cause surprises by performing control actions in ways that contradict the dispatcher's mental traffic plan. Bainbridge (1983) called it the irony of automation, that when workload is highest, automation is of the least assistance. To avoid automation surprises, train dispatchers are often forced to take full control by inhibiting all automatic functions in the 'disturbed' area and solve the disturbed situation 'manually'. Billings (1991) reports that the probability of human failure in monitoring automation increases when operators are not alert to the state of the automation. Automatic functions can be implemented locally, in the centralised control systems or even as a separate complex automatic control system. Including the train dispatcher, there are up to four levels of more or less autonomous functions that try to partially solve the same problems. Interactions between these levels of decision making and execution are complex. Furthermore the paper graph only supports timetable planning and there is no tool for planning the track usage. Planning of track usage is merely a mental process and the dispatcher must remember the plan without support.

The traffic control system does not provide the train dispatcher with adequate tools to perform efficient traffic control during severe disturbances. As a result of this, intense manual control and oral communications induce high workload. The high cognitive workload and the time pressure make it impossible to perform optimal planning and efficient control of the traffic. The result is that the dispatcher is focused on finding and performing merely workable solutions.

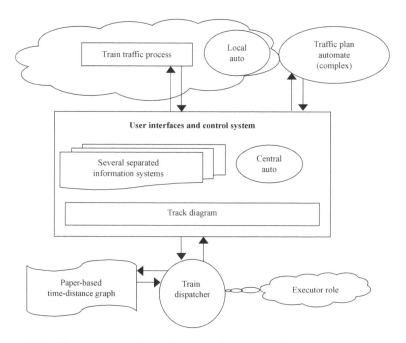

**Figure 27.1   The structure of traffic control system in Sweden today**

In our research we have identified many problems of today's control strategy and in the design of the traffic control systems. Summarised below are some of the main issues:

- lack of overview in the time/distance domain causes sub-optimisation;
- fragmented information as the control system is implemented as several separate information systems. Necessary information can be difficult to obtain;
- lack of precision in data, for example concerning exact train position and speed, etc.;
- information used in the decision-making process can be outdated;

- time and event dependent complexity caused by autonomous automatic functions;
- interlocking routes have train-id as a prerequisite;
- time consuming phone or radio communication with train drivers and others.

## *Future Traffic Control by Re-planning*

To achieve efficient control of systems in general, there are a number of things that need to be considered. The operator controlling the system should have a clear *goal* and an accurate *mental model* of how the entire system works under various conditions. The system should provide the operator with good *observeability* as to the systems past, current and predicted future status. Adequate possibilities to interact with and control the system *(controllability)* are also crucial according to Andersson, Sandblad, Hellström, Frej and Gideon (1997).

Today control is focused on the infrastructure, and through the control of the infrastructure the traffic using the infrastructure is managed. We suggest that the goal should not be to control the infrastructure but rather to plan and control for the train traffic operating on it. The key to the proposed control strategy is re-planning. Due to the nature of traffic the original traffic plan easily becomes obsolete. To solve this, the train traffic controller should be provided with the means to continuously reformulate the current traffic plan into a new improved plan. The aim of reformulation of the plan is to turn a plan with conflicts or inefficient traffic into a better, more optimised plan for how the traffic should be carried out. The traffic plan specifies timetable and planned track usage for every train. From this plan the order of which trains are allowed to use the shared track resources can be derived. Provided that the system has access to a valid functional traffic plan, it is possible to implement an automatic function that executes the plan. The main control task will then be to perform continuous real-time re-planning resulting in a valid traffic plan that can be executed. However, unfortunately there are situations where mere planning will not be sufficient. In case of signalling or mechanical failures in the infrastructure, manual control will still be required. Manual control can then be applied completely or partially over a control area. To be able to perform partial manual control and re-planning tasks at the same time it is necessary that the user interface clearly indicate to the operator what the status of the automatic execution function is.

Poor automation design can be directly linked to lack of feedback, monitoring difficulties, passive decision making, poor mental models, and thus result in situation awareness problems. Automatic systems used today are often perceived as unpredictable by the train dispatcher. In comparison to that, the proposed automatic function will only execute exactly what the traffic plan states. It will not compromise situation awareness in the same sense as today's automation since it will always be known by the operator what the automatic function will do next.

The following functionality of the automatic execution and support systems must be available:

- automatic execution of the continuously updated traffic plan;
- ending of the automatic execution function (by the operator or autonomously);
- automatic test of planned train path in due time in order to test the feasibility of the current traffic plan;

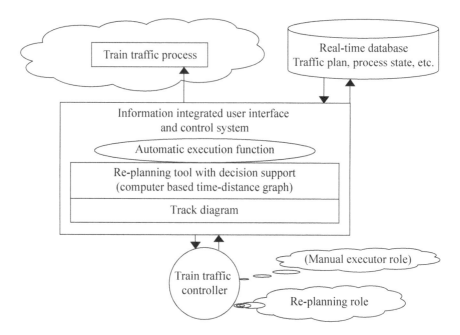

**Figure 27.2   The conceptual structure of the new proposed control strategy**

- automatic interlocking of tested train path according to the plan and train signalling orders;
- automatic functions are made predictable, easy to understand and usable also during severe disturbances;
- automatic functions are for execution, not for changing plans;
- automatic functions *do not* autonomously change track usage or train order.

Endsley (1988) defined *situation awareness* as 'the perception of elements in the environment within a volume of time and space, the comprehension of their meaning and the projection of their status in the near future'. Achieving situation awareness is central to maintaining good decision making and human performance. By showing not only the current status of the traffic process but also showing predictions of what will happen in the near future situation awareness is improved. Today the extensive use of separate information systems induces heavy workload because of limitations of the human short-term memory. Integrating information into one unified user interface will support the situation awareness of the train traffic controller. All decision relevant information should be easily obtained from the user interface. Existing technology can now be used to deliver decision relevant information of higher resolution to the operator, for example it would be possible to deliver exact positioning of trains, their current speed and even acceleration etc. By always having access to accurate information with high resolution the situation awareness of the operators will be improved.

## One Way to Achieve Control by Re-planning

An important aspect in achieving control by re-planning is the design and implementation of a tool that allows the operator to interact with the traffic plan in some way. One way to reach this goal is to provide the traffic controller with an *interactive* computerised time-distance graph. Prototypes of new user interfaces that support the new control strategy have been designed, implemented and preliminary tested in a laboratory environment at Uppsala University. The interface is designed to integrate all decision relevant information into one unified interface and to support continuous awareness of the dynamic development of the traffic process. There are alternative ways to implement necessary re-planning actions. By incorporating some intelligence into the system it could be possible to perform re-planning simply by prioritising certain trains. However, initially we have chosen to focus on a more manual re-planning approach. Considering this approach there are a number of things that the train traffic controller must be able to perform:

- add, remove and change duration of stops for reasons to suit the traffic flow;
- add, remove and change duration of stops for reasons of passenger or freight exchange etc.;
- alter planned track usage;
- add, remove and change duration of resource blocking (construction work etc.);
- turn on or off automatic execution functions partially or completely;
- input information about restrictions concerning specific resources (reduced maximum speed allowed on track etc.);
- change planned speed of a specific train.

The traffic controllers' most important method to manage the process is to maintain system awareness and plan the traffic so that available resources are optimally utilised. To meet expected and unexpected events, the traffic controllers need to have large amounts of decision relevant data visible simultaneously. To manage, plan and control the traffic, the controller needs to know what resources are available at each time interval, and what their attributes and actual states are. Most important is track usage and train characteristics. Information is grouped in a structure that corresponds to train controllers' mental models of track system and traffic processes. Data are, when possible, shown in fixed positions so that they form a well-known pattern that can be interpreted by the traffic controller with minimal cognitive workload. The structure of the user interface is shown in Figure 27.3. and an example of the interface design in Figure 27.4.

## Experiments on Control by Re-planning

Advanced prototypes have been implemented. By connecting user interface prototypes to a train traffic simulator (Sandblad *et al.*, 2000) it is now possible to perform experiments with the design of new user interfaces and decision support tools, and to test new control strategies for the train traffic control operators. In a laboratory at Uppsala University a control room environment has been set up where

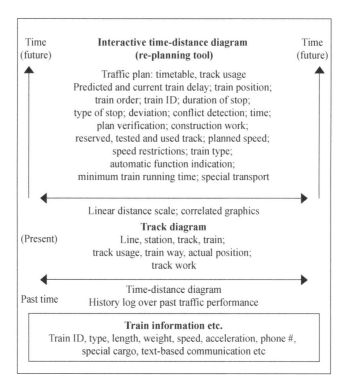

| Time (future) | **Interactive time-distance diagram (re-planning tool)** | Time (future) |

Traffic plan: timetable, track usage
Predicted and current train delay; train position;
train order; train ID; duration of stop;
type of stop; deviation; conflict detection; time;
plan verification; construction work;
reserved, tested and used track; planned speed;
speed restrictions; train type;
automatic function indication;
minimum train running time; special transport

Linear distance scale; correlated graphics

**Track diagram**
(Present) Line, station, track, train;
track usage, train way, actual position;
track work

Time-distance diagram
Past time   History log over past traffic performance

**Train information etc.**
Train ID, type, length, weight, speed, acceleration, phone #,
special cargo, text-based communication etc

**Figure 27.3   The information structure of the new integrated user interface**

it is possible to evaluate the prototypes. It is also possible to conduct more limited experiments on site, at traffic control centres, using a portable computer.

To begin investigating whether or not re-planning in a computerised time-distance graph results in reasonable solutions an experiment was conducted. Twelve professional train dispatchers were presented with the same preconditions and they all got to solve the same disturbed traffic situation. Eight out of the 12 dispatchers used the computerised graph to reformulate a new valid traffic plan. The other four dispatchers performed, as they are used to, the re-planning task in a paper-based time-distance graph. On a logical level there was no significant difference between the solutions produced with the computerised and with the paper-based graph. Basically two different variants of solutions were produced. The majority of the train dispatchers chose an efficient but risky solution, resulting in minimised delays. The others chose a softer approach in solving the problem that resulted in many more delays but probably could be considered to be a more robust solution. Initial indications were that there is no significant difference in solutions produced through planning with a computerised graph in comparison to the paper-based graph used today. In connection with this experiment, by interviewing and using questionnaires, we also got an opportunity to see how the train dispatchers experienced the basic concepts of the control strategy. Reactions were very positive to the concepts and

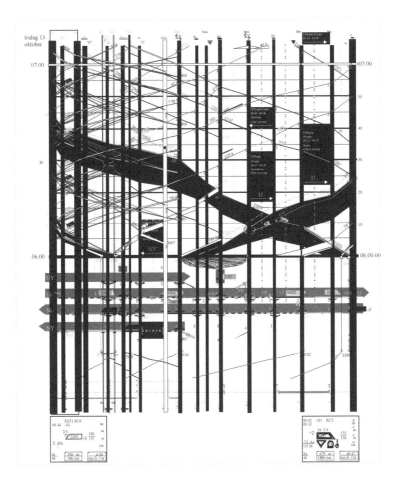

**Figure 27.4 The detailed design of the new user interface**

ideas, even though the experiment only allowed for limited interaction (changing the duration of train stops) in the computer based time-distance graph.

Further testing and evaluation of the proposed concepts and interface design is imperative and is being performed.

## Discussion

Efficient performance requires that the train traffic controllers make good decisions. Situation awareness is crucial in order to make these good decisions. By providing all decision relevant information in an easy to obtain manner system awareness can be enhanced and cognitive workload decreased. The computer based time-distance graph is designed in such a way that it visually supports the operators' comprehension of the current status and the projection of its implications in the future (situation awareness level 3, Endsley (1996)). The user interface, with its planning view, can support early detection of upcoming conflicts, identify possible re-planning alternatives and their predicted effects. The proposed new control strategy has potential to better support the traffic controller's ability to handle continuous re-planning, with the goal to always have a functional traffic plan at hand. This plan can be automatically executed except when a technical malfunction hinders this. The automatic support function executing the traffic plan is predictable and easy to understand. Sensible interface design concerning the automation helps to keep the *human in-the-loop* and to avoid automation surprises. By performing re-planning the operator is deciding what the automatic function will do and at what time. Hence, the operator is continuously in control.

Initial indications during experiments with the new user interface prototypes are promising. Experiments on re-planning (with limited interaction) have been conducted with satisfying results. The majority of the participating train dispatchers have been very positive about the new concepts and ideas. More detailed experiments will be performed on re-planning and execution. Many of the important aspects can be investigated in the simulated environment at Uppsala University, but some issues must be evaluated in an operative environment. The Swedish National Rail Administration is currently performing a feasibility study for the near future implementation of a full scale operative demonstrator system based on the concepts and ideas presented here.

## Acknowledgement

This project has been financially supported by the Swedish National Rail Administration and Vinnova, Swedish Agency for Innovation Systems.

## References

Andersson A.W., Sandblad B., Hellström P., Frej I. and Gideon A. (1997), 'A Systems Analysis Approach to Modeling Train Traffic Control', Proceedings of WCRR 1997, Florence.

Andersson, A.W., Sandblad, B., Nilsson, A. (1998), 'Improving Interface Usability for Train Dispatchers in Future Traffic Control Systems', Proceedings of COMPRAIL 1998, Lisbon.

Bainbridge, L. (1983), 'Ironies of Automation', *Automatica*, 19: 775–9.

Billings C.E. (1991), *Human-centered Aircraft Automation: A Concept and Guidelines*, NASA Technical Memorandum 103885, Moffet Field, CA: NASA Ames Research Center.

Endsley, M.R. (1988), 'Design and Evaluation for Situation Awareness Enhancement', Proceedings of the Human Factors Society 32nd Annual Meeting, Santa Monica, CA.

Endsley M.R. (1996), 'Automation and Situation Awareness', in R. Parasuraman and M. Mouloua (eds), *Automation and Human Performance: Theory and Applications*, Mahwah, NJ: Lawrence Erlbaum: 163–81.

Sandblad B., Andersson A.W., Frej I. and Gideon A (1997), 'The Role of Human Computer Interaction in Design of new train traffic control systems', Proceedings of WCRR 1997, Florence.

Sandblad, B., Andersson, A., Jonsson, K.-E., Hellström, P., Lindström, P., Rudolf, J., Storck, J. and Wahlborg, M. (2000), 'A Train Traffic Operation and Planning Simulator', Proceedings of COMPRAIL 2000, Bologna.

# PART 9
# WORKLOAD IN SIGNALLING

Chapter 28

# Understanding of Mental Workload in the Railways

John R. Wilson, Laura Pickup, Beverley J. Norris, Sarah Nichols
and Lucy Mitchell

## Introduction

This chapter describes the background to programmes of research to understand mental workload in train driving and signalling. It gives more detail on the development of signalling workload tools in particular. It also sets background to the individual chapters by Pickup, Mitchell and Morrisroe in this book.

Two programmes of research have been carried out over the past few years which have explored the nature of mental workload (MWL) and tools for its measurement and assessment: in signalling (for Network Rail) and driving (for Rail Safety and Standards Board). The research teams involved are from academia and industry – the Institute for Occupational Ergonomics (IOE), University of Nottingham and CCD Design and Ergonomics Ltd. This chapter is mainly concerned with the earlier research on signaller workload, but with reference to driving workload where appropriate

While workload has been studied so intensively and for so long in other industries including transport (for example in aviation) there is a dearth of contributions on workload and the railways. For drivers there have been some reports of psychophysiological measurements of workload (for example Caban, 1993 and Myrtek, 1994). Recently, Hamilton *et al.* (in this book) developed and proposed ATLAS as an analytical tool to assess train driver workload, from an information processing angle. For signalling, there have been one or two contributions. Reid *et al.* (2000) attempted to predict railway signaller workload with technical systems change (NX to VDU with Automatic Route Setting) by using task analysis and the NASA TLX workload assessment tool (Hart and Staveland, 1988); McDonald (2001) started with NASA TLX and adapted this to make it more suitable for the work of rail network controllers.

A number of techniques were employed in our own early signalling workload studies (for example Bristol and Nichols, 2001; Nichols *et al.*, 2001; Wilson *et al.*, 2001). The full NASA-TLX was used as well as a cut down version of three of its scales – 'planning and deciding', 'time available' and 'physically tired'. The US Air Force Flight Test Centre (AFFTC) seven point scale (Ames and George, 1993) – a single rating of workload combining perceived demands from activity level, the system, time and safety concerns (see O'Brien and Charlton, 1996: 194) – was adapted.

A measure of work underload, with items for 'under-challenged', 'disinterested', 'bored', 'difficult to concentrate' etc. was developed. Static and dynamic load measures were related to subsequent analysis of video recorded activities and the 'objective' self-report ratings.

## Requirements for Rail Workload Assessment

The industry appears to recognise workload as related to any of:

- the number, complexity and interaction of tasks performed by someone, over a period of time or at one point in time;
- the tasks to be performed in a certain time compared to the capacity of the individual(s) to complete them (successfully);
- the load or effort perceived as experienced by someone, over a period time or at one point in time;
- the number of functions that can be performed successfully by a person or a team in different situations and scenarios;
- the compatibility of work systems such as computer interfaces, job designs or procedures with the functions that need to be completed, with respect to influence on someone's performance and well-being.

To assess such parameters the industry variously require tools for use in-depth by HF experts and/or by managers as an outline check, to be applied concurrently and/or retrospectively, and to tap operators' experiences and views and/or provide an analytical or direct assessment.

Network Rail and RSSB have suggested potential practical uses of MWL assessment tools as:

- safety case or other approvals for new systems;
- follow up general enquiries or complaints from staff or managers;
- assess the consequences of the introduction of increases or changes in information and communications technologies – for example GSM–R, various forms of ERTMS;
- assess the consequences of the introduction of equipment requiring new ways of working – for example reminder appliances, axle counters;
- assess the consequences of increases in information available from the infrastructure or in network centres;
- assess the consequences of changes in operating procedures;
- assess the consequences of changes in type or balance of lineside/cab information;
- assess the consequences of different signage and signalling siting systems;
- support investigation of driver loading for in-cab and out-of-cab activities;
- provide the basis for a 'function complexity index' for signalling;
- identify perceived levels of workload that are unacceptable;
- assess the maximum number of inputs/activities feasible for a signaller to handle simultaneously or over a defined period of time;

- allow identification of the number of trains or amount of track (or other parameters) an individual signaller can cope with.

## Nature of Mental Workload

One of the most widely used and debated concepts in ergonomics/human factors is that of mental workload (MWL). The notion has widespread acceptance in assessing the impact of new tasks, comparing the effects of different job or interface designs and understanding the consequences of different levels of automation. However, mental workload is a seductive concept, apparently meaningful to specialist and lay person alike, and which can be adapted to fit many contexts. As a result, MWL – or often just workload – is used to denote often quite different factors or situations, and has a plethora of measurement methods associated with it. Loose and non-operationalised use of the term, and consequent variations in how MWL is assessed, can leave clients of ergonomics bemused. Another difficulty is that researchers and practitioners may employ a well known measurement method, in the hope of setting their results against some norms, but often may be trying to use a measure created for an entirely different purpose in entirely different circumstances (Pickup *et al.*, 2005).

There are no universally accepted definitions for mental workload in general, nor is there agreement on any one measurement approach or tool:

> There has been much discussion of an acceptable definition of MWL ... As with other areas of ergonomics this has proven difficult ... consider the wide array of definitions of MWL [in] the first significant collection of papers ... [Moray, 1979] ... and things have not got that much better ... The problem is the proverbial one of the chicken and the egg ... [are we] trying to use our [imperfect] knowledge of psychological processes to derive measures of MWL ... or using MWL measures to investigate ... sychological processes ... A bit of both. (Megaw, 2005)

A major distinction is between workload as imposed by the system, somehow measured independently of any individual, and workload as rated by those individuals carrying out the 'work' that is 'loaded' on them, whether this reflects perceived task difficulty, or level of effort exerted etc (see Hart and Staveland, 1988). Expanding on this in the particular context of rail, it seems apparent that MWL is seen by industrial stakeholders in three broad different ways. First it is seen in terms of 'the work that loads' (or the work that creates demand); this leads to questions such as how many functions can one or more signallers handle (under normal and perturbed conditions), will a new technology in the signal control centre or cab increase the demands on the operator beyond their capacity to cope safely and effectively, or how will design and sequence of information inputs (signs or signals) load the driver? Taking this view, there might seem to be need for an analytical tool or an expert assessment method, which can identify levels and limits to the 'work that loads', and considers the actual demands of the work in the way it is organised. This allows for an understanding of the contribution of time pressures to workload; that is, having to carry out certain tasks in a certain way within a particular time limit, and measures may be highly related to information processing theories such as multiple resource theory (Wickens, 2002).

In the second broad view, MWL is seen as 'the experience of load' by the operator. In this case, industry need is for 'workload' to be measurable through operator reports, ratings etc. This approach recognises that ratings and reports may be explicitly or implicitly of the different factors: the many external and internal *sources* of load; the different personal experiences of *'demand'*; the level of *effort* the operator is prepared to, actually does, and perceives that they put in; and the *effects* on them and on the system of the demands and of their efforts.

To make some sense of the different conceptions of MWL we have produced a model embracing several of the important ideas, including imposed, internal and perceived load, demands, effort and effects (Pickup *et al.*, 2005). This was done originally for signalling, and was adapted to underpin the research on driving. In parallel we have clearly proposed that the different MWL notions and the different rail industry needs require a suite of MWL tools which together will build up a total picture of workload for a function, system or scenario. There has been considerable debate down the years over appropriate workload measures. Large numbers of these exist, none produced for the rail industry, and a number of them have shown problems with the appropriateness of wording and form. The majority of the methods have been developed for a North American military or aerospace population, usually in simulators or else highly physically constrained settings, and may not transfer to best effect within fieldwork (or even control room simulators) with a European and civilian workforce. Although we had no desire to develop new workload assessment tools if none are needed, we were not convinced that this was the case.

In a third, related, rail view MWL may be seen more as an actual decrement in performance, as the measurable effects of changes in operator behaviour and their influence on the system. In this case we would assess MWL as implied by changes in actual measured performance. But this is fraught with difficulty. Any relationships in any models adopted are rarely straightforward. As task demands increase there are not necessarily associated increases in operator self-reported load nor decreases in performance. The reasons are many. MWL is influenced by perceptions of the task by the operator and by their skill, experience and training as well as by task demands. Dissociation between workload and performance frequently occurs. Operators may modify their workload on the basis of explicit or implicit feedback, changing strategies or motivation. Performance outcomes themselves can modify the tasks. Different types of performance at different times may necessitate different MWL measurement approaches: differences which might be influential are controlled or automatic processes, data-limited or resource-limited performance, use of single or multiple resources, and task management which includes task switching – an aspect of efficient performance that may bring its own load.

**Choice Across Broad Measurement Approaches**

Before describing the MWL tools that have been developed or adapted during these programmes of research, a general description of measurement is given. Following Megaw's (2005) use of Hill's (1987) framework we can divide MWL measurement approaches into two broad categories, analytical and empirical. Several different tools and techniques might be used within the approaches, none of which is ideal in all circumstances but all of which may have some merit at times.

## Analytical Approaches

*Checklists and walkthroughs* can be used by rail and/or HF experts, and might allow a group of experts to jointly decide on the existence, levels or impact of potential loading factors in the environment or system. They can take the form of structured decision flow charts as well as (weighted) lists. SME (subject matter expert) commentary can be used as they review performance in the field, in a simulator, or from video tape. Existing MWL tools to aid expert judgement include Pro-SWAT and SWORD.

*Task analysis (usually with timeline analysis)* has been used frequently in other domains, usually with a form of timeline analysis. The analysis will be of what tasks at a fairly detailed level (for example recognise EPS sign) must be carried out, when, and with what sequences and what simultaneities. The more complex approaches require a detailed cognitive task analysis and often build upon multiple resource theory (as a very simple example, two visual tasks will conflict and cause more load than a visual and an auditory task). Examples of models and tools include TLAP, VCAP and W/INDEX (which was examined and tentatively proposed for test use in signalling). In many cases these task analytic tools may be classified as computational ones (see later section). One difficulty with these methods is the complexity in their underpinning theory and construction and the effort required to use them.

*Simulation*, abstract computation and/or graphical representations of tasks, may have detail on sequence, time-base and simultaneous performance. As with task analysis we are talking of simulations with an information processing theory basis, and MicroSaint is possibly the best known. It may sometimes be difficult to see how simulations arrive at a workload estimate and they also potentially suffer from the 'rubbish-in rubbish-out' problem. It is probably harder to simulate signalling in this way than driving.

*Computational methods* are very similar to the family of task analytic methods, and attempt to provide quantitative (usually time-based) data to describe the difficulty of tasks (and hence load). Existing theory and data from human information processing (how people sense, perceive, process and act on information from their environment) are used to build up analyses of how long tasks will take and whether they are beyond reasonable capability limits. ATLAS (Hamilton, this volume) for instance is built around multiple resource theory and uses data from well known ergonomics laws (for example Fitts Law, Hicks Law etc.) and from the Model Human Processor to assess the feasibility of completing certain tasks in a certain time and the consequences of simultaneous tasks. Problems for this approach include the sometimes simplistic assumptions made (that there is an 'ideal' person, context, environment etc.) and that the data as are available rarely fit with real complex work situations. Nonetheless, the approach has potential for prediction prior to a prototype system being available.

*Empirical Approaches*

*Performance measures,* determined to result from changes in level of workload, may appear to be a sensible approach. Primary, secondary and alternate task performance measurement has been largely within an experimental context in the laboratory or simulator, but one possible technique for use in the field is the embedded task method (that is, using a real non-critical task for the secondary performance measure – perhaps employing radio or other telecommunications). Unfortunately there appears to be considerable evidence of dissociation of MWL measures and performance, for instance where there is a good deal of automatic processing (for example for driving) and if when loaded the operator just increases their effort to maintain performance levels.

*Psychophysiological methods* such as measures of heart rate (and HR variability), brain activity (event-related brain potential), eye blink rate or galvanic skin response do show promise to allow diagnosticity in MWL assessment. In experimental settings psychophysiological methods tend to relate well to subjective self-report measures (we have shown this in our own work using them for load and stress effects in virtual environments). However, psychophysiological methods can be limited for application in the field (being somewhat intrusive) are relatively resource heavy, and interpretation of what any measures actually mean is always a tricky art!

*Operator reports* are a popular route to obtain workload measurements, for obvious reasons of simplicity and convenience, and the high face validity given by trying to represent the knowledge, views and reactions of the person most effected. As with all self-report methods, assessment of the validity of the tool or the measures will usually require comparison with use of other methods in parallel, as well as knowledge of what events and performance have actually taken place and might have given rise to the reports. The multidimensional nature of workload comes into play here: what in practice is each individual actually rating or otherwise reporting? A key distinction is between self-report which is made as the events occur, or at the end of a defined period of work, or by review of earlier work performance captured on video. A large number of self-report and workload rating tools are available from other domains (for example NASA-TLX, SWAT, DRAWS, ISA, SWORD, Modified Cooper Harper, Bedford Scale, C-SAW). Care must be taken with self-report to ensure that there is minimum influence of the act of reporting itself on task performance or on the feelings of load (although any such confounding would err on the side of caution in any consequent recommendations). As we have seen, subjective reports may not always reflect, or be reflected by, performance; this is especially in instances of task underload, when the subjective measures are saturated if resources are already fully invested, where motivation is a strong factor, and where there is time sharing – even for easier concurrent tasks this may lead to higher ratings.

**Signalling Workload Assessment Tools**

Network Rail have asked:

1   Can we measure peaks and troughs of the input load on signallers and their (perceived) efforts to meet this load, in the field and in a simulator?
2   Can we distinguish high and low loading signalling sites or situations?
3   What types of tool can be developed for prediction of workload at new sites or jobs, and for assessment of potential for workload producing conditions?

A suite of new or adapted tools for use in rail signalling workload assessment has been defined and developed (see Pickup and Wilson, 2004 for full details). A suite of tests was thought the most effective way of assessing the number of relevant dimensions; one tool would not be capable of assessing all variables relevant to the signaller (Brookhuis and de Waard 2002; Williges and Wierwille, 1979).

An *integrated workload scale (IWS)* of nine points has been produced, inspired by the ISA (instantaneous self-assessment – Brennan, 1992) Scale. This is reported on elsewhere in this volume (Mitchell, Pickup) and is discussed no further here.

Simulator trials have been run with planned and extended scenarios completed by a number of signallers. The IWS at one minute intervals was complemented and tracked by use of a *timeline analysis* based on direct observation of activities classified at the level of monitoring, communicating etc and by concurrent and *retrospective subject matter expert commentary* on strategies adopted and likely performance outcomes. This proved a highly interesting and successful approach, with some validation of workload measurement possible through such application of multiple methods.

An *operational demand checklist* (ODEC) has been produced, based on field observations, signaller interviews and use of repertory grid with SMEs. The ODEC's purpose is to document all entities within the signalling system relevant to signaller workload, and to enable assessment of the system itself for its contribution to workload. As its name suggests ODEC comprises a structured and weighted checklist of those entities, factors and characteristics in the signallers' operational environment which might create different levels of load – for instance train movements per unit time, terminating stations, controlled signals, number of points traversed. The original purpose was primarily for staff levels, but more recently the ODEC has been used as an analytical tool before and during a visit to signalling sites.

To provide a more systematic approach to appreciate the loading factors that relate to perceived and imposed loading factors, *probe tools* in the form of decision ladders (for example Brabazon and Conlin, 2001) have been developed to identify sources of effort and demand that impact upon the signaller's functions. The probes, employed by HF specialists and SMEs, provide a rich picture of loading factors specific to the signalling system and signal box or control centre environment.

In cases where a relatively simple assessment of potential for high (or low) workload is needed, a checklist of weighted *workload principles* has been produced. The items in the checklist do not measure workload *per se*, but give an indication that the conditions and circumstances of work may give rise to a workload problem (or not) and for what reasons. The principles allow an analyst (or in principle a member of the signalling team) to assess a signaller's ability to complete functions and to identify working arrangements that may be detrimental for performance or well-being. The principles have also been used to provide an end point to the probes (decision ladders), to assess what the output of the ladders means in practice. Use in practice by Network Rail Ergonomics Team at signal control sites suggests that the greatest value of the principles checklist is as a convenient first check on workload issues.

The widely accepted modified Cooper Harper tool has had adaptations made for relevance to the rail industry and to account for underload. This *Rail Adapted MCH* is used to try to systematically judge a profile of workload obtained during any workload assessment and identify unacceptable conditions.

One component of workload is the degree of well-being of job holders. The *well-being assessment* scales of the REQUEST questionnaire (Wilson *et al.*, 2001) can be used to support overall assessment, as can in principle records on sickness and recruitment.

## Extension into Driving

The Network Rail sponsored research on workload in signalling has been paralleled by an RSSB examination of driving workload and development of tools. This ensured that both sets of workload tools are based on the same fundamental understanding of workload, establishing any coherences in theory, framework, method and tools.

RSSB have asked:

- Can we understand how changes in signalling or signage systems will influence driving efficiency and reliability?
- Can we, in principle, predict levels of workload which will occur with new cab information and control systems?
- Do certain operational requirements and driving regimes reduce the load so much that inattention or boredom become a risk factor?

A number of tools have been proposed and developed for driving workload assessment, including an IWS, timeline analysis, a checklist of principles and a probe tool. Testing has taken place in a variety of ways, formal use in a simulator, trials in a train cab during actual journeys and less formal examinations and discussions in driver workshops.

## Criteria on Which to Judge the Value of MWL Tools

One major criterion for both Network Rail and RSSB has been that any MWL tools available to the industry are relevant to use in the railways; this is not as obvious a point as it may seem since some well known existing tools can appear to be suitable and then betray their military or aerospace North American origins only in practical use. Interviews with human factors experts suggested that many tools currently available did not meet the needs of rail signalling at least, some had been used but with alterations in wording or method of administration, and one or two had potential but would need thorough trialling and possible adaptation.

Criteria for tools have emerged from Network Rail and RSSB requirements to include: support diagnosis and/or prediction; be applied concurrently and/or retrospectively; support direct assessment and be analytical; take an analytical and/

or direct measurement approach; be qualitative and/or quantitative in form; permit instantaneous and/or continuous measurement; allow tracking of peaks and troughs of load; be 'chronic' (that is, make assessment of load over time) and/or 'acute' (that is, make assessment at a point in time); and be informative about performance consequences. These are in addition to the normal requirements for validity, reliability, acceptability, sensitivity. No one tool or measure will meet more than a few of these criteria. As the tools are defined and used, different levels and types of validation will have to be carried out. Many of the criteria that have been given in fact can be seen as spectra and there is rarely any best or worst point on these spectra. For instance, diagnosis can be at different levels; it need not mean only the identification of causal factors and a simple workload assessment – even ratings of probable/possible/no – is a diagnosis of sorts. Likewise prediction does not have to be quantitative and such a measure therefore does not necessarily have to have computational power; for instance we can predict that a new cab system will not increase workload via application of expert assessment checklists or self-report scales in a simulation.

## Next Steps

It would be useful to have some commonality of workload tools across different rail network areas and functions, but this will not be easy to achieve and may not always be appropriate. In maintenance work for instance, where the nature and settings of work are very different, we are probably looking at very different approaches and tools to those suitable for the process control tasks of signalling and driving.

## Acknowledgements

Most of the work reported here was funded by Network Rail and RSSB. Later funding was from Rail Research UK. We are grateful to all these bodies for the opportunity to explore these important issues.

## References

Ames, L.L. and George, E.J. (1993), 'Revision and Verification of a Seven-point Workload Estimate Scale', *Technical Information Manual. Air Force Slight Test Center*, Edwards Air Force Base, California.

Brabazon, P.G. and Conlin, H. (2000), 'Assessing the Safety of Staffing Arrangements for Process Operations in the Chemical and Allied Industries', HSE Contract Research Report 348/2001, Sudbury: HSE Books, HMSO.

Brennan, S.D. (1992), 'An Experimental Report on Rating Scale Descriptor Sets for the Instantaneous Self Assessment (ISA) Recorder', DRA Technical Memorandum (CAD5) 92017, Portsmouth: DRA Maritime Command and Control Division.

Bristol, N. and Nichols, S. (2001), 'Impact of New Technology on Operator Workload: Development of a Measure of Underload', in M.A. Hanson (ed.), *Contemporary Ergonomics* 2001, London: Taylor and Francis: 251–6.

Brookhuis, K.A. and de Waard, D. (2002), 'On the Assessment of (Mental) Workload and Other Subjective Qualifications', *Ergonomics*, 45 (14): 1026–31.

Caban, Ph, Coblentz, A., Mollard, R. and Fouillot, J.P. (1993), 'Human Vigilance in Railway and Long-haul Operation', *Ergonomics*, 36: 1019–33.

Hart, S.G. and Staveland, L.E. (1988), 'Development of NASA-TLX (Task Load Index): Results of Experimental and Theoretical Research', in P.A. Hancock and N. Meshkati (eds), *Human Mental Workload 1988*, New York: North Holland: 139–78.

Hill, S.G. (1987), 'Analytic Techniques for the Assessment of Operator Workload', in Proceedings of the 31st Annual Meeting of the Human Factors Society, Santa Monica, CA: HFS: 368–72.

McDonald, W. (2001), 'Train Controllers Interface Design and Mental Workload', in J. Noyes and M. Bransby (eds), *People in Control*, London: The Institution of Electrical Engineers: 239–58.

Megaw, E.D. (2005), 'The Definition and Measurement of Mental Workload', in J.R. Wilson and E.N. Corlett (eds), *Evaluation of Human Work*, 3rd edn, London: Taylor and Francis.

Moray, N. (1982), 'Subjective Mental Workload', *Human Factors*, 24: 25–40.

Myrtek, M., Deutschmann-Janicke, E., Strohmaier, H., Zimmermann, W., Lawerenz, S., Brügner, G. and Müller, W. (1994), 'Physical, Mental, Emotional and Subjective Workload Components in Train Drivers', *Ergonomics*, 37: 1195–203.

Nichols, S., Bristol, N. and Wilson, J.R. (2001), 'Workload Assessment in Railway Control', in D. Harris (ed.), *Engineering Psychology and Cognitive Ergonomics*, Vol. 5, Aldershot: Ashgate: 463–70.

O'Brien, T.G. and Charlton, S.G. (1996), *Handbook of Human Factors Testing and Evaluation*, Mahwah, NJ: Lawrence Erlbaum.

Pickup, L. and Wilson, J.R. (2004), 'Signalling Workload Tools Guidance Notes', IOE/ RAIL/03/13 to IOE/RAIL/03/20, The Institute for Occupational Ergonomics, The University of Nottingham.

Pickup, L., Nichols, S.C., Clarke, T and Wilson, J.R. (2005), 'Fundamental Examinations of Mental Workload In the Rail Industry', *Theoretical Issues in Ergonomics Science*, in press.

Reid, M., Ryan, M., Clark, M., Brierley, N. and Bales, P., (2000), 'Case Study – Predicting Signaller Workload', in Proceedings of the IRSE Younger Members Conference 2000.

Wickens, C.D. (2002), 'Multiple Resources and Performance Prediction', *Theoretical Issues in Ergonomic Science*, 3 (2): 159–77.

Williges, R.C. and Wierwille, W.W. (1979), 'Behavioural Measures of Aircrew Mental Workload', *Human Factors*, 21: 549–74.

Wilson, J.R., Cordiner, L.A., Nichols, S.C., Norton, L., Bristol, N., Clarke, T. and Roberts, S. (2001), 'On the Right Track: Systematic Implementation of Ergonomics in Railway Network Control', *Cognition, Technology and Work*, 3: 238–52.

Chapter 29

# A Conceptual Framework of Mental Workload and the Development of a Self-reporting Integrated Workload Scale for Railway Signallers

Laura Pickup, John R. Wilson, Sarah Nichols and Stuart Smith

## Introduction

This chapter presents a conceptual framework produced for the rail industry to provide a common language and understanding of the multiple dimensions of workload. The framework has acted to direct the development of workload measurements capable of assessing the dimensions most relevant to railway signallers and the aspects of workload with greatest priority for Network Rail. A self-report assessment tool was identified as necessary to capture signaller's perception of the peaks and troughs in their workload. This chapter justifies the need for and the development of the nine point integrated workload scale (IWS) for signallers.

The previous chapter from Wilson *et al.* provides the background to this chapter. Here we will firstly summarise the exploratory work that was completed for Network Rail to understand the concept of mental workload and its assessment in the context of the rail industry, focusing on the railway signaller. This is presented within a conceptual framework, which has been used to identify and focus the need for measurement tools (reported in greater detail in Pickup *et al.*, 2005). The second part of the chapter justifies the choice and development of a self-reporting integrated workload scale for the assessment of railway signaller mental workload, to meet the need for: a) tools to be used on- or off-line; and b) tools to tap operator experiences and views.

## Fundamentals of the Concept of Workload

The theoretical question of what actually constitutes workload is much debated. This work took a pragmatic approach to understanding workload and related concepts specific to the railway industry. It is well established that mental workload is a multidimensional concept, with components drawn from consideration of factors such as time, mental tasks, physical tasks, and stress (Vicente *et al.*, 1987; Wickens, 1992; Xie and Salvendy, 2000); these different dimensions are relevant to how the concept is understood and assessed in any particular context. However different authors emphasise the importance of different dimensions and therefore incorporate different dimensions within the workload assessment tools they develop (Reid and Nygren, 1988; Roscoe, 1987; Hart and Staveland, 1988). In order to consider the dimensions relevant to signaller mental workload, and how these should be presented, a conceptual framework (Figure 29.1) of relevant dimensions of workload for the rail industry was produced to support a practical approach to workload measurement usable in real situations (see also Vicente *et al.*, 1987).

This conceptual framework has been built up gradually, expanded and refined in the light of relevant literature and also, critically, our own studies of and with signallers. As a consequence, the evolving framework was determined through a mix of empirical findings and theoretical interpretations, and this mix is reflected in how the framework is explained. It should be emphasised that this framework is not necessarily an operational model. It is proposed in order to develop and position a toolkit of methods to understand and assess mental workload. Thus it is explanatory of routes to measurement rather than of the mechanisms by which MWL is caused. The framework also reflects the very different ways in which workload is conceived, and therefore measured, by different investigators.

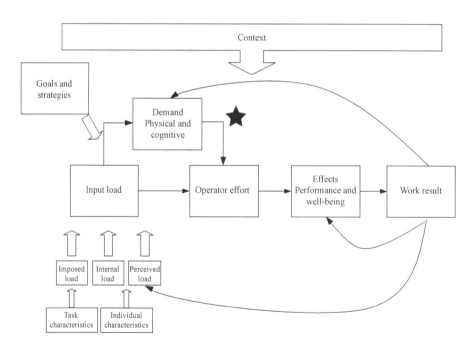

**Figure 29.1    Developed conceptual framework of mental workload; the star marks the part of the framework to be targeted for measurement by the integrated workload scale**

## The Operator's Perception of Workload

Gaining an insight into the operator's views and experiences, capturing peaks and troughs in the load upon an individual and their perception of effort to meet this load are of interest for Network Rail. This type of assessment can capture the dynamic nature of workload determined by the scenario, context, goals, functions and tasks to be completed. The star in Figure 29.1 marks the operator's perception of their workload, the part of the conceptual framework initially targeted for measurement. The benefit is in providing insight into the workload perceived by individuals working a specific signalling system and controlling a particular area of the infrastructure. Other uses envisaged by the Network Rail were for comparison in new technology and procedures implementation. The client was keen that any tool adopted was suitable both for application within a simulator (off-line) and the field (on-line).

## Self-reporting Assessment of Workload

Self-report of workload is widely accepted as capturing the operator's perception of workload or effort. Some authors believe that subjective ratings are the most sensitive and accurate reflection of mental workload (Hart and Staveland, 1988). Others suggest that they can reflect the actual effects on performance better than measures of task demands (Jensen *et al.*, 1994) and benefit from the operator's insight into an increase in effort prior to performance degradation (Muckler and Seven, 1992). When an individual is rating themselves they may well be considering how well they are coping, the resources they are using and the amount in reserve, previous experiences and their level of motivation (Muckler and Seven, 1992).

This type of workload measure falls into two categories of multidimensional and unidimensional workload scales. Multidimensional scales explicitly represent the dimensions of workload and allow a rating to be obtained for each dimension (for example NASA TLX- Hart and Staveland (1988), and SWAT – Reid and Nygren (1988)). Unidimensional scales represent the concept of workload as one continuum (Bedford Scale – Roscoe (1987), Modified Cooper Harper (MCH) – Wierwille and Cascali (1983), ISA – Brennan (1992), and AFFTC – Ames and George (1993). Muckler and Seven (1992) stress the need to select a measure based on the information required. However, Nygren (1991) proposes that multidimensional scales should be used where possible as they have a greater capability to be diagnostic and therefore have greater sensitivity. This presumes that diagnosticity is the main goal and the negative aspect is they require time to present and complete. Hendy *et al.* (1993) present evidence to dispute the claims that multidimensional scales have greater sensitivity and consider unidimensional scales to be better in providing a global rating of workload whilst being quicker and easier to administer.

The goal of the measurement tool to be developed for Network Rail was to enable peaks and troughs in individual effort to be captured on- and off-line. This required an assessment of the fluctuations in the global workload experienced, with a priority for the scale to be quickly administered and be minimally intrusive if it is to be acceptable within the field. A unidimensional rating scale was considered more suitable to capture information under these circumstances.

## Existing Unidimensional Workload Scales

A review of existing unidimensional workload scales assessed suitable terminology, practicability and capability to meet the client's goals (Pickup and Wilson, 2003).

- The modified Cooper Harper (Wierwille and Cascali, 1983) focuses on the load perceived as a consequence of the system being operated. This didn't appear to fit with the functional approach we have adopted.
- The Bedford Scale (Roscoe, 1987) uses the terminology of spare capacity, emphasising the information-processing dimension of workload. This is recognised as an essential workload dimension and is meaningful to the human factors community but does not always appear to have an intuitive meaning for everyone (Brennan, 1992).
- Instantaneous self-assessment (ISA) (Brennan, 1992) is simple in its presentation and application as used concurrently with performance. Discussions with human factors experts in air traffic control suggest the tool had been very useful even on-line within the field. The scale has a five point scale, which had caused some concerns over a central tendency effect being noted which fundamentally reduced the scale to three points. Increasing the scale to seven points was being considered by the industry. The ISA scale descriptors explicitly use the term workload but insight into the culture of signallers has suggested that the term workload is directly and negatively associated with pay and grading.
- The Air Force Flight Test Centre (AFFTC) (Ames and George, 1993) despite being a unidimensional scale retains the multidimensional nature of the workload concept by incorporating between 2–4 dimensions within each scale descriptor. The developers suggest that raters are able to integrate the various dimensions into a single rating and previous use with signallers supported this (Cordiner *et al.*, 2000). However, the scale required minor changes in terminology for it to be acceptable to signallers.

There was no desire to develop a new unidimensional scale if proven scales were suitable. A key concern was terminology, which should represent the concept of workload in the most intuitive way to signalers, thus increasing validity. Cainesmith and Kendall (1963) consider dimensions defined in terminology suggested by end-users to provide a higher degree of content validity. Furthermore Brennan (1992) concluded from experiments during the development of descriptors for ISA that a descriptor set that was more reflective about how people feel about their workload levels may prove to have greater meaning and suggested this could be a 'promising pointer for future investigations'. We decided to develop a signaler-specific unidimensional scale that could achieve the practicality and simplicity of ISA but used descriptors that reflect how signallers themselves feel about their workload levels, in an absolute graphic rating scale.

Construct validity of a workload scale requires the descriptors to consider workload theories (Nygren, 1991) and psychometric theories of measurement (O'Donnell and Eggemeier, 1986; Kantowitz, 1992). The psychometric properties of any workload tool are critical for a tool to measure the relevant dimensions of workload in the context required (Nygren, 1991). Basing tool development on relevant theory, and

on the perceptions and language used by the signallers, would enhance the validity
– face, content and construct – of our integrated workload scale.

## Psychometric Measurement – Psychophysical Scaling

Psychophysical scaling as a part of psychometric measurement is suggested by
Gopher and Braune (1984) to provide a powerful measurement approach to quantify
the subjective experience of workload. Several approaches to psychophysical scaling
were considered and the Thurstone's approach of method of equal-appearing intervals
(Thurstone, 1927; Oppenheim, 1992) was adopted, as it appeared to balance a robust
development process of a scale with equal appearing intervals with efficiency in
its application. Each stage of the Thurstone's approach taken will be explained in
greater detail to justify any modifications to the approach made and to describe the
development of the final nine point integrated workload scale for signallers.

### Development of Scale Descriptors

The conceptual framework (Figure 29.1) and theories of workload gave dimensions
considered as the core psychometric properties to be reflected within the workload
measure, to ensure construct validity and produce a tool capable of capturing the
necessary workload dimensions (Nygren, 1991) – namely load, demand, effort and
effect.

Terms more specific to the signaller's workload dimensions were collected through
interviews and observations completed during workload investigations at a sample of
signal boxes. The interview data were coded based on the basic conceptual framework
of Figure 29.1 and followed best practice in qualitative enquiry (for example Miles
and Huberman, 1994; Robson, 1993). Systematic analysis of interview data extracted
perceived sources of load, causes of effort and demand and types of effects, when
the signaller performs the key functions within their work: obtaining awareness of
the situation, making decisions and planning strategies, and acting to implement
plans. This provided key workload dimensions in terminology familiar to signallers.
Examples are; *load* – amount of work, jobs, tasks, situations, responsibilities, problems,
time available; *demand and effort* – concentration, focus of attention, busy, effort,
demanding; and *effect* – pressure (time and individual), frustration, struggling, spare
time, managing. It was rare that only one term or dimension was used to describe the
workload experienced by signallers. This supported the view of Ames and George
(1993) that raters using the AFFTC are capable of integrating dimensions in interpreting
the term workload and in fact this appeared the more natural approach to expressing
perceptions of workload experienced. Therefore descriptors were developed to present
dimensions in combination to be included in the pool to be judged by signallers to
establish those considered most relevant to describing their workload.

A total of 47 descriptors were developed which drew from typical phrases transcribed
from the interviews for example 'struggling to keep up', 'feeling under pressure'. The
literature varies on the recommendations for number of phrases; Oppenheim (1992)
suggests there could be as many as 100–150 statements. The main aim was to ensure
that the descriptors were in a language familiar to signallers and reflected the most

relevant workload dimensions. The descriptors aimed to provide good 'rating scale cues' to allow sufficient anchor points that could act as mileposts along the workload continuum to facilitate the raters' (signallers') judgement (Guildford, 1954).

Therefore the final scale can be considered as unidimensional as the continuum represents the variable of workload. Although workload is still considered as multidimensional, the integration of the dimensions judged most relevant by the end users and placed along the workload continuum is proposed as a more practical reflection of how the concept of workload is viewed, and can be rated, by signallers.

*Provide Descriptors to Judges*

Two researchers agreed that the final 47 descriptors had clarity and relevance to signaller workload. The number of descriptors were considered manageable and could potentially be considered thoroughly by signallers in a short period of time. It was not possible to gather a large number of 'judges' to complete the card sorting activity as required by the Thurstone technique. Therefore, in a minor adaptation, a questionnaire was designed to administer the card sorting method. The descriptors were numbered and presented in random order and signallers were requested to judge each statement. The judgement requested was not for how favourable the term was, as is traditionally the case in the Thurstone technique, but whether the term appropriately described high or low levels of workload. The workload continuum was represented with a line from high to low workload. The judges (signallers) were asked to put each descriptor in one of eleven boxes that represented points along the continuum. An additional box was titled 'confusing' for any obscure or undesirable statements.

The questionnaire was piloted with signalling SMEs and human factors experts. Over 130 questionnaires were distributed by hand, to allow a verbal explanation to be provided and encouragement to complete the questionnaire. The locations were chosen as representative of the main types of signalling systems, lever frame signal boxes, NX-panel systems and an IECC system. Response rate was 23 per cent (30 out of 130), not good but sufficient (a minimum of 20 judges has been suggested by Dane, 1990).

*Calculation of Spread and Median of Statements*

The responses were initially analysed to exclude two judges who placed more than one third of their responses in the same box. Therefore the final analysis was completed with 28 judges. The next stage of analysis was to consider the spread of judgements made on each statement; the wider the spread the more ambiguous the statement. Therefore the median and semi-interquartile range was calculated for each statement. Any method of judging spread can be used including mean and standard deviation, but the median and semi-interquartile range is considered the simplest (Oppenheim, 1992). The median was initially used but where there were two closely positioned phrases the mean was consulted to assist in the final decision.

A frequency distribution was calculated by recording the number and proportion of phrases collected in each box. A cumulative-percentage frequency graph was produced (Figure 29.2), the 11 box categories form the X-axis and the percentage frequencies the Y-axis. This graph suggests the data collected were relatively well distributed due to the nearly linear properties of the graph line.

Cumulative frequency

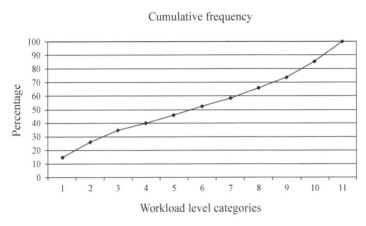

**Figure 29.2 Cumulative-percentage frequency graph of workload descriptors by category**

*Final Descriptors Established*

The optimum number of points for a unidimensional scale is debatable. With too few steps a scale will lack sensitivity and with too many the rater is unable to discriminate to such a fine level. The issue of central tendency highlighted earlier with the ISA tool was a good indication to develop a scale with a greater number of descriptors. Conklin (1923) analysed over 23,000 ratings and suggested that a nine point scale was optimal for bipolar scales. Freeman and Freeman (2000) suggest scales with between 7–12 steps provide the maximum discrimination. Therefore the IWS was to have nine descriptors, with the intention to assess scale length and a possible reduction as part of the subsequent validation process.

Oppenheim (1992) suggests that only descriptors that fall between the first and third quartile of the range of descriptors should be used. However, examination suggested that there was a strong consensus on the descriptors at the extreme points and so a decision was made to include all the data. To move from 11 categories to a nine point scale, a ratio of 1.25 (10 intersections for eight intervals) was applied. Therefore the descriptors were reviewed to extract those that had medians closest to 1, 2.25, 3.5, 4.75, 6, 7.25, 8.5, 9.75 and 11, presented as the final IWS (Figure 29.3).

**The IWS Tool**

For usability in the presentation of the scale, the key workload dimension within each descriptor was highlighted and presented to the left of the full descriptor (Figure

29.3). At a glance the key dimensions can act as a prompt to a signaller familiar with the scale from prior use, with the scale presented as colours rather than numbers to subtly stress rating by description. The output from the IWS tool is a graphical trace of workload experienced by the operator during a particular scenario, task, event or period within a shift (Figure 29.4). Although a mean workload value over time can be calculated, the greatest value of the IWS is to track the peaks and troughs either in a simulator or field environment at intervals to suit needs (for example one, two,

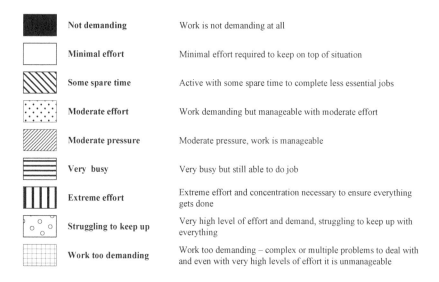

| | | |
|---|---|---|
| | **Not demanding** | Work is not demanding at all |
| | **Minimal effort** | Minimal effort required to keep on top of situation |
| | **Some spare time** | Active with some spare time to complete less essential jobs |
| | **Moderate effort** | Work demanding but manageable with moderate effort |
| | **Moderate pressure** | Moderate pressure, work is manageable |
| | **Very busy** | Very busy but still able to do job |
| | **Extreme effort** | Extreme effort and concentration necessary to ensure everything gets done |
| | **Struggling to keep up** | Very high level of effort and demand, struggling to keep up with everything |
| | **Work too demanding** | Work too demanding – complex or multiple problems to deal with and even with very high levels of effort it is unmanageable |

**Figure 29.3   The integrated workload scale (IWS) for signallers (boxes are colour-coded in practice)**

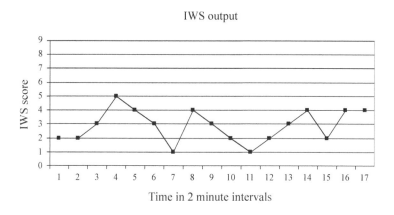

**Figure 29.4   Typical IWS graphical output**

five, 15 minutes etc). The chapter by Mitchell *et al.* will discuss the validation and testing completed with the IWS.

The scale is currently packaged to allow several modes of application. The simplest method of application, for the investigator, is by pen and paper. The scale is presented, in colour, in a position agreed to be visible yet not obtrusive to the signaller and the system of control. The investigator uses a stopwatch to time and prompt the signaller at agreed rating intervals, and then records the rating on paper. The advantage of this approach is its simplicity and no problems of obstructive or malfunctioning technology. The disadvantage is that to create the graphical output the data have to be entered into an excel spreadsheet at a later date.

An electronic device was therefore considered to streamline the data collection, analysis and presentation processes whilst minimising intrusiveness to the signaller's environment. Technology that involved any form of radio waves, for example wireless or bluetooth technology, was quickly dismissed due to the potential for interference with control equipment. Equally, voice activated devices were considered unsuitable due to surrounding noise levels (in some control areas) and the need to transfer between multiple users.

A touchscreen laptop computer was settled on at first, as an interface familiar to most signallers and easily transportable and interchangeable between users and sites. A Visual Basic software programme was designed to prompt the signaller at pre-set intervals. An audible alarm (distinct from other control system alarms) alerts the signaller to press the colour button next to the descriptor most applicable at that time. The software records the rating received, the prompt time and the time the rating was actually received. Two graphs are instantly produced at the end of the session to represent prompt time elapsed and time of actual scoring. Due to space being limited within some control areas a further interface was designed. This was based on the interface used to administer the ISA workload tool by Eurocontrol (Hering and Coatleven, 1996). The input device is a nine-point keypad that can be attached onto the signalling system in a position selected by the signaller. The keypad attaches via a port into a standard laptop computer. The keypad provides an audible alert local to the signaller's position and when the keypad is pressed the rating is recorded using the same software as described previously. An alternative input device has been trialed in the field. The Actiwatch has an interface similar to a watch that requires a button to be pressed a set number of times to reflect the rating descriptor being recorded. The data are then downloaded into an Excel spreadsheet.

The latest software version, IWS 3.0, has been programmed to prompt at each time interval irrespective of a previous rating being entered or not. However, the signaller is provided with a detailed briefing before starting to use the IWS tool to emphasise that supplying a rating is their lowest priority. This is to ensure that when used on-line the signaller is clear that the IWS tool should not have a priority above any other signalling activity, so aiming to limit the intrusiveness of the tool.

The tool is capable of highlighting fluctuations in workload, however does not record the reason for the fluctuations. Concurrent video and observational recordings by human factors experts or SMEs can provide valuable information to support and explore the reason behind workload fluctuations (see the chapter by Mitchell *et al.*).

## Summary

The development of the IWS drew upon knowledge from use of other self-reporting workload scales and built on these to develop a scale specific to the rail industry, in particular the signalling domain. Psychometric theory of measurement, workload theories and qualitative data from the field of signalling provided a rich foundation to ensure the process of scale development facilitated face, content and construct validity. The scale and its methods of applications have been trialed and used both on- and off-line, alongside other tools developed in parallel (see Morrisroe and Wilson *et al.* in this book, and Pickup *et al.*, 2005). The IWS appears to provide a meaningful approach to capturing real time signaller perception of the demands and effort experienced as a consequence of their work, and has not been considered to be intrusive.

## Acknowledgements

This work has required support from the Ergonomics Group at Network Rail and a number of signallers who have provided essential information. We would like to acknowledge and thank all those who have been involved for their time and enthusiasm.

## References

Ames, L.L. and George, E.J. (1993), 'Revision and Verification of a Seven-Point Workload Estimate Scale', *Technical Information Manual*, Air Force Flight Test Center, Edwards Air Force Base, California.

Brennen, S.D. (1992), 'An Experimental Report on Rating Scale Descriptor Sets for the Instantaneous Self Assessment (ISA) Recorder', *DRA Technical Memorandum (CAD5) 92017*, DRA Maritime Command and Control Division, Portsmouth.

Cainesmith, P. and Kendall, L.M. (1963), 'Retransition of Expectations: An Approach to the Construction of Unambiguous Anchors for Rating Scales', *Journal of Applied Psychology*, 47 (2): 149–55.

Conklin, E.G. (1923), in J.P. Guildford (1954), *Psychometric Methods*, London: McGraw Hill: 290.

Cordiner, L., Nichols, S. and Wilson, J. (2000), 'RELOAD: Summary of Workload Assessment from the Field Studies', *Report IOE/99/012*, Nottingham: The Institute for Occupational Ergonomics, University of Nottingham.

Dane, F.C. (1990), *Research Methods*, Belmont, CA: Brooks-Cole Publishing Company.

Freeman, R. and Freeman, J.S (2000), 'A Metrics Toolbox – A Scoring System to Help You Evaluate Proposals and Proposal Processes', *APMP* (28).

Gopher, D. and Braune, R. (1984), 'On the Psychophysics of Workload: Why Bother with Subjective Measures?', *Human Factors*, 26 (5): 519–32.

Guildford, J.P. (1954), *Psychometric Methods*, London: McGraw Hill.

Hart, S.G. and Staveland, L.E. (1988), Development of NASA-TLX (Task Load Index): Results of Experimental and Theoretical Research', in P.A. Hancock and N. Meshkati (eds), *Human Mental Workload*, New York: North Holland: 139–78.

Hendy, K.C., Hamilton, K.M. and Landry, L.N. (1993), 'Measuring Subjective Workload: When is One Scale Better than Many?', *Human Factors*, 35 (4): 579–601.

Hering, H. and Coatleven, G. (1996), 'Ergo (Version 2) for Instantaneous Self-assessment of Workload in a Real-time ATC Simulation Environment', *EEC Note No. 10/96*, European Organisation for the Safety of Air Navigation, France.

Jensen, S.E, Tudor, S.G and Adams, M.I (1994), 'Pilot Workload in Single-Seat TIALD Operation', *Technical Report: DRA/AS/MMI/TR94045/1*, Unclassified DERA Report.

Kantowitz, B.H. (1992), 'Selecting Measures for Human Factors Research', *Human Factors*, 34: 387–98.

Miles, M.B. and Huberman, A.M. (1994), *Qualitative Data Analysis: an Expanded Sourcebook*, Thousand Oaks, CA: Sage Publications.

Muckler, F.A. and Seven, S.A. (1992), 'Selecting Performance Measures: "Objective" Versus "Subjective" Measurement', *Human Factors*, 34 (4): 441–55.

Nygren, T.E. (1991), 'Psychometric Properties of Subjective Workload Measurement Techniques: Implications for their Use in the Assessment of Perceived Mental Workload', *Human Factors*, 33 (1): 17–33.

O'Donnell, R.D. and Eggemeier, F.T. (1986), 'Workload Assessment Methodology', in K. Boff, L. Kaufman and J. Thomas (eds), *Handbook of Perception and Human Performance*, Vol. 2, New York: J. Wiley.

Oppenheim, A.N. (1992), *Questionnaire Design, Interviewing and Attitude Measurement*, London: Pinter Publishers.

Pickup, L. and Wilson, J. (2003), 'Workload Assessment Tools', *Report IOE/RAIL/03/02*, University of Nottingham, Nottingham.

Pickup, L., Wilson, J.R., Nichols, S., Norris, B., Clarke, T. and Young, M.S. (2005), 'Fundamental Examinations of Mental Workload in the Rail Industry', *Theoretical Issues in Ergonomics Science* (in press).

Reid, G.B. and Nygren,T.E. (1988), 'The Subjective Workload Assessment Technique: A Scaling Procedure for Measuring Mental Workload', in P.A. Hancock and N. Meshkati (eds), *Human Mental Workload*, New York: North Holland: 185–218.

Robson, C. (1993), *Real World Research: A Resource for Social Scientists and Practitioner-Researchers*, Oxford: Blackwell.

Roscoe, A.H. (1987), 'In-flight Assessment of Workload Using Pilot Ratings and Heart Rate', *The Practical Assessment of Pilot Workload*, Neuilly sur Seine, France.

Thurstone, L.L. (1927), A Law of Comparative Judgement. *Psychological Review*, 34, 273–286.

Tsang, P. and Wilson, G.F. (1997), 'Mental Workload', in G. Salvendy (ed.), *Handbook of Human Factors and Ergonomics*, Ontario, Canada: John Wiley and Sons: 417–49.

Vicente, K.J., Thornton, D.C. and Moray, N. (1987), 'Spectral Analysis of Sinus Arrythmia: A Measurement of Mental Effort', *Human Factors*, 29 (2): 171–82.

Wickens, C.D. (1992), *Engineering Psychology and Human Performance*, 2nd edn, New York: HarperCollins.

Wierwille, W. and Cascali, J.G. (1983), 'A Validated Rating Scale for Global Mental Workload Measurement Application's, *Proceedings of the Human Factors Society 27th Annual Meeting*: 129–32.

Xie, B. and Salvendy, G. (2000), 'Review and Reappraisal of Modelling and predicting Mental Workload in Single- and Multi-task Environments', *Work and Stress*, 14 (1): 74–99.

# Chapter 30

# Assessment of New Workload Tools for the Rail Industry

Lucy Mitchell, Laura Pickup, Shelley Thomas and David Watts

## Introduction

This chapter reports the results of two trials conducted to evaluate new workload assessment tools developed for the rail industry. The tools assessed were the integrated workload scale developed for rail by the Institute of Occupational Ergonomics and reported in Pickup *et al.* (another chapter in this volume) and a software-based timeline analysis tool configured for the rail environment by CCD Design and Ergonomics Ltd.

The chapter details the first trials with the tools in the signalling environment and provides the reader with up-to-date information regarding current practice for workload assessment in the rail industry.

The trials were conducted in two separate simulator environments: an Integrated Electrical Control Centre simulator at York, and an NX Panel simulator at Crewe. The trials were conducted by CCD Design and Ergonomics Ltd, the Institute of Occupational Ergonomics at Nottingham University, and Network Rail HQ Ergonomics Team.

The tools were assessed for: sensitivity; usability; and intrusiveness to signalling tasks. The tools showed positive results on all assessments. The research also investigated different hardware solutions for the tools.

### Workload Assessment in the Rail Industry

Previously there has been no workload tool specific to the rail industry. A specific tool is required for the following reasons:

*   to ensure that the assessment scale has meaning and reflects interpretations of workload in the industry;
*   because the signalling environment is dynamic and immediate rating is important to avoid loss of accuracy in perceptions;
*   because other techniques that have been used in rail have been regarded as being too disruptive or providing distorted results.

*The Integrated Workload Scale*[1]

An integrated workload scale (IWS), specific to signaller workload has been developed by the Institute of Occupational Ergonomics (IOE), at Nottingham University. This scale was designed for use real time, alongside signalling activities. The scale incorporates nine points, designed to appear equal. A software version of the tool enables the scale to be presented on a touch screen using a standard laptop computer.

*Timeline Analysis*

This study coupled the IWS with a timeline recording methodology created using Noldus Observer software. Timeline data were recorded from direct observations of signaller activity whilst undertaking simulated scenarios. A set of pre-determined activities were developed and coded into the Timeline programme by CCD Design and Ergonomics Ltd (CCD). These activities covered: paperwork, communications, monitoring activity, and actions.

*Combined Picture*

The IWS output and the timeline analysis output were downloaded into Excel and analysed to produce a combined graph – overlaying IWS scores onto Timeline Data. The timeline methodology has been developed to interface with IWS recordings for data collection, analysis and presentation.

*The Trials*

The combined tool was trialled in two separate simulator environments: an Integrated Electrical Control Centre (IECC) simulator at York and an Entry/Exit (NX) Panel Simulator at Crewe. The trials were conducted by CCD, IOE, and Network Rail HQ Ergonomics Team (NR).

**York IECC Simulator Trial**

This trial was run in April 2003 using the York IECC simulator running scenarios developed by Network Rail operational experts. A simulated environment was used to avoid potential disruption to live operations and to allow a degree of control for scenario generation.

*Facilitators*

There were five facilitators at the trial:

- two Network Rail HQ operational experts;
- one simulator facilitator – responsible for running the scenarios;
- one IOE facilitator – responsible for IWS recording and trial administration;

- one CCD facilitator – responsible for timeline recording and trial administration.

*Participants*

Seven participants took part in this trial. These participants were all taken from the roster on duty in the York IECC on trial days. Ages ranged from 24–50 years with a mean age of 39 years. Experience at York IECC ranged from three months to 14 years with a mean of 4.5 years. All participants were familiar with York IECC and the simulated areas of control presented in scenarios.

Participants were told they may take breaks between scenarios or leave at any time if they wished and they would not be penalised in any way. They were given the opportunity to ask any questions before the trial began. They were further informed that the results of the trial would be used for the purpose of the study and by IOE/CCD only. Participants remain anonymous and participation did not carry any implications for NR personnel involved.

*Scenarios*

A realistic scenario was designed for the simulation by the NR HQ operational experts. All participants were familiar with the area and the geography used. The timetable was not one that signallers had used before, however, and they were given time to familiarise themselves with it. Although signallers normally worked this area with automatic route setting (ARS), this was switched off for the trial.

The scenario was adapted into two types:

- Scenario 1 was designed to represent 'normal' working conditions, involving the kinds of tasks a signaller expects to perform on a day to day basis;
- Scenario 2 was designed to represent a higher workload situation, where the signaller must respond to 'abnormal' events.

Both scenarios were 15 minutes long. The two scenarios were identical except that an (wrong side) off side (cess side) Hot Axle Box Detector (HABD) was triggered four minutes into Scenario 2. No audible alert was provided – only a visual cue. All but one signaller had to be prompted to respond to this alert by a facilitator acting as Duty Signalling Manager.

*Trial Procedure*

1 Signallers were briefed regarding the purpose of the trial on arrival.
2 IWS instruction was given. Participants were told that they would be alerted with a tone at two minute intervals. Once prompted they should choose the coloured (touch) key on the screen associated with the terms that best described how they were feeling at that time. It was stressed that this was their lowest priority task and that missed responses were not detrimental to the trial.
3 Participants were given an operational brief providing the context for the scenarios they were about to undertake. They were asked to:

- undertake each scenario in accordance with the rule book;
- complete the IWS ratings when required;
- complete a short subjective questionnaire at the end of the trial.

4   The scenarios were undertaken in turn: two with the IWS, two without the IWS. The order of presentation was randomised using a latin square design to counterbalance learning effects.

5   Throughout the scenario a facilitator recorded activity using the Noldus Observer Tool.

6   After each scenario the participant was asked to complete a short questionnaire assessing the usability and intrusiveness of the IWS.

7   Finally the participant was asked to retrospectively complete an adapted SWAT (subjective workload assessment technique) questionnaire.

*Results*

*Sensitivity*

*Sensitivity to changing demands in the Scenario*   The graphs showed peaks and troughs in workload across the recording intervals. An example is shown in Figure 30.1.

The graph shown in Figure 30.1 shows timeline analysis data overlaid onto the IWS output. The timeline analysis goes some way to provide objective and explanatory data behind the IWS scores.

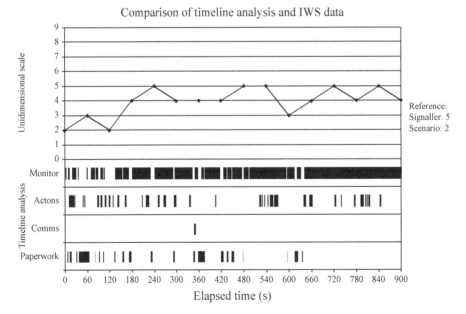

**Figure 30.1   Example graph output showing peaks and troughs over time**

*Comparison of IWS scores with scores on an adapted SWAT* The IWS and SWAT scores were correlated for both 'normal' and 'abnormal events' scenarios. A positive correlation was found in the 'normal' scenario (0.69). Only a weak correlation was found between scores in the abnormal scenario (0.34). It is likely that this weaker correlation was due to a high incidence of missed ratings in the 'abnormal events' scenario. The correlations suggest that SWAT and IWS are both measuring the same thing (workload).

*Responses to the questionnaire* Participants reported that they found the tool reflected how hard they were working and gave an accurate representation of their workload.

*Intrusiveness* The tool was not found to be intrusive. Intrusiveness was assessed via the questionnaire and via comparison of SWAT data collected with and without the IWS. There were no questionnaire responses to indicate that signallers had problems performing their operational duties whilst scoring the IWS. Strong correlations were found between SWAT scores for normal and abnormal scenarios (0.68, 0.97). If the tool was intrusive it is likely that SWAT scores would increase in scenarios where the tools were used.

## Usability

The tool was found to be usable. Usability was evaluated on a five point scale from 'very usable' to 'not usable at all'. All participants responded positively to the questions regarding usability. From a total of seven assessments, five reported the tool to be 'very usable' and the other two reported that it was 'usable'.

## Crewe NX Panel Simulator Trial

A second trial was conducted in August 2003 using an NX panel. The aims of this trial included:

- to develop a scenario that generates a higher workload experience for operators to test the higher end of the rating scale. In the York trial none of the participants scored above five on the IWS;
- to further satisfy Network Rail that the tool will not have any negative effects on operations (intrusiveness);
- to comment on the hardware solution of the IWS currently offered and to provide guidance on the future direction of this element of the design.

A simulated environment was used again to allow controlled scenario generation and to avoid disruption to operational areas until the tool had its unobtrusiveness confirmed. This trial undertook an examination of individual participant data in detail, rather than averaging across participants. This approach was taken to allow individual tracking of score in of the tool in response to the demands of the scenario and the individual decision-making process.

*Facilitators*

There were five facilitators present at the trial:

- two NR operational experts from the NR HQ Ergonomics Group. These two facilitators were responsible for briefing participants, de-briefing participants and providing the subject matter expert (SME) commentary;
- two CCD facilitators – project team members responsible for timeline recording, IWS recording, and trial administration;
- one Crewe NX simulator facilitator – familiar with the NX panel simulator and responsible for briefing participants about the panel, running the simulation, and providing the voice communications input.

*Participants*

Ten participants took part in this trial. Nine participants were trained signallers; one participant was an experienced signalling engineer. the participants were provided by NR from various locations throughout the country. All participants were familiar with NX Panel operations although none of them were familiar with the Crewe Simulator panel before the trial. Experience ranged from 5–35 years, with a mean of 18.5 years. Experience with NX panel operations ranged from 2.5 to 20 years, with a mean of nine years. Two participants did not complete the trial. These two participants struggled with the scenario from the outset and the NR facilitator decided it was not in their best interests to continue.

All participants were briefed regarding the purpose of the trial and were assured that:

- they would remain anonymous;
- the data would be used only by IOE/CCD;
- the trial was not a personal assessment and that there would be no implications for them as NR employees;
- should they feel uncomfortable with the trial at any point and they indicated this to the NR facilitator, the trial would be stopped if they wished.

*Scenario*

Another realistic scenario was developed by NR operational experts based on the NX Simulator Panel at Crewe. The panel control area included a four platform station, a station siding, and a factory siding. Both sidings required phone calls for shunting in and out. A branch line dictated a slot request via phone from the fringe signal box. A simplified timetable was provided.

The scenario lasted approximately 90 minutes and included periods of both high and low workload. Initially a low workload period was simulated, with lower traffic levels and less frequent train annunciations. Forty minutes into the scenario the workload was increased – with a higher traffic level and the incorporation of two failures: a points failure approximately 45 minutes into the scenario, followed by a track circuit failure approximately 60 minutes into the scenario. The points failure was simulated at the entry into the station sidings, and the track circuit failure was simulated on the track circuit for platform four at the station.

Relevant communications were controlled from the Observations/Control Room and were 'role played' by the simulation facilitator. These communications were held as consistent as possible between participants.

*Trial Procedure*

1   On arrival, participants were briefed regarding the trial and given the opportunity to ask any questions.
2   Participants were instructed in the use of the IWS and that they must respond to an audible prompt by selecting the appropriate term to describe their workload at that point on the touch screen.
3   An operational brief was given, providing scenario context and involving familiarisation with the panel.
4   Participants undertook a 90 minute scenario.
5   Throughout the scenario the participant was prompted to respond on the IWS scale every two minutes.
6   Activity was recorded using Noldus Observer and by hand by a NR operational expert.
7   At the end of the scenario the participant was asked to complete some questionnaires regarding intrusiveness and usability.

*Results*

*Sensitivity*

Figure 30.2 shows an example of data output. Some general observations were found across participants indicating the sensitivity of the tool. These included:

•   a peak of activity at the beginning of the trial, related to initial route setting (and, possibly, unfamiliarity);
•   a general pattern rising in accordance with traffic increases around 40 minutes (2400–3000 seconds);
•   a further rise in scoring at around 45 minutes (2700) when points fail;
•   another increase at 60 minutes when the track circuit fails.

The York trial scenario had not elicited scores above 5 ('moderate pressure'). The more demanding scenario in this trial elicited higher scores and showed that the tool is sensitive at the higher end as well as within the lower range values.

*Timeline analysis data*

The timeline analysis again proved useful for identifyin which types of task are the most significant contributors to increased scores on the IWS. IWS scores increased in relation to the level of communications and the frequency with which actions are required to be performed.

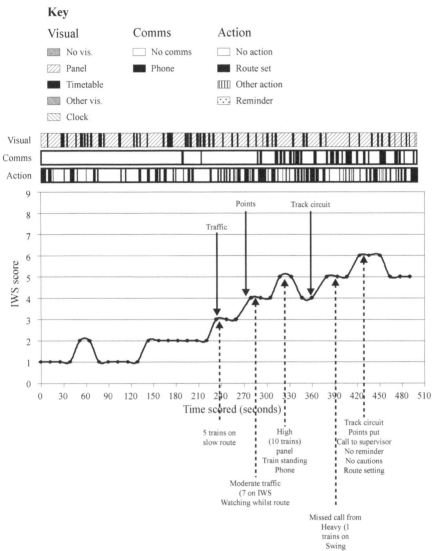

**Figure 30.2  Participant data output example**[3]

Note: the SME notes are incomplete due to graphics production for publication here.

*SME commentary*

Increased workload is also reflected in the SME commentary. SMEs were able to observe the effect that high workload scores had on task output. The SME commentary revealed that performance began to deteriorate for many participants at scores around 5 ('moderate pressure'). This was shown in a variety of ways: increased delay, missed telephone calls, omitted reminder appliances, failure to clear signals, route cancellations, and missed failure indications.

*Consistency across participants*

The consistency of results is shown in Figure 30.3 which shows the output from all participants on one graph. This figure illustrates the similarity in scoring trends and also relates the scores to changing scenario demands over time.

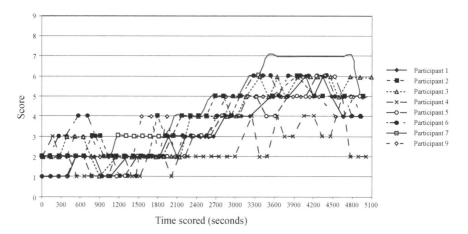

**Figure 30.3   Collective view of all participants' responses over time**

*Questionnaire*

Participant responses to the questionnaire confirm the sensitivity of the scale. All participants believed it produced an accurate reflection of how hard they were working across the scenario.

*Usability*

Again, questionnaire responses showed that participants found the tool easy to use. All participants reported that they found it easy to find the appropriate describer for their experience of workload at any given time.

*Intrusiveness*

The majority of respondents to the questionnaire (all but one) appeared to have no problems using the tool whilst performing their signalling duties. The one suggestion that there may be an issue with intrusiveness was directly related to busier locations (should the tool be used in the future) and it is possible that different hardware implementations may address any concerns.

*Practicality*

Most participants missed ratings at times and had to be verbally prompted to respond. Many were unaware that they had missed ratings. Where participants recognised that they had missed scores they attributed this to the fact that they either could not hear the prompt or that they were concentrating so hard that they did not notice it.

Increased response times were also shown at later points in the scenario and consistently with higher workload scores. A number of reasons were identified for this increased response time:

1   as participants get busier, and their workload increases, they miss the auditory alert due to concentration on the panel;
2   they are already involved in actions/communications and it takes them longer to fit the IWS response into their work;
3   workload shedding by experienced staff.

Some participants reported that the touch screen was not very sensitive. Observational data indicated that participants were unfamiliar with this technology.

This information carries implications for the practicality and usefulness of the tool at busy periods. Particular hardware implementations may be more appropriate for some participants and some environments. The simplest method would be pencil and paper recording with verbal prompts. A different solution would be the Actiwatch (http://www.camntech.co.uk).

**Summary**

The tools assessed have been demonstrated, in two separate trials, to be valid, sensitive, and usable methods for assessing workload in the rail industry. It is also assessed that the tools do not intrude significantly to the job of the signaller in a simulated environment.

The tools show a combined picture that responds to trends and manipulations in the scenario in unison. The IWS identifies peaks and troughs in workload whilst the timeline gives a level of diagnosticity to the picture, indicating the actions associated with higher workload. The timeline indicates that IWS scores increase in relation to the level of communications and task frequency and switching. This information is useful to the industry for assessing/comparing design solutions and for exploring the causes of workload in the field.

It is likely that particular hardware implementations will be best for particular environments. There is also scope for integrating the IWS, timeline analysis, and SME commentary within one software-based tool which would reduce the number of assessors required and could, potentially, aid analysis.

More recently, the tool has been taken into the signalling environment and used in live situations, with promising results (see Morrisroe *et al.* in this volume).

## Acknowledgements

Thanks are given to the Simulator Facilitator at Crewe, the CORUS Training Centre at Crewe, to York IECC, the Signalling Manager at York, to all the signallers that took part, and to Network Rail Ergonomics Team for their help, participation, cooperation, support, and hospitality throughout the trials.

## Notes

1   The integrated workload scale was previously referred to as the uni-dimensional scale (UDS).
2   The graphical output previously shown (Figure 30.1) is not achievable for a 90 minute scenario.

Chapter 31

# Application of Industrial Strength Workload Assessment Tools

Ged Morrisroe

## Introduction

This chapter addresses the experiences gained through administering the mental workload toolset, developed for railway use by the Institute for Occupational Ergonomics at University of Nottingham, on behalf of Network Rail. The tools have been applied to the signalling tasks for locations which have been referred to the assessment team. The assessment team has comprised operational experts and an ergonomist. The efficacy of these tools and representativeness of the results is discussed.

## Background

The subject of occupational workload has a long and distinguished reputation as a theoretical and practical tar pit. Researchers that stray into this field emerge many years later with either another grand unified theory on the subject (subtly different to the ones already in the expanding list) or with significant reservations and qualification of the subject area. In fact, the subject seems to be stuck and progress over the years seems to be ephemeral.

In industry, workload is manifest and recognisable. Logically its effects on the individual and on the performance of the overall systems at work cannot be ignored even if there is lack of a single, sharp, pithy and universally agreed definition and theoretical underpinning for the subject.

This chapter sidesteps the embroiled theory and concentrates on reporting the experience thus far of applying a set of workload assessment tools developed for the rail industry and reported elsewhere in this book in chapters by Wilson *et al.*, Pickup *et al.* and Mitchell *et al.*

This chapter lists the toolset elements that have been applied to date, and for each reports the experience of applying them within signalling centres operated by Network Rail. Any refinements that we have made are described and a resume of the observations about the tools is presented.

The title of this chapter refers to 'industrial strength tools' for workload assessment (at higher levels of workload – underload has not been addressed in these particular assessments). The requirements for such industrial strength or industrial suitability can be stated thus:

- detect workload and indicate whether it is excessive;
- identify the sources of demand and the drivers that are associated with the perception of excessive effort;
- can be administered within a half day site visit.

The success of the tools in meeting these requirements will be evaluated, using the experiences and outcomes of real use made by a team that does not include any of the original tool set developers.

## The Network Rail Workload Assessment Toolset

The context in which these tools have been applied has been to address the question of whether or not a workload problem exists for a signaller. If it is deemed to be the case, then can the tools determine the sources and factors associated with the workload?

*Process*

To date, the source of requests for a workload assessment have arisen as a result of staff complaint, signalling managers' requests or by referral from a third party (for example an Investigation). In each case a small team from the Ergonomics Group at Network Rail, comprising at least one operational specialist and an ergonomist, have carried out the assessment.

The team arrange a mutually convenient time to undertake the assessment and plan to be present during a significant time period that coincides with that of the perceived workload. On the day of the assessment, after introducing ourselves and the methods, we administer all or part of the workload toolset. We make it clear to the signallers participating that the assessment can be terminated at any time and for any reason at their request. Likewise, if the operations specialist deems that the assessment should be terminated for operational reasons then we leave. This report is based on about ten assessments, none have been terminated prematurely, and only a small number have required a follow-up visit to check on the reliability of the findings by reviewing them with a different signaller.

*The Toolset*

The following tools have been used in real workload assessments:

1  operational demand evaluation checklist (ODEC);
2  integrated workload scale (IWS) and timeline analysis;
3  workload principles;
4  SWAT (subjective workload assessment technique);
5  rail-adapted modified Cooper Harper (RMCH);
6  workload probes (decision ladders).

Tools 1, 2, 3 and 6 are new, and the other two are modified versions of well known tools. The new tools have been described in Pickup *et al.* and Wilson *et al.* in this volume and especially in Pickup *et al.* (2005)and so no additional information is given here regarding their derivation or development.

*Operational Demand Evaluation Checklist (ODEC)*

ODEC has been administered in all instances of the workload assessments. The checklist collects a set of data that reflects the 'objective' demand on the signaller, for instance number of track miles controlled. The assessors classify each answer into a high, medium or low category; a simple scoring system is used ('3, 2, 1' for high, medium or low respectively). Some high salience items in the checklist have been identified as scoring double.

Although the data can be collected from a number of sources, the assessors try to access the data in the way that is least disruptive to the signaller. Data are collected from the signal box specific Sectional Appendices (for example track miles and line speed), the signalling display (controlled and automatic signals), and the simplifier (timetable/ number and frequency of planned traffic). If there is a shift manager, they can provide some of the answers that are not directly available to the assessors (for example short notice traffic movements and number of incidents in the last week). In the beginning, some of the categorisation of the data into high, medium or low had to be carried out by the operational specialist(s) based on general experience. This issue is addressed in a later section.

The valuable output of ODEC is the list of high scoring 'demand' items; these are useful indications of the demand factors that may underlie the workload 'problem'. An overall summed ODEC score can be obtained. This score in isolation is of limited value, but when signal boxes are compared it is interesting to note the ordering of the total scores.

*Results and Revision of the ODEC Tool*

The experience of using ODEC has lead to a modified form of the checklist being developed by the assessors. The modifications and the reasons for them are as follows:

• the length of the checklist has reduced from 31 to 26 items. Items that have not helped discriminate between locations or appeared to have less relevance have been removed (for example tunnels and bridges);
• ranges of scores for an item have been added, for example the number of points traversed by a *simple* and a *complex* train movement;
• a fourth category of 'not applicable' has been added for the assessments, because at some sites there are none of that item. This helps to distinguish between sites that have low and no instances of the item;
• notes and prompts are provided to indicate the type of data required. It was noted that when the data from a number of assessments were compared, there was a difference in interpretation of some of the entities to be assessed; for example for sidings, sometimes the data provided were the number of sidings whereas what is required is train movements in and out per day;

- a comprehensive scoring system has been developed for the checklist items. As we gathered more data, it became easier to be categorical about what constitutes high, medium and low for an item;
- automatic scoring of the items in the spreadsheet has now been provided. Since we were entering the data into a spreadsheet, and because the scoring scheme was now defined, it was easy to construct the on-line version of ODEC.

Experience has made us concerned about including an overall score in the feedback reports. Its presence encourages some readers to attach too much weight to it as an absolute indication of the workload level.

This is evident when the scores from a mechanical lever frame are compared with a NX panel. The NX panel scores much higher (for instance, 96 versus 49). If these scores are taken at face value, a false impression of demand is given. There are genuine reasons for the differences: lever frames control small sections of route, have fewer signals, and many of the ODEC items do not exist at these sites (for example hot axle box detectors). In fact, the mechanical signal box that scored 49 was deemed to be a demanding signalling work position. The implication of this is that there needs to be a separate scoring system for mechanical lever frame sites. It is likely to be based on a multiplier scoring system. It is clear that we are not in a position to specify a 'red line' level based on the overall ODEC score.

Finally, a feature of ODEC that was not at first apparent to the assessors is that signallers can construe it to be a job evaluation method. Some signallers asked directly, 'Will this lead to a regrading of the job?' It is important that the signallers are briefed at the outset to establish that this is not about job evaluation, however based on some of the outcomes and recommendations of these assessments it is not possible to say that the two are unrelated.

*IWS and Timeline Analysis*

These two tools are reported together to reflect that in the field they are administered concurrently. The timeline and IWS are typically recorded for a 60-minute period that coincides with the predicted peak workload. The results from the two tools allow the assessors to see a 'trace' of increases and decreases in reported levels of demand and effort and correlate with the activities represented in the timeline.

It became usual for the operational specialist to record the timeline information manually. A recording sheet divided into minutes was used to record the activity and give some indication of start and end if it ran to more than 30 seconds. It was usual for this duration to be exceeded for telephone calls, but less usual for any other signalling task, for example, route setting or data entry for train description. The timeline also recorded walking around the panel and watching a particular event. The activity was video recorded, but on no occasion have we needed to refer to it to amplify the timeline record.

The IWS was prompted every five minutes. Simulator based trials (see Mitchell in this volume) were undertaken with shorter inter-stimulus intervals, but in the live signalling situation a five-minute interval appeared to give the correct trade-off between number of data points and interrupting the signaller to ask for a rating.

Two methods of capturing the IWS data were used. Initially, after briefing the signaller, the assessors set up the response options sheet next to the signaller's work area and then prompted (every five minutes) for a verbal rating. The IWS was designed with nine semantic levels of demand/effort without a numerical rating scale. For reasons of ease of use, we added a numerical value for each of the points.

Another method used was a customised watch (Actiwatch score) that can be programmed to prompt at the specified intervals and record the rating that the signaller makes (by pressing a button several times until the appropriate number is shown). This method also worked well and has the advantage that the signaller can respond on the move. Here the scale was inverted so that ratings at times of highest load required less button presses from the signaller.

*Results*

The results gained show that the IWS ratings vary over the recording period and give a good indication of the relative demand and effort relating to the tasks recorded in the timeline.

The respondents appear to adopt a response set. For the initial series of ratings, the signallers refer to the response descriptions, but once they settle into a pattern of rating they appear to vary about their chosen 'default' level without referring to the scale description.

Prior to using the IWS there was concern that the 9-point scale may be unwieldy. In practice, it does not appear to be a problem. All levels of the scale have been recorded at different times (but not by the same person). It may be that the 'response set' effect described above makes the scale practical and easy to use.

*Workload Principles, SWAT, Rail-adapted Modified Cooper Harper (RMCH) and Probes (Decision Ladders)*

The Principles, SWAT and RMCH are short tools that can be used to address whether a workload assessment is necessary, or else are used at the initial stage of the assessment to determine whether workload may be a problem at the outset.

The probes (in the form of decision ladders) comprise an in-depth, diagnostic tool that surveys a broad area comprising receiving information, giving information and evaluating and decision making. When a problem is detected at a general level, the tool is used to drill down with a sequence of related, detailed questions.

*Results*

These analytical tools have not always been administered during the assessment. There are two main reasons for this:

1   as explained earlier, the conduct of the assessment works best when signaller intervention is minimised. These tools rely on face-to-face interviewing of the signaller and require varying amounts of time. If the signal box is staffed by a single signaller, then it is difficult to administer. If there are a number of signallers or a relief signaller is present, then it becomes easier to administer these tools;

2    as in life, when visiting a signal box there comes a time when to remain is to out-
     stay your welcome. After concluding the ODEC, IWS and timeline activities the
     assessors usually have been on site for two hours. At this point in the assessment,
     the assessors begin to drill down into areas that have emerged as the particular
     set of demands that may be related to workload. This activity is interspersed
     with statements that summarise the understanding of the assessors that the
     signaller is invited to confirm. This phase seems to be critical for the assessors to
     establish that they have detected the sources of workload and that they have an
     understanding of their effects.

On a number of occasions the assessors have used impressions and perceptions gained
during the assessment to complete some of these tools off-line after leaving the site.
The value added by this is not compelling, so it is now rarely completed in this way.
    Within the 12 principles, five are primary and are essential attributes of the
signalling task in that compromising these is likely to lead to an erosion of safety
and performance. On occasions these five primary principles alone have been
administered to signallers to provide a pass/fail view of the signalling task. However,
it is difficult to arrive at a categorical yes/no response. The answers recorded will be
different if one prefaces the questions with either:

*   'In the main ...', or;
*   'In every instance ...'

SWAT and RMCH have been least used of the tools. In addition to the reasons
that have been explained above, the sequence in which an assessment takes place
means that short, summary tools such as these seem out of place at the end of an in
depth assessment. ODEC is the preferred starting point for the assessment since the
data gathering is non-threatening and allows the signallers to become 'comfortable'
with the assessors; it also enables the assessors to familiarise themselves with the
signalling site characteristics. So, those tools that are directed towards detecting, at
an early stage, whether there is workload problem, are somewhat redundant after a
more in-depth assessment has already taken place.
    The probes have been administered on one occasion only. The tool comprises three
parts and part 1 – receiving information required 60 minutes to administer. Part 2
– giving information was conducted with a second signaller and was completed in 15
minutes. Part 3 – evaluating and deciding was less well developed and was curtailed
after ten minutes.
    It was clear that the probes were effective in identifying the major sources of demand,
understanding alternative strategies for addressing the demand, and establishing the
safety, performance and signaller well-being consequences. Ranking of the factors
was not performed on this occasion.
    A continuation assessment visit is required when, after the first visit, the assessors
have not identified whether or not there is a workload problem, or if there is a workload
problem but they have failed to identify the contributing factors. On account of its
extended time requirement and the need for a captive signaller for up to two hours,
it was decided that the probes would be best used when a continuation assessment
visit was required.

The experience of administering this tool has established that it is a worthwhile, in-depth and diagnostic tool. A reworked and repackaged version is being developed.

### Detection of Workload and Sources of Demand

In this section, the performance of the tools is evaluated and discussed with respect to:

1   detecting workload; and
2   diagnosis of the sources of task demands that cause workload problems for signallers.

Based on the results gained from the initial assessments it cannot be claimed that when the tools are applied they categorically indicate the presence of undesirable levels of workload. What the tools do provide is a standard and signaller-relevant set of data that the assessors must weigh to decide whether there is a workload problem.

In particular, ODEC classifies the physical infrastructure elements of the signalling control area of responsibility into high, medium or low. A score is derived from the tool but, for reasons already expounded, it is inadvisable to consider this score as a categorical level of demand.

Similarly, the IWS scores indicate a rating of the load experienced and effort required by the signaller, but caution is needed in taking the average, minimum and maximum values as absolute evidence of workload demand. The IWS and the Timeline are most useful together in associating the activity being undertaken and the ratings of effort over a period of time.

So if these results are indicative rather than categorical, then what is the process that the assessors use to determine an answer to the two basic questions?

As already stated the assessors have not used the summary tools (SWAT, principles and RMCH) as the starting point for the assessment. Instead, the order of tool application is normally ODEC then IWS and timeline. Towards the end of the assessment, the assessors are able to draw on this detailed information to address the detection of workload question. To make the judgement, they need to be confident that they have understood the level of workload and sources of demand on the signaller. If they are confident, then the workload judgement is easy to make; if not, the workload assessment needs to be adjourned and further examination undertaken on a later continuation visit.

At the outset, the assessors had a concern that the workload assessment toolset might not discriminate between high and excessive workload, and would thereby indicate high workload in all sites visited.

Because of the method of referral for assessment, nearly all sites have been found to have high workload. In more than half of these sites, the reasons attributed for elevated levels of workload have been short-term deficiencies either created by the infrastructure or by the signaller's equipment. In these cases, the assessors consider that once the deficiencies are addressed and remedied, the level of workload will return to an acceptable level. In less than half the assessments the recommendation for extra staffing resource has been made during particular periods where the demand on the signaller is particularly high. Regrading of the job has not been recommended on any of these occasions.

## Representativeness of the Results

For these assessments, there are a number of concerns about whether the workload measure is:

- representative of all the individual signallers and the range of task demands that confront them;
- manipulated by the respondents;
- restricted in generality due to the short period of time that the assessments were conducted.

Each of the assessments carried out have addressed and observed a single signaller's experience of workload and are not necessarily a consensus view of the signalling team. On account of this, on occasions when we have suspected that the signaller studied is less representative of the rest of the staff, we have followed up the visit to meet other staff and validate the findings with these signallers' perceptions. An instance of this was observed in a signal box where a highly experienced signaller participated in the assessment, but the other staff were relatively new to signalling.

The assessors are aware of the possibility that some manipulation of results might take place, especially when the staff are already specifically complaining about workload. This risk has been reduced by having an operational specialist on the assessment team. Also, the bias may be in the opposite direction. A few signallers have commented, after completing the IWS session, that they find it difficult to give ratings at the higher levels of effort and that it 'goes against the grain'. Therefore, there is a risk that signallers will downplay their perceived effort.

The timing and duration of the assessments is potentially troublesome. Since the assessment rarely exceeds three hours, there may be some doubt about the generality of the results based on this short window. However in practice, the remit of the assessment often specifically directs us to consider a particular period, for example the morning or evening rush, the periods of possession management, the combination of two areas of control for the night shift, and requests for line crossing by the public (for example ramblers, holiday makers). In the Workload Assessment Report, the prevailing state of the railway during the assessment is always described, and it is emphasised that the conclusions and recommendations are made on the basis of what was observed and recorded on the day. This provides a useful platform to consider variations in demand on the overall workload.

In summary, the three points of concern raised have, in practice, not had a significant effect on the generality of the results.

## Conclusion

This chapter presents the experience of applying a set of tools for workload assessment. The tools have been developed with the rail industry in mind and target the signalling task in particular.

Criteria for assessing the industrial efficacy of the tools have been stated in this chapter. The conclusion, based on the assessments conducted to date, is that the tools generate the necessary information for the assessors to determine whether there is a workload problem, they are effective in identifying the sources of demand, and they can be administered within a half day.

## Acknowledgements

This chapter is based on the application of the toolset in a number of the signalling locations assessed. The assessors that collected these reported data were part of the Ergonomics Group at Network Rail and I gratefully acknowledge their support in preparing this chapter. Finally, thanks are due to the signallers who agreed to participate in and were supportive of these assessments.

## References

Pickup, L., Wilson, J.R., Nichols, S., Norris, B., Clarke, T. and Young, M.S. (2005), 'Fundamental Examinations of Mental Workload in the Railroad Industry', *Theoretical Issues in Ergonomics Science* (in press).

# PART 10
# HUMAN RELIABILITY AND SAFETY CULTURE

Chapter 32

# Human Error Risk Management Methodology for Safety Audit of a Large Railway Organisation

P.C. Cacciabue

## Introduction

This chapter considers the application of the HERMES (human error risk management for engineering systems) methodology for safety assessment studies. The application concerns the recurrent safety audits (RSA) of a large organisation in the domain of railway transportation systems. The objective of such study was the identification of the most relevant areas of intervention for improving safety and reliability of the service.

The methodology has been applied to the whole organisation and its working processes. Specific attention was dedicated to train drivers. A number of indicators of safety (IoS) and recurrent safety audit matrices (RSA-matrices) were identified, which enabled the assessment of the safety level of an organisation and the generation of safety recommendations.

The application of HERMES to this case study shows that the methodology is applicable in practice and can give valuable and significant results.

The need to include human factors (HF) considerations in the design and safety assessment processes of technological systems is nowadays widely recognised by almost all stakeholders of technology, from end-users to providers and regulators.

The critical role assigned to HF in design and safety assessment is further enhanced by the common-sense appreciation that it is impossible to conceive a plant that is totally 'human-error free', which must be considered an intrinsic characteristic of any technological system. Therefore, the improvement of the safety level of a system can only be achieved through the implementation of appropriate measures that exploit the enormous power of both human skills and automation potential for preventing or recovering from human errors, and for mitigating the consequences of those errors that still occur and cannot be recovered. This represents a process of human error management (HEM).

From a HF perspective, the constituent elements of complex technologies can be identified in the presence and interconnection of organisational and cultural traits, working conditions, defences, barriers and safeguards, and personal and external factors (Reason, 1997). These factors have to be evaluated in order to examine the state of a system/organisation with respect to safety/risk conditions. Moreover, the intrinsic dynamic nature of socio-technical organisations demands that these data are regularly evaluated with adequate approaches that emphasise the cognitive ergonomics aspects of Human Machine Interactions (HMI) (Wilson *et al.*, 2001).

Recurrent safety audits (RSA) of organisations attempt to evaluate the safety state (level) of an organisation with respect to a variety of safety indicators and markers associated with the current state of the elements of a system.

The recurrent assessment of the safety level of an organisation requires a methodological framework where different methods and approaches are combined and integrated for considering HMI. RSAs of organisations are critical and essential key processes for preserving systems integrity and for preventing and protecting from accidents.

In this chapter, a methodological framework of reference, called human error risk management for engineering systems (HERMES), is presented (Cacciabue, 2003). HERMES offers: a) a 'roadmap' for selecting and applying coherently and consistently the most appropriate HF approaches and methods for the specific problem at hand, including the identification and definition of data; and b) a body of possible methods, models and techniques to deal with the essential issues of human error and accident management (HEAM) approaches.

Firstly, the points of view to be taken into consideration by HF analysts before starting any analysis or safety study will be reviewed. Then, models and methods for HMI will be discussed focusing on their integration in a logical and temporal sequence within the overall design and safety assessment process: this represents the methodological framework HERMES. Finally, focusing on safety assessment, the specific application of HERMES for the safety audit of a railway organisation will be discussed in detail.

## Standpoints for Performance of Human Machine Studies

A number of standpoints should be considered when studying a Human machine system (HMS). Five standpoints are to be set at the beginning of any study (Figure 32.1):

1   definition of the goals of the hms under study and/or development;
2   concept and scope of the hms under study and/or development;
3   type of analysis to be performed;
4   area of application; and
5   indicators of safety level.

The first standpoint demands that the goals of the systems under study must be clearly identified and constantly accounted for during the whole process. As an example, the Defences, Barriers and Safeguards (DBS) represent all structures and components, either physical or social, that are designed, programmed, and inserted in the human-machine system with the objective of making more efficient and safe the management of a plant, in normal and emergency situations. DBSs should tackle one or more of the objectives of HEAM, namely: a) prevention; b) recovery from human errors; and c) containment of the consequences that result from their occurrence.

Control systems are directly related to some forms of performance, either appropriate

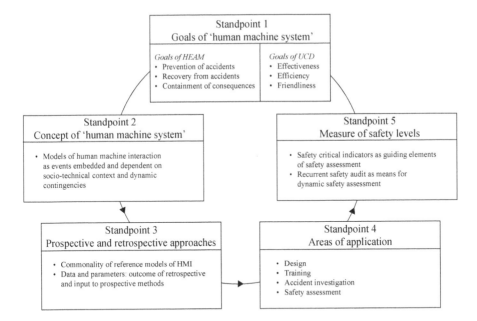

**Figure 32.1  Standpoints for HMS studies**

Note: HEAM = Human Error and Accident Management; UCD = User Central Design.

or erroneous. It is therefore important to develop a clear understanding of human performances or behaviours and their dependence on specific dynamic context (contingencies) and socio-technical environment in which they are embedded. This is a second standpoint for the development of effective HEM measures.

A variety of models and methods are necessary for the development of HMSs. In general, they can be structured in an integrated framework that considers two types of analysis, that is, retrospective and prospective studies, which are complementary to each other and contribute equally to the development and safety assessment of HEM measures. These analyses rest on common empirical and theoretical platforms: the evaluation of socio-technical context, and the model of HMI and related taxonomies.

When performing a process of design or assessment of a HMS, it is essential that the specific area of application of the system under study is well delineated. Four areas of application must be considered, namely, *design, training, safety assessment,* and *accident investigation.* Each of these four areas of application encompasses specific types of assessment. The fourth standpoint for the development of effective HMS and HEM measures lies in the appreciation of the fact that there are four possible areas of application, and in the links existing between different areas.

The fifth standpoint is related to the definition of appropriate safety levels of a plant. In all type of analyses, and for all areas of application, it is essential that adequate *indicators* be identified that allow the estimation or measurement of the safety level of a system.

## A Methodological Framework

A methodology that aims at supporting designers and analysts in developing and evaluating safety measures must consider the basic viewpoints discussed above. In particular, a stepwise procedure may be considered according to each area of application and to the key objective of the safety measure under analysis, namely prevention, recovery or protection. In addition, a set of methods and models can be considered as the functional content supporting the application of procedures.

From the procedural viewpoint, it is necessary to ensure consistency and integration between prospective and retrospective approaches. The methodology described in Figure 32.2 is called human error risk management for engineering systems (HERMES) (Cacciabue, 2003) and can be applied for all safety studies that require HF analysis.

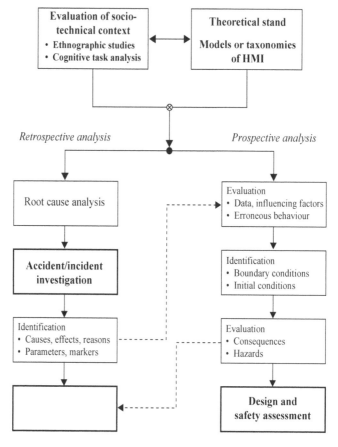

**Figure 32.2   Human error risk management for engineering systems (HERMES)**

HERMES must be supported by existing models and specific methods for performing each step of the procedure. The steps to be carried out in the application of HERMES are the following:

- firstly, it is necessary to select a common theoretical platform for both retrospective and prospective types of analysis;
- at the same time, data and parameters typical of the system are derived from ethnographic studies and cognitive task analysis;
- then a set of data, influencing factors and erroneous behaviours are evaluated by investigation of past events based on root cause analysis (RCA). This process leads to the identification of parameters and markers of cognitive behaviour;
- then, for a complete prospective study the analyst and designer needs to apply his/her experience and creativity for identifying boundary and initial conditions for performing predictive safety studies, and for evaluating unwanted consequences and hazards;
- in this way, HMI methods can be consistently and coherently applied for design, safety assessment and accident investigation as well as for tailored training.

## Application of HERMES for the RSA of a Large Railway Organisation

*Problem Statement and Boundaries of Study*

The top management of a large European railway company, which will be called ERC for convenience in the remaining part of this chapter, felt the need to carry out a study of the human factors and safety related issues existing within the organisation, with particular attention to the population of train drivers (TDs). The objective of such study was the identification of the most relevant areas of intervention for improving safety and reliability of the service, and in general to ascertain the current state and possible needs for human error and accident management measures.

The HERMES methodology turned out to be the most appropriate approach to apply for a safety audit, as it offers a wide variety of methods and models amongst which to choose the best ones for solving the variety of problems presented during the audit, while paying attention to the integration and balance between the different approaches.

The ERC organisation consisted of more than 100,000 employees, engaged every day in the management of the railway traffic, with a population of more than 70 per cent train drivers. The technology and means available at ERC consisted of a high number of 'trains' presenting a wide variety of technological solutions on board, from the most modern fully automatic controlled, high speed machines, to very conventional locomotives, based practically on full manual control. At the time of the study, the railway network covered almost 20,000km over a vast territory and a variety of terrains. Approximately 10,000 trains per day ensured the movement of several hundreds of thousand passengers.

The application of HERMES was limited to the identification of safety critical factors, or indicators of safety (IoS), and their organisation into recurrent safety audit matrices (RSA-matrices) that serve the purpose of defining the existing level of safety within the organisation and defining the reference measures for future audits.

The work was performed in three coordinated and distributed phases.

*Phase 1*  This phase included the setting up of the team of human factors experts and working groups supporting the study, and covered the initial steps of HERMES:
   a)   acquisition of information and knowledge in the field about the working environments and practices of train driving (ethnographic studies);
   b)   study of work requirements by task and job analysis; and
   c)   identification of theoretical models and techniques to apply.

*Phase 2*  The second phase of work was dedicated to the extensive field assessment and thus to the collection of the most important sources of information and data through questionnaires and interviews. The analysis of all collected data aimed at identifying possible areas of concern.

Moreover, the annual reports of ERC on incidents and accidents were made available and could be studied from a retrospective point of view.

*Phase 3*  The third phase was totally dedicated to the core development of the safety audit and preparation of recommendations.

The identification of safety indicators and matrices, as well as the recommendations on safety improvements were performed with reference to the results of Phase 2 and exploiting the experience and creativity of the analysts.

*Phase 1: Evaluation of Socio-technical Context of ERC and Theoretical Stand*

This activity involved some staff members and managers at different levels within the organisation. The focus of this phase was to select a consistent number of train drivers, or the most representative depots, that would represent the reference populations of TDs and depots for further detailed interviews and data collection.

Training procedures and books with norms, regulations and standards were also made available by ERC, as part of the theoretical documentation.

Phase 1 consisted of five main steps, namely: creation of a human factors team and steering committee; preliminary selection of model and taxonomy; preparation of ethnographic studies; task analysis and initial ethnographic studies; and adjustment of model and taxonomy.

An essential outcome of this process was the identification of a model for representing the socio-technical interactions, and the workflows and procedures applied within the organisation (Figure 32.3). The model SHELL (Edwards, 1988) was selected as a reference for the analysis of the behaviour of train drivers and their working context and environment. This model is a consolidated framework and has been adopted in other domains strongly affected by human factors issues. The chart of human factors relationships was particularly important for identifying an initial set of factors that affect performance and behaviour of TDs, called performance influencing factors (PIFs); and for preparing the guidelines for the interviews and field observations.

The following step of development focused on the definition of the *tasks* of a TD. The combination of task analysis and SHELL helped in refining important PIFs.

The theoretical instruments selected to represent formally the tasks of TDs were the hierarchical task analysis (HTA) and the time line analysis (TLA) (Kirwan and Ainsworth, 1992). HTA allows us to describe in a simple structured format (tree-like) the sequence of tasks to be carried out by TDs. TLA focuses on dynamic aspects.

*Phase 2: Field Studies within ERC and Identification of Data, Causes and Parameters*

The work of data analysis was performed with the support of a limited selected number of experts of the company, including train drivers and managers.

The activity involved substantial field studies, data collection and analysis. It required the application of well-established tools and methods, such as software specifically dedicated to statistical analysis. It consisted of four main steps: cab rides and timeline analysis; detailed interviews; definition of error analysis method; analysis of data collected from questionnaires and interviews.

The cab rides were carried out immediately before and after the first set of interviews and workshops. They helped the human factors team in refining the TA method and in preparing the study of the workload associated with the performance of the TD's task, especially with respect to available times for performing the required duties.

The second step was to conduct many interviews and collect questionnaire data. Three main depots had been identified as representative of the organisation, and extensive data collection in these depots was carried out. More than 300 TDs were interviewed. The questionnaire was distributed to the entire population of TDs of the three depots, that is, approximately 2,500 TDs. Altogether, 710 questionnaires were colleted and the data were stored in a database.

The selection of a technique for representing error generation and management (step 3) during the activity of a TD has been strongly influenced by the fact that the human factors team aimed at selecting a simple theoretical configuration of human machine interaction. The goal was to favour the discussion with TDs for the identification of PIFs in association with possible *error types* and *error modes.*

With these objectives in mind, the theoretical framework called THERP (*T*echnique for *H*uman *E*rror *R*ate *P*rediction) (Swain and Guttmann, 1983) was selected. THERP is normally utilised in human reliability assessment to calculate probabilities of human errors associated with the performance of tasks. However, in this case, only the basic theoretical structure and graphical representation of the method was utilised, leaving behind the quantification part.

The combination of the results of the data collected with the questionnaires and the information retrieved from the interviews led to the identification of eight areas of influence and eight generic PIFs that may lead to different types of errors, such as mistakes due to lack of knowledge, violations or simple slips that may endanger seriously the safe performance of the service (Table 32.1).

*Phase 3: Definition of IoS and RSA-matrices for the Safety Audit of ERC*

This phase of work merged two essential components, namely the information and experience gained by the HF team during the field studies, and the outcome of task analyses and study of reports on incidents and accidents from past operating

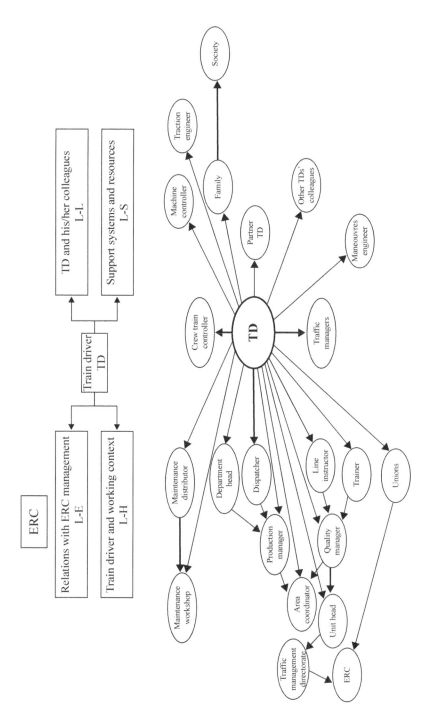

**Figure 32.3   SHELL model and workflows for the organisation ERC**

experience. The objective of this last phase of work was for the development of the IoSs and RSA-matrices to be applied for assessing the state of the organisation.

This phase was conducted in two subsequent steps: development of THERP-trees; and generations of IoS and RSA-matrices, and safety recommendations.

A variety of scenarios have been studied, and the associated THERP-trees have been developed. For each scenario, the success or failure of the mission has been evaluated in consideration of PIFs, workloads, working processes and expected performances of TDs.

The final step of this study focused on the generation of a set of IoSs and RSA-matrices that would allow ERC to perform successive evaluations and audits of its own safety level. The IoS were also the guiding elements for safety recommendations.

PIFs and possible types and modes of errors identified during the previous phases of work have been considered for identifying relevant IoSs. In Table 32.2 the PIFs have been associated with possible causes/effects on TDs' behaviour and consequently to corresponding families of IoSs. In summary, a total of 13 IoSs were utilised for performing the safety audit: two IoSs related to organisational processes (OP), two related to personal and external factors (PEF), five concerning local working conditions (LWC) and four associated with defences, barriers, safeguards (DBS).

Safety recommendations have been developed with reference to the PIFs and *IoS*s. Specific attention was dedicated to the identification of areas of concern and definition of possible interventions or improvements.

A number of *RSA-matrices* could be derived from the table of IoSs. An example of the RSA-matrix relative to the DBS is shown in Table 32.3. Each IoS is associated with three elements: specific safety objectives, that is, prevention, recovery or protection of accidents/incidents; approaches for evaluating the IoSs, for example, *field assessment, medical checks, data analysis, questionnaire/interviews,* and *quality and engineering checks*; and frequency ($\varphi$) of performance of the specific audit.

## Conclusion

This chapter considered the application of the HERMES (human error risk management for engineering systems) methodology for safety assessment studies.

The application concerned the RSA of a large organisation in the domain of railway transportation systems. The objective of such study was the identification of most relevant areas of intervention for improving safety and reliability of the service.

This case study showed that a substantial application of the HERMES methodology to large systems is feasible and can provide important and useful results. Figure 32.4 shows the summary of methods and techniques utilised.

However, a real application of HERMES requires considerable effort from the human factors specialists, especially for the collection of the relevant information and data. Lacking this critical part of the work, the application of the methodology may lead to trivial and possibly misleading results.

**Table 32.1  Final set of PIFs affecting TD performance**

| PIF | SHELL reference – error type | Error modes |
|---|---|---|
| **Communication within ERC (PIF 1)** | Serious problems encountered in the contacting managers for discussing rules and standards (L–E) | Errors and violations of rules |
| **Communication means (PIF 2)** | Inadequate communication technology between several actors, i.e, TDs with train crew controller, traffic manager, line instructors, etc. (L–L) | Catastrophic errors in emergency conditions and high workloads |
| **Technological interfaces (PIF 3)** | Difficult ergonomics of signals in the cabin and inconsistent displacement of signals on the railway (L–H) | Possible occurrence of passing signals at danger |
| **Maintenance of trains/railway (PIF 4)** | Inadequate maintenance of trains and railway reduce reliability of system and increase difficulty of train driving (L–H) | Induced errors of traffic management and violations |
| **Comfort of working contexts (PIF 5)** | Inadequacy of working contexts and poor logistic of rest areas increase stress and workload on long distance shifts (L–E) | Errors due to stress |
| **Roster and shifts planning (PIF 6)** | Too little involvement of TDs in the definition of shifts; the 'fixed couple' strategy may not contribute to safety (L–E) | Errors due to stress or complacency |
| **Rules and regulations (PIF 7)** | Too many complex rules and regulations, sometimes contrasting with each other (L–S) | Errors and violations of rules |
| **Training methods and simulators (PIF 8)** | Existing training is inadequate to cope with advanced automation and complexity of tasks (L–S) | Errors due to lack of knowledge and training |

**Table 32.2 Identification of IoSs with reference to PIFs**

| PIF | Causes/effects | | IoS | IoS |
|---|---|---|---|---|
| PIF 1 | • Serious problems encountered in contacting managers for discussing rules and standards | ✓ | $OP_1$ | Unwritten rules; reporting systems |
| | • Uncertainty about future – low morale | ✓ | $PEF_1$ | Mental conditions |
| | • Unions as unique channel for communicating with top management level | ✓ | $OP_2$ | Role of unions vs management |
| PIF 2 | • Obsolete technology for communication | ✓ | $LWC_3$ | Quality of tools |
| | • Inadequate maintenance of means | ✓ | $LWC_3$ | Maintenance of tools |
| | • Unclear rules for communication | ✓ | $DBS_3$ | Policies, standards |
| PIF 3 | • Poor ergonomics of interfaces of train cabins | ✓ | $LWC_1$ | Workplace design |
| | • Problems in understanding/managing automation | ✓ | $LWC_2$ | Automation |
| | • Inconsistency between signals on track/cabin | ✓ | $LWC_4$ | Signals: track and cabin |
| PIF 4 | • Inadequate and insufficient maintenance of trains and tracks | ✓ | $DBS_1$ | Safety devices |
| | | ✓ | $PEF_2$ | Conditions, stress |
| PIF 5 | • Obsolete technology for communication | ✓ | $LWC_3$ | Quality of tools |
| | • Poor comfort, rest areas and long-haul | ✓ | $LWC_1$ | Workplaces |
| | • Lack of development plans | ✓ | $PEF_1$ | Mental condition, morale |
| PIF 6 | • Heavy and too stiff TD shifts | ✓ | $PEF_2$ | Physical conditions |
| | • Inadequacy of TD team vs safety | ✓ | $LWC_5$ | Job planning |
| PIF 7 | • Excess of rules and regulations | ✓ | $DBS_3$ | Policies, standards |
| | | ✓ | $DBS_4$ | Procedures |
| PIF 8 | • Inadequate training | ✓ | $DBS_2$ | Training standards |
| | • Insufficient expertise of trainers/instructors | ✓ | $OP_2$ | Human relationship |

**Table 32.3 ERC RSA-matrix for the IoS defences, barriers, safeguards**

| IoS – DBS:<br>defences, barriers, safeguards | Prevention | Recovery | Containment |
|---|---|---|---|
| DBS$_1$ = safety devices<br>• *Quality and engineering checks* | $\varphi_{eng}$ | $\varphi_{eng}$ | $\varphi_{eng}$ |
| DBS$_2$ = training<br>• *Questionnaire/interviews*<br>• *Field assessment* | $\varphi_{int}$<br>$\varphi_{fld}$ | $\varphi_{fld}$<br>$\varphi_{int}$ | $\varphi_{int}$<br>$\varphi_{fld}$ |
| DBS$_3$ = policies, standards<br>• *Quality and engineering checks* | $\varphi_{eng}$ | $\varphi_{eng}$ | $\varphi_{eng}$ |
| DBS$_4$ = procedures, instructions,<br>supervision<br>• *Quality and engineering checks*<br>• *Questionnaire/interviews* | $\varphi_{eng}$<br>$\varphi_{int}$ | $\varphi_{int}$<br>$\varphi_{eng}$ | $\varphi_{eng}$<br>$\varphi_{int}$ |

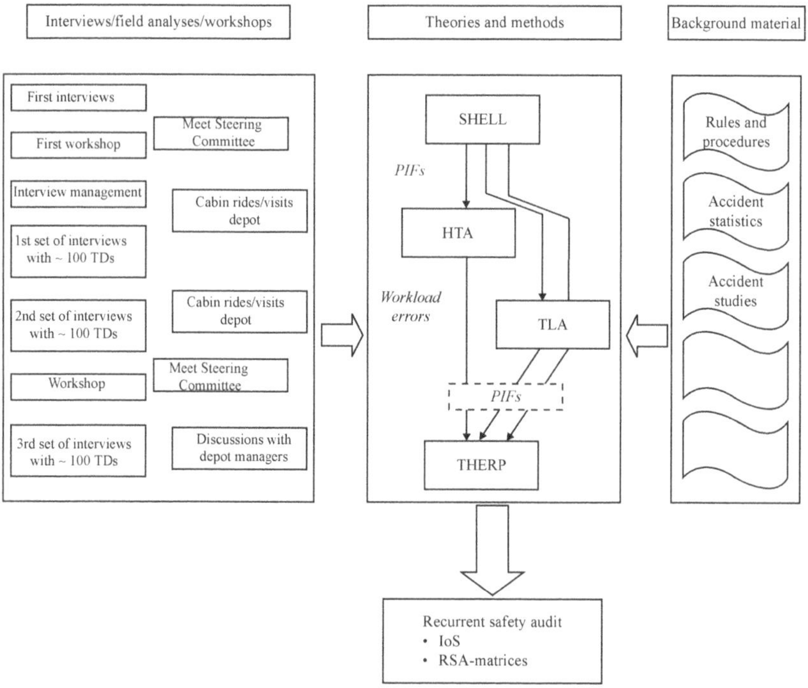

**Figure 32.4  Methods applied for the safety audit of a large railway company**

# References

Cacciabue P.C. (2003), *Guide to Applying Human Factors Methods*, London: Springer-Verlag.

Edwards, E. (1988), 'Introductory Overview', in E.L. Wiener, and D.C. Nagel (eds), *Human Factors in Aviation*, San Diego, CA: Academic Press: 3–25.

Kirwan, B. and Ainsworth, L.K. (1992), *Guide to Task Analysis*, London: Taylor and Francis.

Reason, J. (1997), *Managing the Risks of Organisational Accidents*, Aldershot: Ashgate.

Swain, A.D. and Guttmann, H.E. (1983), 'Handbook on Human Reliability Analysis with Emphasis on Nuclear Power Plant Application', NUREG/CR–1278.

Wilson, J.R., Cordiner, L., Nichols, S., Norton, L., Bristol, N., Clarke, T. and Roberts, S. (2001), 'On the Right Track: Systematic Implementation of Ergonomics in Railway Network Control', *International Journal of Cognition Technology and Work, IJ-CTW*, 3 (4)London: Springer-Verlag.

# Chapter 33

# Competence Management Systems for Rail Engineering Organisations

John Baker

## Introduction

The competence of individuals is an important component in effective health and safety management. Within the rail industry the requirements of the Railways (Safety Critical Work) Regulations are amplified by a Railway Group Standard requiring organisations to have formal management systems in place for assessing staff competence.

This chapter explores some of the issues that railway engineering organisations face when developing competence management systems, using models to illustrate the discussion. The limitations of current methodologies are explored and, from this, the factors that need to be considered if more effective systems are to be delivered are discussed.

The chapter builds upon recent research into current practice in rolling stock organisations, conducted by Risk Solutions for the Rail Safety and Standards Board. Findings of this research will be presented and their relevance to other fields within the industry.

Traditional approaches to safety have focused on engineering and process risks, and sought hardware solutions to them. However, studies show that 'human factors' contribute to up to 80 per cent of workplace accidents and with competence issues being a significant factor in many, if not all, cases. Examples in the transport sector include Clapham Junction, the Kegworth Air Crash, Ladbroke Grove and Chancery Lane.

## What is Competence?

'Competence' is a complex concept, consisting of personal and job-related factors. Personal factors include the thinking and practical skills of the individual as well as their experience, knowledge and attitude. Job-related factors that affect the competent performance of an individual include the provision of a suitable working environment (adequate lighting, heating) as well as the provision of the necessary tooling and equipment. The final job-related factor is the provision of sufficient additional resources (such as technical support and co-workers) to permit the maintenance tasks to be completed without the need to omit elements because of time constraints or for an incorrect diagnosis to be made due to the absence of technical manuals, drawings or expert technical advice.

These factors are all inter-related as depicted in Figure 33.1 – the absence of, or shortfalls in, any one of these elements will mean that competent performance is adversely affected.

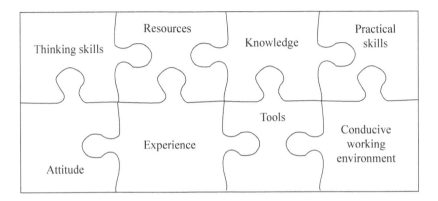

**Figure 33.1  Components of competence**

The significance of competence in maintenance error is now widely recognised and several regulatory standards and guidance documents have been produced to ensure that organisations formally monitor and assess staff competence. A formal competence management system means that there is an intrusive, independent means of checking a practitioner's activities which should result in problems being detected at an early stage and for appropriate corrective action to be taken (such as on-the-spot coaching or by removing the individual from the work activity until competence is developed). From this it can be seen that having a competence management system is about reducing risk; competent practitioners may still make errors but the frequency of occurrence should be significantly reduced and the severity of the consequences far less.

The UK rail industry (in common with practice in the UK generally) focuses on measuring a practitioner's performance against standards of occupational competence (standards defining the competences required to do the job). This approach is well established and is now the accepted way of doing things. However, there are some implicit assumptions with this approach which should be challenged as they do not ensure that practitioners really are competent.

This study explores some of the key findings of recent research funded by the Rail Safety and Standards Board as part of the Rail Safety Research Programme and carried out by Risk Solutions. This research explored how rolling stock maintenance organisations have implemented competence management systems to comply with the Railway Group Standard.

**Factors Affecting the Competence of an Individual**

The target of effort for competence management systems within rail industry organisations is the individual maintenance practitioner (fitter, electrician, technician – Figure 33.2).

The organisation

**Figure 33.2   Target of competence management system efforts**

However, if we consider the points made earlier about competence being made up of both personal and job-related factors it soon becomes apparent that the links to individual competence are more far-reaching, extending out within the employing organisation – and even beyond this.

For example, line managers have an important role in providing a suitable variety of work for practitioners to develop and maintain competence and management style can have a significant effect on individual practitioners' attitudes towards the tasks and the care that is taken in performing them. Engineering specialists within the organisation can affect the competence of individual practitioners through the provision of accurate and intelligible technical documentation, coaching and teaching staff about technical issues. The policies and systems in place in the employing organisation itself can also have a significant impact on individual competence. For example, the policy of only employing individuals that have completed a recognised engineering apprenticeship (or equivalent) should ensure that a base level of competence exists throughout the workforce. Likewise, a rewards policy that recognises and appropriately rewards personal competence may encourage individuals to actively seek opportunities to maintain and develop their competence. Effective communication, through briefing and supported by printed material, should ensure that individual practitioners are aware of changes and technical developments. The range of organisational systems that therefore influence competence of an individual is represented in Figure 33.3.

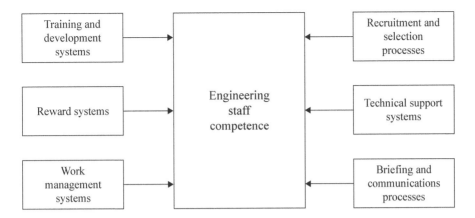

**Figure 33.3   Influences of organisational systems on competence of front line staff**

Looking outside the employing organisation it is apparent that there are factors that can influence the scope and operation of competence management systems, the standards of competence required by individual front line staff and, to a lesser degree, an effect on an individual's competence itself. Regulatory bodies (HSE Railway Division and the Strategic Rail Authority for example) can, through the creation of legislation or national frameworks, affect the scope of competence assessment systems and influence industry-wide initiatives such as national award programmes (NVQs). Engineering professional bodies (such as the Engineering Council and the various Institutions) also affect the competence requirements for practitioners through the setting of generic competence standards and the creation of national registration schemes.

The reliability of the components and the quality of supporting documentation can impact on the competence requirements for individual practitioners – if the quality of delivered spares is variable, then a higher level of competence may be required in the maintenance staff to ensure that defects are identified and rectified.

A further influence comes from the asset owners (rolling stock leasing companies, infrastructure controllers) who may specify maintenance requirements and take an active interest in the competence standards of the organisations carrying out maintenance on their assets.

This network of relationships affecting the competence requirements of individual practitioners is represented in Figure 33.4.

All this means that the individual is fairly passive in the development and maintenance of their own competence – there is little incentive (either 'carrot' or 'stick') for them to take a more active role. The result of this is that organisations are committing significant resources to managing the competence of individual practitioners.

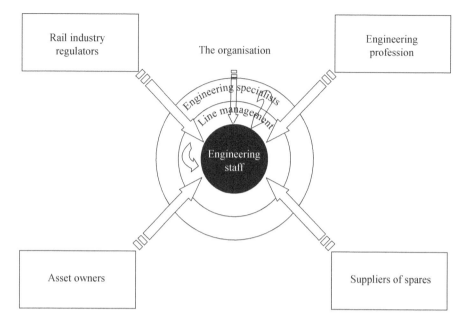

**Figure 33.4  Influences on competence of front line staff**

## Competence as a Variable

A further issue is the way that competence standards are used in the rail industry (as described before, this is typical of the UK in general). The industry's approach is to describe an individual as being either 'competent' or 'not yet ready'. Unfortunately this does not accurately reflect reality; competence is not a binary state, it varies over time as skills, knowledge and experience develop. In a maintenance context we can use the example of a first year apprentice – a 'novice' who is likely to rigidly follow taught rules as their perception of the situation they are operating in is very limited. As the individual's experience, knowledge and skills develop they will be able to see how their actions relate to longer-term goals and outcomes until they reach the point at which their line manager trusts them sufficiently to permit them to work alone – they have become 'competent'. However, the competence of the individual is likely to continue to develop; they will be able to identify deviations from normal patterns, decisions will be made intuitively. In time, the individual will be regarded as something of an 'expert', someone that others go to for advice; someone that is able to track down the most difficult of defects in an asset's systems. The range of this variation in competence is represented in Figure 33.5.

Recognising that there are progressive stages in an individual's competence provides a framework for incentivising individuals to develop and maintain their own competence. This would shift the balance of responsibility away from the employing organisation towards the individual practitioner.

**Figure 33.5   Competence as a variable**

## Competence Standards

There are a number of different competence standards (against which a practitioner's competence is assessed) in use in the industry. Some organisations have adopted the national occupational standards for NVQ Levels 2 and 3 developed by the Rail Industry Training Council (RITC), some organisations have developed their own standards and some standards have been developed by consultancies. Whilst use of the RITC NVQ standards should theoretically be very attractive (for example applicable to any type of traction, easily available), there appear to be a number of drawbacks that have caused them to be rejected by many organisations. One of these factors is NVQ standards use a vague form of language, requiring interpretation by assessors and which may then lead to inconsistent assessments. Another significant factor is that the units and elements of competence are generic in nature and not easily related to the specifics that an individual organisation may wish to focus on. In contrast, competence standards developed in-house can focus on specific asset systems and the context in which they operate. This should lead to a more effective assessment of an individual's competence as this approach should assess key risks applicable to the asset type being maintained and the way that it is used in service. The major difficulty with developing these specific, detailed standards is that they are costly (in terms of both money and time) for an individual organisation.

The outcome of some organisations developing standards independently and others using 'off-the-shelf' standards is that the industry now has many different 'standards' for similar or identical activities! The fact that different approaches have been applied also makes it difficult to compare standards of competent performance across organisations.

## Frequency of Reassessment

During our research we noted considerable variations in the reassessment frequencies across different organisations even where the *same rolling stock* (for example HSTs) were being operated in *similar operating environments*. In such cases the reassessment frequency might reasonably be expected to be the same unless there were other significant factors affecting the overall competence levels (such as high staff turnover rates). Examples of this include the safety-critical systems of AWS where reassessment frequencies varied from 1–3 years and brakes where the frequency varied from 1–5 years. No significant differences were identified and the variations in reassessment frequencies are therefore difficult to justify.

## Conclusion

This chapter has discussed how the competence of individual practitioners is dependent on many other groups of staff, extending beyond the boundaries of the employing organisation. We have also shown that the management of staff competence is interwoven with many organisational management systems. These two points together demand that the management of competence is treated holistically, not as a 'bolt-on' but as an integral part of the organisation's activities.

We have also discussed that there are variations in the rail industry's standards of competence and reassessment frequencies. Whilst this should be expected where there are differences in the types of asset, the operating environment and in the organisation itself, there should be some harmonisation where these factors are very similar or even identical.

## Acknowledgements

The author would like to acknowledge the support, contributions and comments from Lesley Hodsdon, Mick James and Cliff Cork of RSSB in the research project.

## References

Dreyfus, S.E. (1981), 'Four Models of Human Situational Understanding: Inherent Limitations on the Modelling of Business Expertise', USAF Office of Scientific Research, Ref. F49620–79–C–0063.

HSE (2002), 'Developing and Maintaining Staff Competence', Railway Safety Principles and Guidance, Part 3 Section A, London: HSE.

Rail Industry Training Council, National Vocational Qualification Standards (Levels 2 and 3) in Engineering Maintenance.

Risk Solution (2003), 'Competence Management – Rolling Stock Organisations', research report for RSSB, April.

RSSB (1998a), Competence Management for Safety Critical Work', Railway Safety Railway Group Standard GO/RT 3260, Issue 2, August.

RSSB (1998b), 'Code of Practice: Competence Assessment' Railway Safety Code of Practice GO/RC 3560, Issue 1, August.

Chapter 34

# The Analysis of Human Communication Errors During Track Maintenance

W.H. Gibson, E.D. Megaw, Mark Young and E. Lowe

## Introduction

This study investigated human communication errors related to track maintenance. The analysis was based on communication errors captured from the analysis of voice recordings between signallers and trackside personnel. The collected recordings were transcribed and then classified in relation to communication topic and error types. These data were collected with the aim of better understanding the communication process such that improvements can be made and also to provide human reliability data to be used in risk assessments. The study has collected novel human error probability data for use in risk assessments and will be added to the cross-industry CORE-DATA database of human error probabilities.

*Communication and Track Maintenance*

Safety systems are required on the railway when maintenance personnel work on or near the railway line. They are primarily used to reduce the risks from personnel coming into contact with moving trains and they also help to ensure that trains do not travel over railway infrastructure which is not safe due to ongoing maintenance. The systems, of which verbal communication is a prime component, are defined as 'protection of the line' in this chapter. The procedures underpinning these systems are primarily specified nationally in the railway industry Rule Book (Rail Safety and Standards Board, 2003). The scope of the analysis has covered the following types of protection:

- Rule Book Section T(ii), 'Protection of the line for engineering work';
- Rule Book Section T(iii), 'Protection of engineering work when the engineer takes possession of the line';
- local operating instructions related to protection of the line at stations, for personnel working on the line to carry out tasks such as train detachments.

Most of the time, a signaller has the key responsibility of controlling train movements and maintaining safety for an area of railway line. This control occurs remote from the line at a control centre called a signalbox. However, during maintenance, another person takes responsibility for the area of the line in which maintenance is being undertaken. This person is normally on the ground (that is, on or near the railway line) and they are normally termed the person in charge of possession (PICOP) or controller of site safety (COSS). They are referred to as the PICOP/COSS in this chapter.

The communications analysed have been those occurring between the PICOP/ COSS and signaller. The signaller and PICOP/COSS must communicate to ensure that the handovers of responsibility for the line occur correctly. For example, verbal communication is used to check or define the geographical limits in which it is safe for the maintenance to occur and at what time the work can commence. There are clear risks from failures in this communication as it could contribute to personnel being exposed to train movements or trains travelling over infrastructure that is not fit for purpose.

Communication is also recognised more generally as a safety-concern in the railway industry (Railway Safety, 2003). Safety-critical communication for protection of the line was therefore selected as a key area where it would be valuable to collect human error probability data.

*CORE-DATA*

CORE-DATA is a database of human error probability data and associated background information. CORE-DATA has two main objectives: firstly, to support and strengthen the consideration of human error in risk assessments by providing human error probability data relevant to safety-critical tasks; secondly, the process of collecting and normalising data on error in a specific industrial task develops an understanding of that task such that, where necessary, suggestions for practical improvements can be made.

The CORE-DATA database was initially developed at the University of Birmingham (Taylor-Adams and Kirwan, 1995) and computerised as a database with the support of the UK Health and Safety Executive (Gibson and Megaw, 1999). CORE-DATA remains at the University of Birmingham and further data collection to populate the database has recently been sponsored by a consortium of industry groups representing nuclear, air traffic control and railway industries. This chapter presents results from the railway component of that data collection process which has been sponsored by the body now called the Rail Safety and Standards Board.

*Transcript Analysis*

The transcription and classification of communication to identify errors has a long history dating back approximately one hundred years to studies undertaken by Meringer and Freud (Fromkin, 1980). Studies using transcripts of work-related communication have previously been undertaken by a range of researchers in the aviation context (for example Morrow *et al.*, 1993 and Corradini and Cacciari, 2001) and specifically related to human error probability data collection in the aviation context by Gibson (2002). These studies have highlighted that transcript analysis can capture some communication error forms. The data also have the potential to provide human error probabilities because the error frequency can be combined with opportunity for occurrence data such as the number of times that communication occurs or the number of times that a specific communication element occurs (for example communication of a numeric value).

## Method

Voice recordings were collected from audiotapes which are automatically stored when signallers use fixed signal box communication equipment. The recordings were scanned to identify those related to communication for protection of the line.

The sampling strategy for the collection of recordings aimed to minimise bias from specific local conditions. Data from six signalling centres in each of three zones were used. The data included communication related to different safety systems which are defined in the railway industry Rule Book sections T(ii) and T(iii) (Rail Safety and Standards Board, 2003) and also local safety systems defined in local instructions. The recordings were selected for a range of times of day and days of the week.

One hundred and eighty-eight transactions were analysed for the study. A transaction is defined as an exchange between a signaller and a PICOP/COSS. Each transaction can be made up of a number of turns for each participant.

Each transaction was transcribed into a Microsoft Excel spreadsheet and then the communication was classified in relation to:

- failures to implement general communication procedures. Communication procedures are required to be used for messages concerning safety. They are specified in Section A of the railway industry Rule Book (Rail Safety and Standards Board, 2003). Some key examples of the communication procedures are:
  - o the use of standard terms which have defined meanings (for example 'disregard' meaning 'consider that last message as not sent or not applicable');
  - o use of the phonetic alphabet (for example speaking the word 'alpha' to signify the letter 'A');
  - o speaking numbers singly (for example saying 'one four zero' when speaking the number 140, rather than 'one hundred and forty');
  - o identifying oneself consistently (who you are, your location, etc.);
  - o repeat back messages to ensure they have been properly understood (termed 'readback' in this chapter);
  - o the requirement that the signaller has lead responsibility in ensuring that communication procedures are used;
- deviations in information content. These are errors which may have direct safety consequences because incorrect information is being communicated or there is a misunderstanding between participants. For example, a readback may be inconsistent with the initial message presented (for example read back signal identifier 'sierra zulu one two five' as 'sierra zulu one five two'). If not recovered, misunderstanding a signal number could lead directly to personnel being exposed to train movements. Another example is where the 'station side' of a set of points is being discussed, but one participant misunderstands which station is being referred to. It should be noted that all deviations in information content cannot be identified from the communication transcripts. For example, someone may

produce a correct readback but still not have understood the information correctly – what they 'say' in the recordings is correct but what they have understood may be something else. It was not possible to capture these error types within the scope of this study. The error rates derived may therefore be seen as conservative. It should be noted that the data communicated were compared with the written data for the task where possible (primarily a document called the Weekly Operating Notice), in order to confirm the veracity of the original instructions;
• all the transcribed data were also classified by the topic being communicated (for example statement of name, readback of blocking points, query of telephone number, etc.).

The classifications were stored in the Microsoft Excel spreadsheet along with the transcripts and the data were analysed to identify the frequency of occurrence for the errors and their opportunity for occurrence where possible. Tests for inter-rater reliability were undertaken for applying the classification to the transcripts and acceptable levels of reliability were obtained. Eight hundred and thirty-nine classifications were made independently by two assessors, with a reliability likelihood of 0.91 being achieved. This is greater than the acceptability criteria set at 0.90.

### Results and Commentary

It should be noted that this chapter presents some of the key results contained in Gibson (2003).

*Failures to Implement General Communication Procedures – Results*

Failures to implement communication procedures from the Rule Book very frequently occurred across the whole range of procedures. Some key results are presented in Table 34.1. The data are expressed as a percentage of the opportunity for error derived from the 188 transactions analysed.

**Table 34.1 Failures to implement communication procedures**

| Error | Percentage incorrect |
|---|---|
| Omit statement of signaller's signalling centre in a transaction | 35% |
| Omission of job title of PICOP/COSS in a transaction | 79% |
| Omission of location of PICOP/COSS in a transaction | 65% |
| Numerics not stated in single digits | 54% |
| Omission or failures to use the phonetic alphabet | 78% |
| Do not respond to a statement with a readback | 83% |
| Do not use specific terms -'state your message', 'correction', 'readback', 'over', 'out', 'negative', 'disregard', 'not clear' | 100% |
| Do not use specific terms – 'say again' | 78% |

It should be noted that while these data present relatively high levels of failure to implement communication procedures, definition of failure was based on a strict interpretation of the Rule Book procedures. For example, the protection method T(ii) is universally communicated as a 'tee two' in the industry, but in this study this was deemed as incorrect, as the procedures require the use of the phonetic alphabet, and this should therefore be defined as a 'tango two'. Also, a strict interpretation of readback was applied, such that each safety-critical statement (for example signaller communication of their signal box) required a readback.

It should also be noted that these types of failure do not necessarily lead to critical misunderstandings as non-procedural methods can be used as an alternative. For example, the omission of the job title of a PICOP/COSS would not lead to a misunderstanding if the PICOP/COSS has communicated their name and the signaller has additional information (for example from a previous conversation or written documentation) that the person with that name is acting in the role of PICOP/COSS.

*Failures to Implement General Communication Procedures – Commentary*

This section aims to consider why relatively low levels of compliance have been achieved in relation to the communication procedures.

Firstly, the communication procedures have been implemented for a relatively short period of time in the railway industry. It is difficult to change habits, particularly in communication, which we use both in and outside work situations. The requirements for competency and development of a good communication culture to improve communication quality are beyond the scope of this chapter. It is noted that a number of initiatives are underway in this area, in particular through the Railway Industry's Communications Working Group and the specification of Railway Group Standards.

However, evidence can be derived from the collected data and available literature on the communication procedures, as to how they themselves create difficulties for their users to apply them uniformly. The key issue is that the communication rules are difficult to apply consistently or are not sufficient in all situations. Examples are:

1   how people identify themselves during communication. For example, a signaller is required to state 'signaller at ... signal box'. However, this is not always the form of address which will most reliably identify them to the other participant in certain situations. For example, for protection of the line, it may be necessary for the signaller to be identifying themselves in terms of the panel within the signal box they are working on. However, because the rules do not state how this should be achieved various methods are used (for example, the use of signaller name or panel number). This leads to the adoption of a range of possible methods through which the signaller may identify himself or herself. Similar problems occur for the PICOP or COSS (who may use their name, location or item number to define who they are). Personnel, often for reasonable operational objectives, have to deviate from the procedure prescribed in the Rule Book, but are not provided with support as to how best to do this. This brings the rules themselves into disrepute and is considered to increase the likelihood of failure to apply the rules;

2   the railway communication procedures are modelled on those applied in Air
    Traffic Control for communication between Air Traffic Controllers and pilots.
    In Air Traffic Control communication, numeric data are largely limited to one,
    two or three digits when presented singly (Civil Aviation Authority, 2001). The
    time is generally only presented in terms of minutes and therefore hours are
    not required; longer numerics are not required to have numbers spoken singly
    (for example 2,500 is communicated as 'two thousand five hundred'). Presenting
    numbers singly is required for all numbers based on the railway Rule Book,
    but this is difficult to achieve and potentially confusing for longer strings of
    numbers, and the time (for example midnight communicated as 'one two zero
    zero'). This again leads to situations where it is not sensible to apply the rules,
    and is also expected to contribute to failures to apply the rules;
3   standard terms are frequently not useful. For example the terms 'over' and 'out'
    are not applicable to telephone communications and the term 'state your message'
    is largely not required during telephone conversations, as natural conversational
    processes are used to signal a request for a message. Telephone conversations are
    the primary means of communication between signallers and PICOPs/COSSes.
    Again, difficulties with applying these rules are expected to contribute to the
    overall low levels of rule application;
4   communication for protection of the line has a significant problem solving
    component (data collected for this study have shown it accounts for 29 per cent
    of communication undertaken between signaller and PICOP/COSS). In contrast,
    Air Traffic Control communication contains very few problem solving activities.
    This is not because problems do not occur but because controllers and pilots
    have separate responsibilities for resolving problems which do not require
    communication. Problem solving requires flexibility and may not be suitable for
    the application of formal Air Traffic Control type communication procedures.
    Personnel are not given guidance on how to switch between problem solving
    and the more standardised forms of communication which again may lead to
    difficulties in the application of the rules.

In summary, the data collected highlight that communication for protection of the
line is a mixture of routine communications and problem solving. Communication
procedures are difficult to apply during problem solving and therefore may not be so
necessary in this context. For routine communications, this study suggests that the
current Rule Book does not provide complete support as to how these communications
should be carried out and provides rules which are difficult to apply.

*Deviations in Information Content – Results*

Deviations in information content are errors which may potentially have direct safety
consequences, because incorrect information is being communicated or there is a
misunderstanding between participants. The key human error probability data are
summarised in Table 34.2. It should be noted that error recovery can occur when
either the signaller or PICOP/COSS identifies and corrects a deviation in information
content during a transaction. The key critical errors are those which are unrecovered.
Also, the rates presented in Table 34.2 are a potential underestimate of human error

likelihood because, as noted in the method section, not all forms of deviation can be identified from reviewing transcripts.

**Table 34.2  Human error probability data for deviations in information content**

| Error | Error frequency | Opportunity for occurrence | Human error probability |
|---|---|---|---|
| All slip type errors per transaction | 30 | 188 | 0.16 |
| Unrecovered critical slip type errors per transaction | 4 | 188 | 0.02 |
| All slip type errors in numerics, per numeric communicated | 25 | 1058 | 0.02 |
| Unrecovered critical slip type errors in numerics, per numeric communicated | 4 | 1058 | 0.004 |

*Deviations in Information Content – Commentary*

The key discrepancies identified were primarily in the transmission of numeric data. These are predominantly errors which are a naturally occurring feature of human behaviour to which language is particularly susceptible. From the data collected for this study they are predominantly the exchange of one number for another (for example saying 'one five four' when 'one six four' was intended). The non-numeric data also tended to be confused within the same class of words (for example 'up' used instead of 'down' in relation to line descriptions). Such error forms are a function of basic cognitive processes. They occur particularly where units from the same class of word are fairly interchangeable. For example, for a signal identifier, the sequence of numbers is to some extent random and therefore we do not have strong rules as to the sequence required. Similarly, a line may almost equally be 'down' as much as 'up'. It is much less likely that errors of the type 'message state your' for 'state your message' will occur because the units have to conform to a constrained pattern. Based on the collected data, tentative conclusions have been made that such errors seem to be fairly well distributed between different numeric types. These findings suggest that the errors and associated error rates are a function of natural human variability in relation to the communication of numerics.

An interesting comparison can be made with Air Traffic Control where error rates have been collected in pilot readbacks. These errors are also likely to be dominated by slips which are predominantly related to numerics. The Air Traffic Control data provide unrecovered error rates of 0.002 (Cardosi, 1993; Cardosi, 1994; Burki-Cohen, 1995; Morrow *et al.*, 1993; Prinzo, 1996; and Cardosi *et al.*, 1996) compared with 0.004 for data collected in this study. In reliability terms these data from different industries are certainly in the same ballpark. That these rates are approximately equivalent, could also be used to support the view that errors of these type occur at a certain rate due to inherent human variability, particularly for numeric data, regardless of the individuals involved or operational context.

Increasing the use of readbacks may be one method for reducing the impact of these errors, and training personnel to be aware of such variability may provide some improvements. However, human variability is a function of human performance which to a certain extent cannot be changed. If humans are to be relied upon to transmit large amounts of numeric information, then such errors should be expected and probably at similar rates to those presented in the sample. This will probably occur regardless of the quality of communication procedures implemented. To improve these rates, this study suggests that consideration may be required for the use of alternative communication methods (for example, use of text-based communication using future technologies such as GSM-R) or redesign of the information communicated.

**Conclusion**

The study has successfully collected human error probability data in safety-critical communications for use in risk assessments and will be added to the cross-industry CORE-DATA database of human error probabilities. It should be noted that none of the data collected led to a reportable incident and therefore risk modelling and investigation will be required to fully relate the collected data with outcomes such as possession irregularities or trackside injuries/fatalities.

The following summarises some of the key safety performance implications identified from the data.

Failures to implement communication procedures from the Rule Book very frequently occurred across the whole range of procedures. Initiatives currently underway within the industry (particularly through the Communications Working Group) aim to reduce these errors through improved competencies and development of a better culture for compliance. However, the detailed analysis has highlighted that another problem, which has to be tackled, is the usability of the communication procedures within the Rule Book. They are not optimal for human performance, do not all provide proven safety benefits (Gibson, 2003) and are difficult to apply consistently in the real world.

The review of deviations in information content has highlighted the significant risks which arise from the requirement for the communication of large amounts of safety-critical data, particularly numerics such as signal numbers and times. Research has identified that this form of communication will not be reliable due to inherent features of human linguistic performance. These errors will therefore not be successfully removed through improving personnel competencies. There is therefore a requirement for a more fully defined and widely applied readback process to recover from these errors, and consideration of alternatives to verbal communication, such as text, for transmission of these data.

Despite the high frequency of these errors, the data sample did not lead to a reportable incident. This suggests that, while deviations in information content frequently occur, they may be recovered later in the process (for example the PICOP/ COSS identifies the error following the communication and implements any required remedial actions) or do not lead to safety consequences (for example personnel are not properly protected from train movements but there are no train movements in the locality and therefore an incident does not occur). Further investigation into how deviations in information content lead to safety events would be required before further comments could be made on this topic.

A final key finding is not related to error directly but to the significant amount of communication which is required to resolve problems (for example problem solving between COSS and signaller relating to protection limit requirements).

## Further work

The key recommendations for further work are:

1   redesign of communication procedures to improve their usability for railway personnel and increase their impact on improving communication reliability;
2   investigate problem solving in further detail and how the requirement for problem solving can be reduced at the time protection of the line is being undertaken;
3   investigate methods for preventing or mitigating deviations in information content, particularly for slips in the transmission of numeric data;
4   ensure that research into future communication technologies (for example Global System for Mobile Communications – Railways (GSM–R)) considers the potential of that technology to reduce communication error;
5   expand the current study to collect more data: on deviations in information content (these are relatively rare events for which only four critical unrecovered errors were collected for this study); on context factors (so that differences between zones, signal boxes, personnel with different experience, etc. can be explored); on other related error forms (for example form filling/log book entry errors); using signaller, PICOP and COSS debriefs to explore the underlying factors associated with their performance;
6   investigate the link between routine communication errors and safety-related events. It should be noted that this topic is being investigated as a component of the Rail Safety and Standards Board communication research theme.

## References

Burki-Cohen, J. (1995), 'An Analysis of Tower (Ground) Controller–Pilot Voice Communication', DOT/FAA/AR-96/41, Washington DC: US Department of Transportation.

Cardosi, K.M. (1993), 'An Analysis of En-route Controller–Pilot Voice Communications', DOT/FAA/RD-93/11, Washington DC: US Department of Transportation.

Cardosi, K.M. (1994), 'An Analysis of Tower (Local) Controller–Pilot Voice Communications, DOT/FAA/RD-94/11, Washington DC: US Department of Transportation.

Cardosi, K.M., Brett, B. and Han, S. (1996), 'An Analysis of TRACON Controller–Pilot Voice Communications, DOT/FAA/AR-96/66, Washington DC: US Department of Transportation.

Civil Aviation Authority (2001), *Radiotelephony Manual*, CAP413, London: Civil Aviation Authority.

Corradini, P. and Cacciari, C. (2001), 'Shiftwork and Workload: Effects on Air Traffic Controllers Communications', in *People in Control: An International Conference on Human Interfaces in Control Rooms, Cockpits and Command Centres*, London: Institution of Electrical Engineers: 129–34.

Fromkin, V.A. (1980), 'Introduction', in V.A. Fromkin (ed.), *Errors in Linguistic Performance: Slips of the Tongue, Ear, Pen and Hand*, New York: Academic Press: 1–12.

Gibson, W.H. (2002), 'The Analysis of Controller–Pilot Communication and the Implications for Future Systems', confidential PhD thesis.

Gibson, W.H. (2003), *Core-data Project: Analysis of Transcripts of Communication for Protection of the Line*, London: Rail Safety and Standards Board.

Gibson, W.H., and Megaw, E.D. (1999), 'he Implementation of CORE-DATA, a Computerised Human Error Probability Database', HSE Contract Research Report 245/199, Suffolk: Health and Safety Executive.

Morrow, D.G., Lee, A. and Rodvold, M. (1993), 'Analysis of Problems in Routine Controller–Pilot Communication', *International Journal of Aviation Psychology*, 3: 285–302.

Prinzo, O.V. (1996), 'An Analysis of Approach Controller/Pilot Voice Communication', DOT/FAA/AM-96/26, Washington DC: US Department of Transportation.

Rail Safety and Standards Board (2003), *Master Rule Book*, GO/RT3000, London: Rail Safety and Standards Board.

Railway Safety (2003), *Railway Group Safety Plan*, London: Rail Safety and Standards Board.

Taylor-Adams, S.E. and Kirwan, B. (1995), 'Human Reliability Data Requirements', *International Journal of Quality and Reliability Management*, 12 (1): 24–46.

Chapter 35

# Understanding Safety Culture and Strategies for Improvement in Railway Maintenance

Trudi Farrington-Darby, Laura Pickup, John. R. Wilson and Lucy Adams

## Background

This chapter discusses the approach and main findings of qualitative research carried out in collaboration with a UK rail infrastructure maintenance company (IMC). The research was based on interviews (group and individual) with staff working for that IMC. The semi-structured interviews aimed to address the question of 'What makes people unsafe when working out on the track?' A conceptual model derived from safety culture and behaviour literature was utilised for analysis of the interviews and resulted in 40 factors perceived to contribute to unsafe working behaviour. These 40 factors formed the basis of a strategy for the organisation to concentrate their efforts in reducing unsafe act precursors and move in the direction of a positive safety climate.

## Introduction

In common with many other (high risk) industries the rail business has great interest in the notions of safety culture and safety climate. The suggested link between safety culture and major accidents has encouraged high risk industries to reduce their reliance on accident and incident data and to direct health and safety systems towards investigating the culture and climate that may contribute to incidents (Flin *et al.*, 2000). Several of the rail industry partners in the UK have organised or been participants in some reviews of safety culture/climate in the past few years, motivated in part by official reports such as Cullen (2000). Many of the rail studies and analyses remain unpublished for obvious and less obvious reasons, although accounts are available from, for example, Clarke (1998), Hale *et al.* (2003), Lawton (1998).

A UK infrastructure maintenance company (IMC) was concerned about death and injury to staff on the track and wanted deeper information about safe and unsafe working following the results of a recently applied safety climate questionnaire survey. They approached the Centre for Rail Human Factors at the University of Nottingham to work with them. Whilst there is great value in the use of safety climate tools for problem detection and for measuring changes over time and after the implementation of change, there are also limitations to their use. Some of these include: a lack of explanation (the IMC required more in-depth information regarding the issues that the questionnaire had highlighted); a fear of reprisals when writing responses (of particular relevance in the UK rail industry at the time); overuse of questionnaires with little or no feedback; possible misinterpretation of what a respondent intended, and the use of written words for expression in an industry that is traditionally driven by verbal communication. Additionally, the questionnaire results had been sometimes contrary to what the organisation suspected really was happening in terms of working safely. Accordingly they employed an independent group to collect data directly from their staff. Their underpinning question was: 'Why do our experienced staff (sometimes) work unsafely?'

This chapter provides a summary of the work. It involves a qualitative investigation into safety culture, violations and the reasons for those amongst track workers, which led into a comprehensive company strategy to improve safe working on the track and the organisational systems that influence this. This chapter describes the study, with explanation of the qualitative approach, sources and analysis of the data, findings and how these were used to inform a programme of strategic change. The basic remit was to discover why track workers sometimes behave unsafely and what factors within the IMC could be contributing to a negative safety culture and unsafe behaviour. The approach taken, as much as the specific findings for this situation, are considered of possible value to organisations wishing to manage the issue of unsafe working behaviour in the context of safety culture.

## Safety Culture and Safety Climate

The term safety culture generally explains how safety is placed as a priority within an organisation. This is reflected in the decision and policies of that organisation and filters down through these into every aspect of operational performance. A single definition for the term 'safety culture' has yet to be agreed upon. This is in fact one of the stumbling blocks in working with this potentially very valuable, but also rather vague, concept. A lack of consensus on definition can result in an uncertain framework on which to base research or practice.

According to the Health and Safety Executive:

> The safety culture of an organisation is the product of individual and group values, attitudes, perceptions, competencies, and patterns of behaviour that determine the commitment to, and the style and proficiency of, an organisation's health and safety management. Organisations with a positive safety culture are characterised by communications founded on mutual trust, by shared perceptions of the importance of safety, and by confidence in the efficacy of preventative measures. (HSE, 1999: 45)

Safety climate is described as 'the workforce's perception of the organisational atmosphere' (Gonzalez-Roma *et al.*, 1999). Climate is seen as the 'surface features' of a culture (Schein, 1990) and if measured can provide an understanding of the workforce's attitudes and perceptions at the time of measurement; that is, it is a snapshot, generally obtained thorough surveys (Cox and Flin, 1998) such as the

Safety Climate Questionnaire (HSE, 1997a). At present one tool does not exist to measure all features relating to culture, and interviews and focus groups for example are suggested as the most effective ways of understanding culture and the underlying factors that influence it (Flin *et al.*, 2000; Cullen, 2001; Cox and Cheyne, 2000).

Several factors have been identified as supporting the development of a positive safety culture within various industries, including management (DePasquale and Geller, 1999), immediate supervisors (O'Dea and Flin, 2001), individual and behavioural (workforce) factors (Cooper, 2000), reporting systems (Brown *et al.*, 2000), rules and procedures (HSE, 1997b), communication (Brown *et al.*, 2000) and organisational subcultures and subcontractors. Research in rule violation also provides valuable information on the causes of unsafe acts and of negative safety culture. The UK Health and Safety Executive have classified violation types and some of the mechanisms contributing to their occurrence, based on work by Reason *et al* (1990) and HSE (1999). The different types of violations and the mechanisms that drive them provide important clues about the factors that affect unsafe behaviour (Lawton 1998). There is a considerable body of knowledge of work describing safety culture in terms of what makes a good or positive safety culture and what makes a bad or negative safety culture (Cheyne *et al.*, 1998). There are also a large number of studies where the approach to changing unsafe behaviour is through behaviour–based programmes (Lingard and Rowlinson, 1997), goal setting (Duff *et al.*, 1994), and incentives (Haines *et al.*, 2001).

Whilst the literature enables better understanding of safety culture and climate, and can help support systems of reliable measurement, an organisation who think they may have a safety culture problem will want to make sense of their own situation, to determine the best course of action to address their own negative safety culture. Reason (1998) describes managing safety culture through the management of five aspects of culture – learning, reporting, just, flexible and informed culture, which acknowledges the contributions to operator unsafe acts from more distal sources within an organisation. These five aspects of culture might be used as top-level goals, but how this is translated into tactical and strategic actions requires goals to be made tangible, with need to examine what influences the safety related attitudes, beliefs and behaviour of people in an organisation.

**Approach and Methodology**

The investigation process is summarised in Table 35.1. More detail can be found in Farrington-Darby *et al.* (2005). The issue of blame was a concern, so to encourage cooperation and open and honest responses, measures were put in place that included confidentiality agreements, working closely with the trade union representatives throughout, ensuring feedback and providing payment for time and travel. These measures turned out to be critical in several respects: to obtain enough study participants; to support contributions and interest in order to identify key factors and behaviour; to allow subsequent review and commentary by the stakeholder participants on the interpretations and findings of the researchers; and to ensure dissemination and buy-in to the organisational strategies put in place as a result.

**Table 35.1 Outline of investigation process**

| | |
|---|---|
| *Stage 1*: | Background research and building framework |
| *Stage 2*: | Interview preparation and confidence building |
| *Stage 3*: | Semi-structured interviews (6 groups totalling 34 track workers and 8 individual interviews) |
| *Stage 4*: | Data gathering and preparation |
| *Stage 5*: | Data reduction and analysis |
| *Stage 6*: | Data representation |
| *Stage 7*: | Dissemination |
| *Stage 8*: | Action plan and strategy |

Group and individual semi-structured interviews had a primary question 'What makes track workers unsafe?' Follow up questions covered: safety culture, contract staff, the use of the rulebook and procedures in the rail industry, suggestions for improvement of safety and safety culture; these questions were based on the researchers' experience of railway maintenance and rail human factors (for example Wilson *et al.*, 2001) and the relevant literature, and were particularly informed by the trade union safety representatives and IMC management. The questions evolved with each session, which formed a reflective approach to building on the information.

The methods for analysing rich contextual data in an efficient manner were pragmatic in light of the client's needs. The aims were to determine what track workers, their supervisors and managers thought caused unsafe behaviour and to present these factors in a meaningful way to allow development of appropriate strategies. The analysis process underwent several iterations and the main steps are shown in Table 35.2.

**Table 35.2 Process for data analysis**

| | |
|---|---|
| *Stage A*: | Direct transcription of tapes to database |
| *Stage B*: | Initial coding against conceptual model |
| *Stage C*: | Inter- and intra-coder reliability testing |
| *Stage D*: | Final framework for coding from reviewed model |
| *Stage E*: | Completed and updated coding of data |
| *Stage F*: | 100 factors identified through thematic coding using framework |
| *Stage G*: | Dependence classification to reduce categories |
| *Stage H*: | 40 factors identified that influenced unsafe behaviour |

The statements transcribed in stage A were coded against a conceptual model of the grouped themes of factors that were described in the literature; an initial 100 factors were identified from the rich data; stage G ensured that a workable number of factors for managers of the organisation were presented without losing the original themes. Sorting on the basis of dependence and those factors that would strongly influence others gave 40 main factors that influence track workers' safety behaviour. These were presented with quotes to illustrate and expand on them, suggestions for ways of dealing with the factors and whether the factors were considered to have a direct and immediate influence on track workers' behaviour (for example weather) or are more medium term (for example supervisors' style of management) or long term and distant (for example contradictory rules) in the proximity of their source to the unsafe event.

## Findings

The final influencing factors are summarised in Table 35.3 (not sorted into any classifications), and represent a range of contributions to safety behaviour. The inclusion of both expected and less expected factors and the detailed explanatory examples given in the detailed report reflect the commitment and insight shown by all participants. The 'blame', and 'us and them' cultures that were evident from what was said in the interviews (and reported by many other rail groups informally over the years) did not prevent track workers and managers throughout the organisation from openly recognising the variety of sources for unsafe behaviour and negative safety culture that existed (increasing the awareness of more distal causes of accidents as described by Reason, 1997).

There was a strong underlying sense that safety was not perceived as just the responsibility of those who perform the work, and that the planners and supervisors had a strong influence on safety and how this was to be achieved. Many staff admitted that they look for an easy and comfortable way of achieving a task goal even if that may involve risk, or do not consider some risks because they have become the norm (routine violation). Further from the sharp end, the senior management and policy makers within the organisation were seen as having considerable influence on how easy or difficult it was to be safe on the track. This was via the culture that was encouraged through examples set, their messages (direct and indirect through actions and statements), and their policies on everything from recruitment and training to dealing with the demands from their client, the railway infrastructure controller.

Subsequently, factors were separated into three categories denoting temporal and distance proximity to unsafe acts at trackside. This was not a mutually exclusive classification, but it did enable the organisation to see how unsafe acts and culture permeate an organisation in its many functions and roles.

Factors thought to affect behaviour at the time of performing a job at the track side include: excessive and poor communication; poor behaviour on the part of safety role models such as controllers of site safety (COSS); inconsistent team members due to poor planning, staff shortages or the heavy reliance upon contract staff. Track workers also reported a lack of systematically performed and regularly performed risk assessments. Various reasons could be found to explain this, that staff become familiar with their jobs and the risks associated with them, that they do not have the skills with which to perform risk assessments or they do not have the resources to perform risk assessments in a dynamic work environment. Other factors influencing unsafe behaviour at the time of a job were reported as the perception of safety and risk; an operator's knowledge; the equipment available, the condition it was in and the appropriateness of the equipment for that job; operator fatigue and loss of concentration; individual competence to perform a task safely; certification of

**Table 35.3  40 factors influencing safe and unsafe behaviour on the track**

| | | | |
|---|---|---|---|
| Communication on the job (excessive and poor quality) | Poor and underused real time risk assessment skills | Individual perception of what safe is | Track workers knowledge and understanding |
| Inconsistent teams/ subcontractors | Safety role model behaviour | Social pressure of home life | Setting up site safety on the day |
| Rule dissemination | Physical conditions | Peer pressure | Feedback cycle |
| Competence capability and certification | Working hours: different behaviours out of normal hours | Manager's communication methods | Information/communication route clarity |
| Pre-job information dissemination | Planners knowledge for job resourcing | Job feedback to planners | Volume of paperwork |
| Feedback messages from managers | Manager's railway knowledge | Manager's visibility and accessibility | Perceived purpose of paperwork |
| Supervisor's style visibility, communication, representation of staff | Supervisors: (technical competencies and assessment of them) | Supervisor's presence (visibility, leading by , example opportunity for verbal communication) | Fatigue, concentration, ability to function (alcohol) |
| Equipment (condition, appropriateness and availability) | Practical alternatives to rules | Perceived purpose of the rule book | Rule book usability and availability |
| Information pathway flow | Planners competency to plan | Methods for reporting | Information systems use |
| Contradictory rules | Recruitment methods | Training needs analysis | Training methods |

competency and a sense that things are different and the same rules do not apply when you are working out of the normal hours of work.

More medium-term factors were related primarily to the immediate supervisors (line managers) and work planners, who provide a link between the track workers and the organisation. Supervisors are seen to be the first line representation of management, organisational beliefs and policies, regulation and discipline, support and communication. Planners provide a link between operational staff and the programme of work, which forms the organisation's main business. They indirectly influence safety trackside through their competence to plan and their operational knowledge. The planners and the supervisors form part of a communication feedback loop and are particularly relevant to the capacity of the organisation to learn and evidence of operational staff that they were being listened to. Some of the medium-to long-term factors relate to managers more distant from the track workers than their direct supervisors. In particular factors such as management visibility, accessibility, communication methods, feedback and messages provided implicitly through decisions and actions were suggested to affect behaviour at trackside and culture in general.

Some of the factors identified as long term were inherent in the whole organisation and business, for instance paperwork volume, relevance and purpose, rules that are contradictory to each other, what is practical and what is expected, poor rules dissemination, the usability of the rule book (on several levels including format, readability, availability and weight) and the perceived purpose of the rules. There is sceptism regarding the motives behind some of the rules, procedures and paperwork in the rail industry. Rejection on that level may lead to negative culture through a lack of trust. Safety systems that affect the organisation's functioning throughout, such as communication pathways, feedback cycles, and reporting methods, were reported as indirect factors influencing the safety behaviour of staff on the track. Another area of concern highlighted through the study was that of training and recruitment. Staff and skill shortages were seen as directly affecting ability to perform work safely on the day but had their source in long-term organisational policy.

## Strategies for Change

Throughout this programme of research the emphasis had been on collaboration and participation. The whole piece of work was an active collaboration between the outside body University of Nottingham who could show independence and guarantee individual anonymity, and the organisation itself who were actively interested in putting change strategies in place. Needs for quality data collection as well as good practice had meant that the work was participatory, involving the trade union and safety representatives. Therefore it was agreed that the generation, development and implementation of any change strategies as a result would also have the emphasis on a participatory approach. Great care was taken in identification, discussion and development of appropriate strategies. For example, if operating procedures relating to safety are brought in reactively in a panic, this may not actually improve safety at all, and in fact may make the job harder to do, impairing potential effectiveness and performance which then may actually promote a culture of violations.

Dissemination began shortly after the completion of the interviews and during production of the various reports. Interactive briefing sessions were held where real world problems were worked through in groups, and the causal factors identified, with emphasis on the framework from this study. More distal factors have become even more apparent through this mechanism. High quality posters, leaflets and pamphlets were produced to present the findings and gain feedback. These were distributed widely within and outside the company, turning up in unexpected places (for instance on the wall at a Personal Track Safety course attended by one of the authors!) Key individuals were identified by the company, made familiar with the research and the case studies, and provided with training on facilitation techniques, to carry out the dissemination programme to a wider audience.

Risk assessment training was planned to address how some of the study findings could be included as factors in risk assessments at all levels, formally and informally. This was carried out through the company's in-house training school. At the same time the influence of training and training staff on operational staff perceptions and the reinforcement of culture have been highlighted and particular attention was given to their crucial role. They were seen as a key group for moving organisational messages that promote safety. In another training initiative, policies and procedures for COSS training and recruitment were reviewed and revised. Emphasis was to go beyond purely technical skills and onto the provision of those social, decision-making and communication skills that are required to perform the important safety critical role of COSS.

A working party was already reviewing the management of subcontractors and they were informed of the relevant issues from this study and how they relate to safe working practices and safety culture in general. One particular focus was the training of site agents (staff from the IMC who dealt directly with subcontractor staff on site).

Efforts to reduce unnecessary telephone calls to track staff, and appropriate diversion of these communications to supervisors and managers were begun. This interestingly has echoes in current work by the authors for Network Rail looking at the roles of engineering supervisors, and the disturbances to their work from very frequent and often unnecessary communications.

Finally, accident and incident reporting systems were reviewed as part of a second phase of work with the IMC; a reporting culture was seen as essential for the development of informed and learning aspects of safety culture within the IMC.

An overall company strategy has been developed that includes all the above strategic approaches and also review of higher level safety related systems. Safety was considered to be endemic in all systems and the organisation was keen to break down any apparent or real demarcation between safety and performance.

## Conclusion

For many years much of the attention in rail human factors was upon driving and more lately signalling, whether this was to do with effective performance or safety of running the trains. Maintenance has been under the spotlight only much more recently. The behaviours we are studying for track workers can impact on the safety of the network as a whole (for passengers, drivers, the public etc.) as well as on the safety of themselves and their colleagues.

Creating a positive safety culture is a challenge for most organisations. Although the individual does have a part to play, the environment created by the organization largely determines the relevant attitudes, beliefs and perceptions of safety. Behaviour, safe or unsafe (including violations) is determined partly by the attitudes, beliefs and perceptions of individuals, and so the organisational safety culture can be considered to be what ultimately shapes behaviour. The research reported in this chapter concentrated on a specific rail maintenance organisation. Through careful and traceable identification of perceptions of staff, from track workers to senior management, we have determined 40 primary factors that are considered to influence track workers' safety behaviour and the organisation's safety culture. These factors span immediate 'at the time of the job' factors, such as the behaviour of someone in a safety critical role, medium-term factors such as supervisors' visibility and longer term strategic factors such as the quality of organisational accident and incident reporting systems.

Unlike many research projects carried out by a university or consultancy for an industrial organisation, the researchers have stayed in touch with the company, and in fact were involved in giving substance to many of the strategy choices. Many of the strategies and actions implemented as a result of the work appeared to be well received. A formal process of evaluation over time of changes in attitudes, incident reports and behaviours was to be undertaken. However, in common with the whole of the UK network, the IMC has since ceased to exist in the form that it was when the work was performed. Over 2004, maintenance functions were being taken back in house by the infrastructure owner, Network Rail. This will result in an amalgamation of people from different IMCs, under new management.

### Acknowledgements

The authors acknowledge the time and effort of all the IMC staff that participated so willingly in this study, including the safety representatives, trade union representatives and the managers. Their commitment, energy and insight was key to the success of the project and its outcomes.

### References

Brown, K.A., Willis. G.P. and Prussia, G.E. (2000), 'Predicting Safe Employee Behaviour in the Steel Industry: Development and Test of a Sociotechnical Model', *Journal of Operations Management*, 18: 445–65.

Cheyne, A., Cox, S., Oliver, A. and Thomas, J. (1998), 'Modelling Safety Climate in the Prediction of Safety Activity', *Work and Stress*, 12 (3): 255–71.

Clarke, S. (1988), 'Safety Culture on the UK Railway Network', *Work and Stress*, 12 (3), 285–95.

Cooper, M.D. (2000), 'Towards a Model of Safety Culture', *Safety Science*, 36: 111–36.

Cox, S.J. and Cheyne, A.J.T. (2000), 'Assessing Safety Culture in Offshore Environments', *Safety Science*, 34: 111–29.

Cox, S. and Flin, R. (1998), 'Safety Culture: Philosophers' Stone or Man of Straw?', *Work and Stress*, 12 (3): 189–201.

DePasquale, J.P. and Geller, E.S. (1999), 'Critical Success Factors for Behaviour-based Safety: A Study of Twenty Industry-wide Applications', *Journal of Safety Research*, 30 (4): 237–49.

Duff, A., Robertson, R., Phillips, R. and Cooper, M. (1994), 'Improving Safety by the Modification of Behaviour', *Construction Management and Economics*, 12: 7–78.

Farrington-Darby, T., Pickup, L. and Wilson, J.R. (2005), 'Safety Culture in Railway Maintenance', *Safety Science*, 43: 39–60.

Flin, R., Mearns, K., O'Connor, P. and Bryden, R. (2000), 'Measuring Safety Climate: Identifying the Common Features', *Safety Science*, 34: 177–92.

Gonzalez-Roma, V., Peiro, J., Lloret, S. and Zornoza, A. (1999), 'The Validity of Collective Climates', *Journal of Occupational and Organizational Psychology*, 72: 25–40.

Haines III, V., Merrheim, G. and Roy, M. (2001), 'Understanding Reactions to Safety Incentives', *Journal of Safety Research*, 32: 17-30.

Hale, A., Heijer, T. and Koornneeff, F. (2003), 'Management of Safety Rules: The Case of Railways', *Safety Science Monitor*, 7 (1).

Health and Safety Executive (HSE) (1997a), *Health and Safety Climate Survey Culture*, Sudbury: Health and Safety Commission.

Health and Safety Executive (HSE) (1997b), *Successful Health and Safety Management*, Sudbury: Health and Safety Commission.

Health and Safety Executive (HSE) (1999), *Reducing Error and Influencing Behaviour HSG48*, Sudbury: Health and Safety Commission.

Health and Safety Executive (HSE) (2001), *The Ladbroke Grove Inquiry: Part 1* ('The Cullen Report'), Sudbury: HSE Books.

Lawton, R. (1998), 'Not Working to Rule: Understanding Procedural Violations at Work', Safety Science, 28 (2): 77–95.

Lingard, H. and Rowlinson, S. (1997), 'Behaviour-based Safety Management in Hong Kong's Construction Industry', *Journal of Safety Research*, 28 (4): 243–56.

O'Dea, A. and Flin, R., (2001), 'Site Managers and Safety Leadership in the Offshore Oil and Gas Industry', *Safety Science*, 37: 39–57.

Reason, J. (1997), *Managing the Risks of Organizational Accidents*, Aldershot: Ashgate.

Reason, J. (1998), 'Achieving a Safe Culture: Theory and Practice', *Work and Stress*, 12 (3): 293-306.

Reason, J., Manstead, A., Stradling, S., Baxter, J. and Campbell, K. (1990), 'Errors and Violations on the Roads: A Real Distinction?', *Ergonomics*, 33: 1315–32.

Schein , E. (1990), 'Organisational Culture', *American Psychologist*, 45: 109–19.

Wilson, J.R., Cordiner, L.A., Nichols, S.C., Norton, L., Bristol, N., Clarke, T. and Roberts, S. (2001), 'On the Right Track: Systematic Implementation of Ergonomics in Railway Network Control', *Cognition, Technology and Work*, 3: 238–52.

# Chapter 36

# Cross-border Railway Operations: Building Safety at Cultural Interfaces

S.O. Johnsen, I.A. Herrera, J. Vatn and R. Rosness

## Introduction

*Background*

As legislation is introduced to ensure the interoperability of railway systems across Europe [EU–96], the issue of safety at cultural interfaces has become a subject of considerable interest to the rail industry. It is recognised that different cultures exist in organisations that will be increasingly required to interface with each other. Cultural interfaces represent a potential source of safety problems, but also a potential for learning from other cultures. To approach this challenge proactively, UIC (the international organisation for railways see http://www.uic.asso.fr) has initiated a study to develop a methodology to identify and improve safety problems that arise at cultural interfaces. The project, SCAI (Safety Culture at Interfaces), covers the development of a method for managing cultural interfaces and piloting of the method in three railway undertakings.

The UIC study will take operational problems as the starting point for a methodology. The method will be based on a scenario approach where different cross-border scenarios are identified. Such a scenario could be that track work is carried out (near a border crossing), and then trains from different countries are approaching.

The scenarios are analysed by a group of people representing the various railway undertakings, infrastructure managers, and the traffic control. The analysis is based on an event diagram, and an evaluation of differences at interfaces. The method should assist the various railway undertakings, infrastructure managers etc. to implement counteractive measures in order to control the risk at interfaces. Piloting the method will improve the quality and demonstrate whether cultural interfaces can be exploited as a source of impulses to share best practices and thus improve operational safety. Basically the method will act on a bottom-up level where specific safety culture issues are resolved by the involved parties, but the method is also expected to identify more general issues that have to be resolved at for example a European level (harmonising of rules, reporting systems etc.).

This chapter will present the first draft of the method (developed by SINTEF) to identify and manage risk at interfaces.

**Safety Culture at Interfaces**

*Problem Definition*

The following questions need to be explored in order to improve safety culture at interfaces:

- What is safety culture at interfaces, that is, which definition is useful?
- How do we understand safety culture at interfaces?
- How could we improve safety culture at interfaces?

*What is safety culture at interfaces, that is, which definition is useful?*

To set the scene for what is meant by safety culture we adopt the definition:

> The safety culture of an organisation is the product of individual and group values, attitudes, perceptions, competencies and patterns of behaviour that determine commitment to, and the style and proficiency of, an organisation's health and safety management. (HSE, 1993)

This definition is very broad, and in order to narrow the scope of the study, we have chosen to focus on: 'Characteristic interaction patterns when organisations interface each other, i.e., how people collaborate and communicate at interfaces'.

At the same time it is important to specify which of the interfaces between the different stakeholders are most important. We have tried to identify the most important stakeholders and interfaces based on interviews and discussion with the industry, see Figure 36.1.

Based on the deregulation in the EU and the increased competition among railway companies, the primary interfaces to be analysed are between railway undertakings crossing borders within the EU. Also interfaces between infrastructure operators and maintenance operators supporting the railway undertakings should be investigated. The most important interfaces regarding cross-border traffic are seen as the interface between the 'driver' and the 'traffic control' and between 'regulatory authorities' and the 'industry'.

*How do we understand safety culture at interfaces?*

In the present study we are considering safety culture closely related to an explicit qualitative risk model. This risk model is spanned by a set of relevant *scenarios*, and each scenario is influenced by a set of *safety critical functions*.

The risk model, and how it connects to the safety culture is further elaborated in the 'Proposed Method' section. At this point we would, however, emphasise that safety culture is seen as 'something' that affects the performance of the safety critical functions, and in this context a correlation between safety culture (attitudes to job, organisation and management) and accident/incident rate seems reasonable. Such a correlation has been documented by among others Itoh, Andersen and Seiki (2004).

When analysing safety culture we will need to be rather explicit about what safety culture is and we differentiate between two situations:

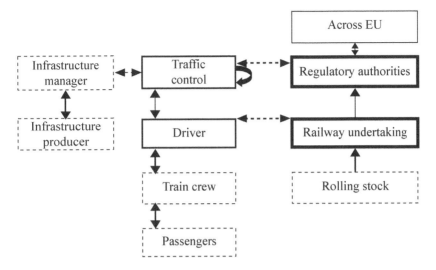

**Figure 36.1   Important interfaces and stakeholders**

- safety culture as a set of properties of an organisation that are unconditionally positive with respect to the safety level (an example is a *reporting culture*);
- safety culture as a pattern of behaviour and commitment to reach an agreed standard, but where 'a best practice' to reach the agreed safety standard does not exist (an example is *problem solving*).

When treating safety culture as a set of properties of an organisation, these properties could be seen as key elements of safety culture. By a literature survey we have identified such common key elements of safety culture and listed them in Table 36.1. We have tried to adjust and elaborate the key elements regarding safety culture at interfaces. This is a preliminary list of key elements and this list will be updated through further study and through our workshops and pilot trials. These elements are building blocks for our proposed methodology.

Regarding problem solving, it could be difficult to treat safety culture as a property, for example there is 'no best way' of solving problems. A study of cultural differences across nations looks upon culture as a collective phenomenon which could influence decision making and (in our case) safety (Hofstede, 1991). For example the management style and the mechanisms to resolve problems have been shown to differ between different national cultures, the 'right' and natural solution in one culture was not seen as the preferred solution in another culture.

As an example Hofstede refers to a conflict between two department heads within a company. This case was presented to students from France, England and Germany to recommend a preferred ('correct') solution to the conflict. The suggested 'correct' solution differed between each country as described:

- the French solution was for the opponents to take the issue to their common boss, who would issue orders for settling such dilemmas in the future;

**Table 36.1 Key elements of safety culture at interfaces**

| Safety culture | Safety culture at interfaces |
|---|---|
| **Management involvement and commitment** | Management involvement and commitment at interfaces regarding safety: Identifying who is involved, establishing clear responsibility, common communication and common understanding across interfaces. |
| **Shared commitment** and level of care for hazards | Shared commitment across interfaces. Commitment may reach outside the border of the company, to other companies or suppliers, when safety is at stake. |
| **Flexible rules** and regulations to reduce risk | Systematic evaluation and adjustment of rules and regulations to reduce risk at interfaces. |
| **Focus on organisational learning** | Focus on learning from participants across border: reporting of relevant incidents, open discussion with participants across border, good co-opting processes. (Meaning cooperation between management and workforce across the different organisations into a meeting arena where ideas and experience can be exchanged and actions can be agreed upon.) |
| **Reporting culture** | A reporting culture, also considering incidents across border or at interfaces. Reporting of specific operational safety problems that may occur at cultural interfaces. |
| **A just culture** | A just culture between interfaces. Suppliers doing out-sourced work not punished when incidents are reported or error committed. Competition across interfaces is not unjustly blamed. |
| **Industry-wide cooperation** and information sharing | Focus on industry wide learning including participants across border and new entrants to the industry, regarding safety. Establishing common competencies across interfaces. |
| **Legislative cooperation** | Cooperation and information sharing between legislative authorities across border. Focus on legislative-wide learning from participants across border. |

- the German solution was to establish specific procedures or routines to be used;
- the British solution was to recommend a management course to the opponents to improve their interpersonal skills.

In this situation it is apparent that the different cultures could not be classified as 'good' or 'bad' with respect to safety; they have different ways of doing things. This fact is of vital importance when aiming at improving safety culture and resolving conflicts related to different cultures. This will be discussed next.

*How Could We Improve Safety Culture at Interfaces?*

The perspective of safety culture in this study is to be able to grasp important elements of safety culture at interfaces and to improve it.

In the literature different concepts for organisational learning and development are found, and in this study we have based our approach on the framework developed by Argyris and Schön (1978), to be able to improve safety culture at interfaces.

Organisational learning could work at many levels; at the operational level we have the specific railway undertakings, and their subcultures, and how they interfere with other railway undertakings and actors in railway operation. On a regulatory level we have the European Commission responsible for the Safety Directive on railway operation and the UIC. Improvement and learning should take place at all levels where organisations interface each other. Therefore our approach has been designed to get involvement and commitment from management and the workforce, to ensure the possibility of organisational improvement and change.

We have chosen a 'scenario-based' approach to ensure good communication, involvement and commitment from the important stakeholders at interfaces.

The scenarios should be analysed by an experienced team of management and operating personnel. Suggested participation is from traffic control, drivers and railway management across interfaces– to ensure that the participating organisations as a whole can learn and develop.

To ensure an approach focusing on relevant scenarios, we are using the experience of the participants based on incidents/near misses and a generic list of scenarios. Experience show that participants feel more comfortable analysing a scenario where they are knowledgeable (Kjellén, 2000).

The involvement of the different stakeholders as drivers, train dispatchers and management is important if change can take place. The stakeholders must be able to change routines, competencies, management directives or even influence laws or regulations.

In our concept *the analysis of scenarios and safety culture* could not be treated independently from the arena of *organisational learning* and vice versa. Therefore these elements will be integrated in 'one method' and will be outlined explicitly next.

**Proposed Method – Called 'the Track to Safety'**

A useful methodology to be used to analyse railway operation must consist of the following parts:

1    establish a *framework to succeed* across different railway undertakings in
     Europe. This must encompass clear and *common goals* regarding safety culture
     at interfaces that could be accepted by the rail community and improved as
     experience are gathered;
2    establish a *method to foster organisational learning* based on management
     commitment and work force participation. This process must be robust
     enough to be used in different national cultures. The method should benefit
     from organisational learning, and take place in an arena where the important
     stakeholders can participate. The method is based on the framework developed
     by Agyris and Schön (1978), and consists of the steps:

     a)   scenario description based on safety critical functions;
     b)   root cause analysis related to deficiencies in safety critical functions and
          influenced by safety culture;
     c)   adjustments of routines, directives and competencies.

*An Approach to Succeed Cross-border*

To succeed cross-border, the railway industry must prioritise work related to safety
culture at interfaces. Each organisation must be involved in identifying risks and
prioritisation of actions.

   An important element is the cooperation between management and workforce
across different organisations where ideas and experience in a meeting arena can
be exchanged. This arena could be the source to share best practice and improve
operational safety and could aid in establishing a continuous learning and improvement
process.

   The participants should consist of stakeholders, as described in Figure 36.1, for
instance:

•    traffic controller from each interface (from each country);
•    driver from each interface (from each country);
•    management from railway undertaking;
•    process leader and scriber/secretary.

   The suggested approach includes the following steps (as described in Figure
36.2).

1    *Develop the framework.* Establish common goals, management commitment, and
     Learning arena. Involve workforce, management and regulation authorities (if
     possible). Collect new scenarios of interest and work to sustain the commitment
     to the process itself.
2    *Describe relevant scenarios* based on safety critical functions.
3    *Scenario analysis* – identify differences and safety challenges at interfaces.
4    *Identify actions* and adjust based on good co-opting processes.

   The scenario approach should foster organisational learning. The method should be
used both proactively and reactively. For example to aid in:

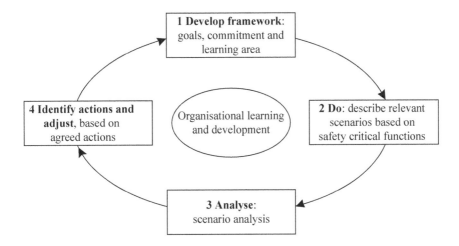

**Figure 36.2   Suggested scenario approach to foster organisational learning**

- establishing good routines when a railway undertaking is going to establish a new international connection (have all risks been addressed?);
- establishing safety culture across organisational interfaces after organisational changes, outsourcing activity;
- identifying root causes of incidents and accidents.

Steps 1–4 are further described next.

*Develop Framework (Step 1)*

As the method is being used, the framework of the method must be continuously developed based on experience from the users and discussions within cross-border authorities such as UIC and EU. A repository of incidents in cross-border traffic should be available as a basis to develop relevant scenarios.

The use of the methodology should not be too resource demanding. The anticipated level of effort is two workdays from the participants.

*Describe Relevant Scenarios Based on Safety Critical Function (Step 2)*

*Description of relevant scenarios*

The scenarios should represent significant areas of concern for the stakeholders. Near misses, a generic list of scenarios (to be developed by the project in cooperation with the industry) or a brain storming process at the start of the meeting could be sources for representative scenarios. Experience from a Norwegian study (Tinmannsvik and Rosness, 2002) indicates that scenarios derived from near misses can give a good generic coverage. It is, however, important to update the scenarios to cover new technology, changing regulations and new operational experience.

The scenarios could be illustrated by a STEP diagram (Hendrick and Benner, 1987): see Figure 36.4 for an example.

*Identification of safety critical functions*

To analyse a scenario we have introduced the concept of safety critical functions (SCF), defined as 'functions of a system for which a malfunction would immediately increase the risk of injury, or damage to health'. The SCFs could be viewed as 'basic events' in a fault tree analysis (FTA), or 'barriers' in an event tree analysis (ETA). However, the scenarios are not described to the level of detail and formalism as is usually done in FTA and ETA.

Combining SCF analysis with STEP analysis has proved fruitful both with respect to getting a good understanding of the scenario being analysed, and to ensuring user commitment. Figure 36.4 gives an example in relation to the safety critical function 1.1: Ensuring that a train does not enter a section that is occupied by another train.

A complete set of safety critical functions would be of value when conducting a scenario analysis. So far we have categorised the safety critical functions into seven areas:

SCF–1   SCFs related to normal operation;
SCF–2   SCFs related to ordinary traffic disturbances;
SCF–3   SCFs related to technical failures in signalling system/central train control (CTC) system;
SCF–4   SCFs related to degenerated infrastructure;
SCF–5   SCFs related to work on the track;
SCF–6   SCFs related to deficiency on rolling stock;
SCF–7   SCFs related to cross-border activity.

Each area is divided into several primary safety critical functions, and these are listed in the appendix. As an example, SCF–1, SCFs related to normal operation contains 1.1: Ensuring that a train does not enter a section that is occupied by another train.

An example will illustrate our method. The example is based on an accident as described in Figure 36.3. Two trains are colliding because of misunderstandings related to where the trains are crossing.

Maintenance is carried out on a track 1 near a border crossing. Train B is instructed by a rail traffic controller in country B to cross to track 2 from station 2 towards station 1. The rail traffic controller in country B informs the rail traffic controller in country A correctly about the crossing.

However, the rail traffic controller in country A understands that train B is going to cross to track 2 from station 3 towards station 2. The rail traffic control in country A allows train A to continue on track 2, from station 1 to station 2. This leads to an incident where train A collides with train B on the track between station 1 and 2.

This accident is illustrated in a STEP diagram, as shown in Figure 36.4. The timeline is along the x-axis and the different actors or stakeholders are listed along the y-axis Each box illustrates an action by one of the actors. The critical action is illustrated by a safety critical function, in this example the SCF is 1.1: Ensuring that a train does not enter a section that is occupied by another train.

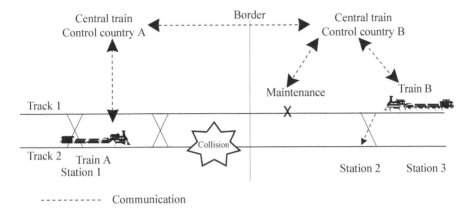

**Figure 36.3   Interactions in cross-border rail traffic (simplified example)**

*Scenario Analysis (Step 3)*

To identify the major safety challenges at interfaces we have to analyse the safety critical functions that have been identified in the preceding scenario analysis.

The analysis is based on identifying the differences across interfaces. Known differences have been named direct influencing factors. Examples of these influencing factors are environmental factors such as language, climate and nature. For each of these differences we must analyse the safety challenges related to collaboration and communication at interfaces. We have explored the necessary steps further. Based on a literature review and our own research, the influencing factors have been identified and structured. We have divided the influencing factors into indirect and direct (tangible) influencing factors. The indirect influencing factors are national culture, organisational culture and professional culture that work through the direct influencing factors:

By analysing the safety critical function, for example 1.1: Ensuring that a train does not enter a section that is occupied by another train, we could identify major differences and safety challenges related to collaboration and communication at interfaces as suggested in Table 36.3.

This must be analysed for each influencing factor. To ensure user involvement it could be important to discuss the challenges based on a visual diagram such as the STEP diagram, as illustrated in Figure 36.4. A graphical illustration of the steps is illustrated in Figure 36.5.

*Identify Actions and Adjust Based on Good Co-opting Processes (Step 4)*

The adjustment of the governing variables must be done in a way that ensures that safety is improved at interfaces. Adjustments and changes must be done in cooperation across interfaces and both management and the workforce must support the changes. Each adjustment must be discussed in a 'co-opting' process with employees and management from the two countries (organisations) present. The 'co-opting' process and the actions are equally important.

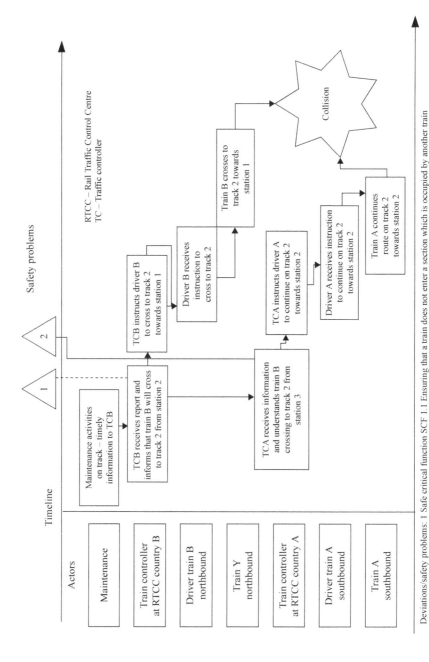

**Figure 36.4 Step diagram of collision between two trains**

Deviations/safety problems: 1 Safe critical function SCF 1.1 Ensuring that a train does not enter a section which is occupied by another train

**Table 36.2  Direct influencing factors**

| Influencing factors | Description/examples |
|---|---|
| Environment | Public opinion, climate/nature, legislation, authorities, language, Regulations, |
| Infrastructure and rolling stock | Tracks, signalling systems, communication equipment, rolling stock, human machine interface |
| Organisation | Structure, goals, strategies, management, cooperation across borders, |
| Safety culture at interfaces | Management involvement, shared commitment, focus on organisational learning, reporting culture, a just culture, industry wide cooperation, legislative cooperation |
| Routines | Work descriptions, contingency and emergency plans |
| Individual and ream | Motivation, risk perception, identity, competence, communication |

**Table 36.3  Analysis of safety critical function – identifying safety challenges**

| Influencing factor | Major differences | Safety challenges related to collaboration and communication at interfaces |
|---|---|---|
| Environment – language | Different language being used | Misunderstanding between traffic control |
| Infrastructure – communication equipment | Different systems being used, different frequencies | Important messages could be delayed in a contingency |

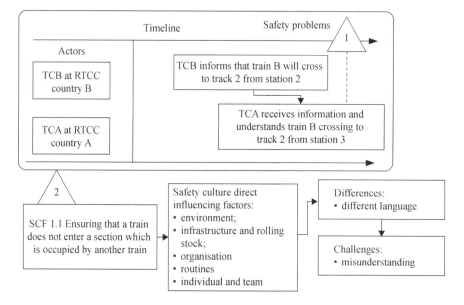

**Figure 36.5   The steps in the methodology – from a SCF to the challenges**

Actions could be implemented by management, or by employees and could consist of changes of routines, training, organisation or other actions, as suggested in Table 36.4. (We have used the preceding example.)

**Table 36.4  Actions based on challenges**

| Major differences | Safety challenges related to collaboration and communication at interfaces | Actions (agreed between participants) |
|---|---|---|
| Different language being used | Misunderstanding between traffic controls | 1 Common language training of traffic control centre (TC) and train driver (TD)<br>2 Common set of phrases being used across interfaces<br>3 Repetition of key information between TC and TD to ensure common understanding<br>4 Long term: common language established in all communication (as in aviation) |

## Conclusion

Building safety culture at interfaces has been seen as a learning process that requires involvement and commitment between organisations. This is a difficult challenge.

One of the first challenges is to motivate and get involvement from the relevant parties in the process. The next challenge is to develop real commitment from the organisations that are involved in railway traffic across borders.

Our approach to these challenges has been to use scenarios where the involved parties feel confident of their knowledge and can be motivated to share experiences. At the same time, we feel that involvement and participation from the workforce and management from the beginning of the learning loop will create ownership and commitment to the problems and their solutions.

An additional challenge is to get the suggested solutions implemented between the involved organisations across interfaces or borders. Our opinion is that the participation and collaboration during problem analysis and selection of agreed solutions would increase the probability of the implementation of the solutions identified.

We would like to point out that the method does not seek to solve all problems but to improve safety culture at interfaces. The utilisation of scenarios based on safety critical functions will facilitate the learning process in an operational way.

It would be a mistake to assume that safety culture is improved once root causes and corrective actions have been identified. It is necessary to implement the chosen solution and to evaluate that the solution has the planned effect. Continuous learning is important to complete the learning loop process as shown in Figure 36.2.

Safety culture should be regarded as a common interest within the railway industry. Competing railway undertakings must learn from each other. Safety culture at interfaces should not be seen as a way to compete at interfaces; open exchange of experience and best practice must take place between the competing firms.

This is our first version of a method to identify problems attached to safety culture at interfaces. The method is based on the combination of existing methods and some new developments to get a methodology that can be used in the railway environment.

It is not possible to isolate the issue of safety culture at interfaces from issues such as environment, infrastructure, organisation, individual and teamwork. Safety culture at interfaces is an integral part of the 'whole picture'.

## Acknowledgements

This research has in part been funded by the UIC Safety Culture at Interfaces project awarded by UIC to SINTEF. We would like to thank to Mr Bill Gall, Mr Teodor Grădinariu and the project team at UIC for their valued support.

# References

Advisory Committee on the Safety of Nuclear Installations (1993), 'Organising for Safety', 3rd report, London: Health and Safety Commission.

Argyris, C. and Schön, D.A. (1978), *Organizational Learning*, Reading MA: Addison-Wesley.

EU (1996), 'A Strategy for Revitalising the Community's Railways', White Paper, COM(96) 421, July.

Greenwood, D.J. and Levin, M. (1998), *Action Research*, London: Sage Publications.

Helmreich, R.L. and Merrit, A.C. (1998), *Culture at Work in Aviation and Medicine: National, Organizational and Professional Influences*, Aldershot: Ashgate.

Hendrick, K. and Benner, L. (1987), *Investigating Accidents with STEP*, New York: Marcel Decker.

Hofstede, G. (1991), *Cultures and Organisations: Software of the Mind*, London: McGraw Hill.

Itoh, K., Andersen, H.B. and Seki, M. (2004), 'Track Maintenance Train Operators' Attitudes to Job, Organisation and Management and their Correlation with Accident/Incident Rate', *Cognition, Technology and Work*, 6 (2): 63–78.

Kjellen, U. (2000), *Prevention of Accidents through Experience Feedback*, London: Taylor and Francis.

Reason, J. (1997), *Managing the Risk of Organizational Accidents*, Aldershot: Ashgate.

Reason, J. (1990), *Human Error*, Cambridge: Cambridge University Press.

Rosness, R. (2002), 'Safety Culture: Yet Another Buzzword to Hide our Confusion?', presentation to the conference *Sikkerhetsdagene* (Safetydays) *2002*, Trondheim, 29–30 October.

Schein, E. (1985), *Organizational Culture and Leadership*, San Francisco: Jossey-Bass.

Schein, E. (1992), *Organizational Culture and Leadership*, 2nd edn, San Francisco: Jossey-Bass.

## Appendix: Safety Critical Functions in the Railway

*Main Categories of Safety Critical Functions*

We will categorise the safety critical functions into seven areas:

SCF–1   SCFs related to normal operation;
SCF–2   SCFs related to ordinary traffic disturbances;
SCF–3   SCFs related to technical failures in signalling system/central Train Control (CTC) system;
SCF–4   SCFs related to degenerated infrastructure;
SCF–5   SCFs related to work on the track;
SCF–6   SCF–6, SCFs related to deficiency on rolling stock;
SCF–7   SCFs related to cross border activity.

*SCF–1: SCFs related to normal operation*

The situation is that all technical systems are functioning sufficiently. The infrastructure is also without any (apparent) deficiency, and the trains are within their schedules. We will assume that the line is either single or double track, and that the line is equipped with a complete signalling system. Six primary safety critical functions are evident in this situation:

*   1.1 Ensuring that a train does not enter a section which is occupied by another train;
*   1.2 Identify and take proper action if a train enters a section which is occupied, or reserved for another train;
*   1.3 Identify any 'system' change that takes the system to a degenerated operation mode;
*   1.4 Ensure the safety of passengers at stations;
*   1.5 Safe operation of level crossings;
*   1.6 Avoid excessive speed.

.

*SCF–2: SCFs related to ordinary traffic disturbances*

The situation is similar to section 0. The situation now is that trains are delayed, cancelled etc.. Hence it is necessary to change the scheduled crossings. A new SCF is thus:

*   2.1 Safe change of crossing.

Note that change of crossing is usually not a problem. However, on lines without a complete signalling system (controlled by train messages) the change of crossings is important. For example on train message (single track) lines in Norway, the locomotive driver shall verify that the passing train has arrived before he enters the next block, even if the train dispatcher has indicated 'green'. Thus, when the scheduled crossing is changed, the procedure needs to be altered as well.

*SCF–3: SCFs related to technical failures in signalling system/Central Train Control (CTC) system*

Relevant SCFs in these situations are:

- 3.1 Diagnose system in order to reveal which functions are trustful in the new degenerated state;
- 3.2 Allow trains to enter sections that could not be confirmed free from other trains;
- 3.3 Issue relevant traffic circular when for example level crossing is defective;
- 3.4 For rolling stock; comply with orders given by traffic circulars, radio messages etc.

*SCF–4: SCFs related to degenerated infrastructure*

When the infrastructure has degenerated or is threatened by extreme weather conditions it would be necessary to impose traffic restrictions. Further repair or maintenance activities must be conducted. Relevant SCFs in this situation are:

- 4.1 Issue relevant traffic circular (speed restrictions, signal out of order due to maintenance etc.);
- 4.2 Comply with instructions in traffic circular.

*SCF–5: SCFs related to work on the track*

When maintenance is conducted on the track, special safety arrangements are necessary. The most important SCFs are:

- 5.1 Issue relevant traffic circular;
- 5.2 Comply with instructions in traffic circular;
- 5.3 Put up signpost with reduced speed information;
- 5.4 Install track coils with reduced speed signature;
- 5.5 Comply with traffic circulars, signposts etc.;
- 5.6 Timely and accurate notification and dissemination of information in contingency situations.

*SCF–6: SCFs related to deficiency on rolling stock*

When there are problems with the rolling stock, it is important that this is detected, and relevant actors are informed. Relevant SCFs are:

- 6.1 Detect deficiency (by train crew, or by infrastructure systems like hot-boxes, stroke detectors);
- 6.2 Timely and accurate notification and dissemination of information in contingency situations;
- 6.3 Take appropriate action (for example stop the train when necessary, and take the train to the nearest station with speed restrictions);

- 6.4 Fetch defective train;
- 6.5 Emergency preparedness in case of accidents.

*SCF–7: SCFs related to cross-border activity*

- 7.1 Ensuring that rolling stock is compatible for cross-border traffic;
- 7.2 Ensuring that rolling stock is maintained adequately;
- 7.3 Ensuring that train crew is qualified and trained for cross-border traffic;
- 7.4 Ensuring that technical systems are reset/configured when passing the border;
- 7.5 Ensuring that dangerous freight is handled properly in relation to cross-border traffic.

# PART 11
# INCIDENT REPORTING

# Chapter 37

# The Development of PRISMA-Rail: A Generic Root Cause Analysis Approach for the Railway Industry

Tjerk W. van der Schaaf and Linda B. Wright

## Introduction

For many domains, incident analysis is a key component of organisational learning and continuous process improvement. Root cause taxonomies should allow for an overview of relative frequencies of the full range of causes in a database of large numbers of incidents, and for identification of their dominant failure patterns. The resulting recommendations for failure prevention should allow managers to take effective and efficient action to increase the safety and other performance indicators of their organisation. This chapter deals with the reasons and criteria for such causal taxonomies, and describes the development of a generic incident analysis approach for the railway industry: PRISMA-Rail.

In recent years structural railway safety problems have become apparent in many European countries, as indicated by a series of incidents at both the accident and near miss level. In this chapter we state that one of the main barriers of effectively and efficiently tackling these problems is the lack of a safety-science based approach for generic incident analysis in this domain. Such an approach should allow identification of the full range of root causes, their classification according to a relatively simple taxonomy of failure factors, and the translation of dominant failure patterns in the resulting database into appropriate structural countermeasures. It should support learning mechanisms at different levels within a certain railway company (that is, train operating team, region, company levels), across different companies in the same country, and across different countries.

There are two basic reasons for using a root cause taxonomy: in the first place it allows an organisation to rise above the traditional qualitative mechanism of learning at the single-incident level. Such 'learning' is usually very sensitive to highly specific, unique aspects of that single incident, and often leads to symptom fighting. Taxonomies on the other hand allow one to build a quantitative causal overview over large numbers of incidents, enabling an organisation to identify its dominant, recurring failure factors. Secondly, when organisations have agreed to use the same standard taxonomy this allows them to directly compare the resulting databases, and learn from each other (for example successes and failures of different railway companies with different interventions aimed at similar causes). Also, by aggregating these databases a causal overview may be created at the higher levels of the entire domain (for example the railway industry), both nationally and internationally.

After listing the criteria for a proper root cause taxonomy, we briefly stress the full set of safety-relevant data types (for example failures, recoveries, and context variables), and the need for distinctly separate incident analysis phases to describe, classify, and interprete incidents and their causes. The historic development of PRISMA is outlined, after which PRISMA-Rail (version 1) is introduced. Finally, we look ahead to sketch a model for learning at different levels in the railway industry.

## Taxonomy Criteria

Wright (2002) provides an overview of the literature regarding criteria for (failure) root cause taxonomies, citing work like Kirwan (1992), Shorrock (2002), and Fahlbruch and Wilpert (1997). These various criteria that have been suggested in the literature are given here in no particular order:

- theoretical validity;
- technical, human, and organisational root causes;
- inter-rater reliability/consensus;
- comprehensive;
- quantitative;
- consequential;
- revealing;
- predictive accuracy;
- auditability;
- resource efficiency (training; useage);
- usability;
- life cycle applicability;
- flexibility;
- context validity;
- hierarchical classification;
- terminology;
- ease of communication.

## Failure, Recovery, and Context

In its report (IOM, 2003) on safety data standards for the healthcare domain the US National Academies' Institute of Medicine proposes an important additional concept to that of failure/error causes (or: failure factors): *recovery factors/causes*, which explain why, in the case of near misses/inconsequential incidents, a series of errors/ failures in the end did *not* result in injuries or worse, but were detected, understood, and corrected in time. Next to the traditional safety management approach of error prevention (based on insight into the dominant failure factors), insight into these recovery factors then offers an alternative way of safety improvement: by introducing conditions and system characteristics which build-in or strengthen opportunities for successful recovery an organisation may protect itself against initial errors/faults turning into actual damage, injury, or other negative consequences.

Another important point in incident analysis (IOM, 2003) is to clearly distinguish between causal factors (failures and recoveries, describing the how and why of the incident) and *contextual variables* (describing the who, what, when, and where of the event), which although not actually causal, still provide useful information pinpointing the details of the relevant circumstances. A combination of causal insight and contextual information often allows organisations to precisely focus their interventions, thus improving the effectiveness and efficiency of such countermeasures.

## Description, Classification, and Countermeasures

Strictly speaking, a causal taxonomy/root cause classification does not function to *describe* the incident or its root causes; nor does it generate advice or suggestions on the appropriate *countermeasures* aimed at influencing these causal factors. It merely assigns the (potentially unlimited range of) identified, incident-specific root causes to a limited set of abstract, non-overlapping, theory-based classes (that is, the taxonomy categories). This classification process then allows one to aggregate the coded root causes of each incident into a database, describing the absolute and relative frequencies of each causal category over a large number of analysed incidents. The result of this is a diagnostic causal overview/summary of the types of processes, systems, and work conditions which generated those incidents.

In other words, a taxonomy needs an *input* describing the incidents, in a complete and neutral way; furthermore, in order for the entire incident analysis approach to function as a safety management tool, the classification's own *output* in the form of the causal pattern arising from the database (also often called 'profile' or 'fingerprint') needs to be translated into grounded links towards effective and efficient countermeasures. Preferably the same theoretical constructs (or empirical relationships) responsible for the basics of the taxonomy should be used to define the links between dominant root causes on one hand and optimal types of interventions on the other.

In this view a classification model thus becomes truely functional only when 'sandwiched' between a proper input phase (created by a tool for describing the development of the incident in sufficient detail) and an intervention-oriented follow-up phase (supported by a tool which maps each causal category of the taxonomy onto a specific type of countermeasure).

## Development of the PRISMA Taxonomy

All of the above considerations have been used to build the original version of the PRISMA approach to incident analysis (van der Schaaf, 1992), and more specifically to construct the classification scheme known as the Eindhoven classification model (ECM).

For the description phase an adapted version of fault trees was developed, 'causal trees', which start at the 'top event' (that is, the observed or reported symptom) and then trace this to the direct causes, then to the indirect causes, untill finally a set of simple stop rules identifies the causes at the most fundamental level: the root causes. In the case of near miss incidents the causal tree, has a (traditional) part describing the failure side ('why things went wrong initially'), plus a separate 'recovery side' explaining how in the end the developing symptoms were detected, understood and corrected in time to avoid negative consequences.

The taxonomy and classification process itself relied heavily on the notions of latent failures (Reason, 1990) and the skill-/rule-/knowledge-based model (Rasmussen, 1986) for the human contribution. A task analysis of the workers in the domain for which the first applications were developed (that is, the chemical process industry) provided subcategories within each of Rasmussen's three behavioural levels. In order to counteract any tendency to mainly (or even only) look at the human component as 'the' failure cause, a flow diagram led the analyst to *first* check for possible technical/ design reasons for errors, then for possible management/organisational characteristics which could have triggered behaviour deviations, before finally 'human/operator error' was considered as a root cause. Later work in the steel industry, civil aviation, and health care extended the set of organisational categories somewhat, and introduced the distinction between internal and external causes: *internal* are those factors which may be directly tackled by the organisation or department in which the resulting incidents surfaced, because that is where they originated, while *external* causes are 'imported' from other departments, or even other organisations (like suppliers, contractors, etc.). From the earliest version an important feature was the decision to allow for a small percentage (maximally 5 per cent) of the identified root causes in a database to remain 'unclassifiable'; this kept the number of classification categories in the taxonomy relative small (that is, 16–20, depending on the version), and therefore easily trainable, transferable, and thus understandable and useable by *all* levels in the organisation. Regular checks on the contents of this 'unclassifiable' category (known as X) made sure that in a new domain no hidden sub-class of causes went unnoticed. For the PRISMA-Medical version for instance a new sub-group of 'patient-related factors' was discovered within the X category, and subsequently added as a separate heading.

For the countermeasures/intervention phase a 'classification/action matrix' was added to provide guidance to managers to decide on appropriate system changes to control or even prevent the dominant failure cause(s) in the resulting database. The matrix distinguished between five fundamentally different groups of interventions (that is, technical/design changes; rules and procedures; skills training; information and communication; and punishment and reward) and linked these with the causal factors for which they would be most effective (or not at all). In this way the most could be made of the previous efforts by the organisation to collect, describe, and analyse the incidents. As one safety manager put it: 'It's good to know what's wrong, but even better to apply the right intervention to it: the one that will prove to have the best cost/benefit on the long term, rather a quick fix of a symptom'.

## Early PRISMA Applications in the Railway Industry

Using the original industrial version of PRISMA (van der Schaaf, 1992) over the last ten years, hundreds of advanced students following one of the safety management courses at Eindhoven University of Technology have applied the failure root cause taxonomy to descriptions of railway accidents. The sources of these descriptions were mostly publicly available databases (usually on websites) of the 'national transportation safety boards' from Canada, USA, and Western European countries as well as Scandinavia. Depending on the particular database at least 90 per cent, and usually over 95 per cent of the identified root causes could be classified.

Geurts (1996) carried out an extensive feasibility study using a slightly simplified version of PRISMA for the analysis of Signals Passed At Danger (SPADs) in the Dutch railways. In a first detailed re-analysis of 13 SPADs, only 2.5 per cent of the root causes were unclassifiable. In a second much larger dataset of 30 SPADs without consequences and 25 SPADs resulting in injuries and/or material damage, the level of classification was 95 per cent.

## PRISMA-Rail Version 1: Failure Factors

Despite the satisfactory classification percentages mentioned above, we have explored the possibility of a more tailor-made extension of the taxonomy for rail along the lines of the similar extension for PRISMA-Medical. Also we sought to provide a broader coverage by including not only railway staff as possible actors in incidents, but also passengers/customers, and even members of the general public (such as passers-by). This resulted in the following proposed set of failure factors:

*High-level Codes:*

| | | |
|---|---|---|
| Technical | (T) | technical/design/workplace factors (including ergonomics). |
| Organisational | (O) | organisational conditions/management decisions. |
| Staff | (S) | railway-employed individual staff |
| Customer-related | (CR) | customers/passengers, or other non-staff users of railway facilities. |
| Public-related | (PR) | members of the general public (for example passers-by), excluding customers. |
| Unclassifiable | (X) | all factors not included in the above (for example natural disasters, extreme weather). |

*Sub-level Codes for T, O, and S Categories:*

| | | |
|---|---|---|
| Technical–External | (T–EX) | all T factors not directly influenced by the organisation. |
| Technical–Design | (TD) | equipment-/software-/workplace design triggers errors (many ergonomics factors). |
| Technical–Construction/ –Maintenance | (TC/M) | original T design itself OK, but built differently/not maintained properly. |
| Technical–Material | (TM) | T factor, not yet clearly classifiable using the above T-codes: more (usually very |

|  |  | time-consuming) expert technical testing needed; so this code is meant as a *temporary* classification, awaiting further results. |
| Organisational–External | (O–EX) | all O factors not directly influenced by the organisation (for example dependent on contractor, supplier, etc.). |
| Organisational–Supervision | (OS) | employee not properly instructed/ supervised (especially with new staff/ contractors). |
| Organisational–Rules/ –Procedures | (OR/P) | Rules/Procedures not existing/not correct or updated/not easily understandable |
| Organisational–Management priorities | (OM) | production/financial/other short-term goals overriding basic safety principles. |
| Organisational–Culture | (OC) | (sub)groups having work practices/ risk perceptions etc. differing from the official organisational policies, by local consensus. |
| Staff–External | (S–EX) | all S individuals not part of the organisation. |
| Staff–KB–Status | (SKS) | staff not aware of process characteristics/ present status (for example not knowing about tracks/trains being maintained). |
| Staff–KB–Goals | (SKG) | staff pursuing other than proper goals (for example making up for lost time by speeding). |
| Staff–RB–License ( | SRL) | staff performing task without proper diploma/licence. |
| Staff–RB–Permit | (SRP) | staff performing high-risk task without proper temporary permits/safeguards. |
| Staff–RB–Check | (SRC) | staff starting job without proper checks on correct/safe status. |
| Staff–RB–Task | (SRT) | staff performing subtasks incorrectly/ incompletely/in wrong order. |
| Staff–RB–Monitoring | (SRM) | staff not monitoring (other) vital system characteristics while performing main task. |
| Staff–RB–Resources | (SRR) | staff not using proper available resources during task (for example tools, information, p.p.e., etc.). |
| Staff–SB–Intentional | (SSI) | staff making wrong intentional movement (for example typing error). |
| Staff–SB–Unintentional | (SSU) | staff accidently making movement (for example tripping/falling/leaning against controls, etc.). |

## PRISMA-Rail : Recovery factors

Table 37.1 shows a provisional, high-level model for classifying recovery factors.

**Table 37.1  High-level model for classifying recovery factors**

|  | Planned recovery | Unplanned recovery |
|---|---|---|
| Technical | T-p | T-up |
| Organisational | O-p | O-up |
| Staff | S-p | S-up |
| Customer-related | – | CR-up |
| Public-related | – | PR-up |
| Unclassifiable | – | X-up |

As the concept of recovery is still relatively new (van der Schaaf and Kanse 2000) the two dimensions above only distinguish between Planned vs Unplanned recovery, and for the moment do not go into more detail than that of the high-level codes introduced earlier for the failure factors.

Planned recovery factors (that is, purposely designed-in barriers/defences against expected, serious errors/faults) may include automatic back-up systems (T-p), or prescribed double-checking of safety-critical decisions (O-p). Unplanned recovery factors, however, consist of not-designed/-prescribed/-trained for aspects: in many cases it will involve *ad hoc*/spontaneous actions by workers 'at the sharp end' (S-up) who are able to detect the developing symptoms, understand their meaning and origin, and correct the ongoing series of events by timely and effective countermeasures, evasive actions etc.

When this scheme is applied to sufficient numbers of incidents (or near-misses) in the railway industry, we expect high concentrations of recovery factors in the T-p, O-p, and S-up categories. In that case further analysis within these three categories will empirically allow more detailed sub-categories to be identified, creating the possibility for companies of more precise focusing of recovery interventions.

## PRISMA-Rail: Context Variables

Many of the context variables may probably be copied straight from the existing SMIS (Safety Management Information System) of the UK Railway Safety and Standards Board (RSSB). For the 'who' category we may think of variables describing the specific company, department, types of staff, work experience, etc. A large set of variables detailing the huge variety of equipment, signals, and all the specific rules and regulations will fall under the 'what' category. The 'when' variables will include date, time, shift, etc., and for the 'where' we may expect variables such as trainline, section, crossing, signal box, station, etc.

As with the recovery factors, experience in applying this approach to a substantial sample of railway incidents will show which additional context variables will be useful enough to include in the coding scheme.

## Learning from Incidents at Different Levels

As mentioned in the introduction, industry-wide standardisation of incident collection mechanisms, and specifically root-cause classification schemes, will not only allow each company to build its own databases as described above, and then compare and aggregate these databases over the different levels of its organisation. It will also allow them to compare themselves to other, similar companies, in terms of benchmarking and associated learning mechnisms.

For the context variables (which allow the capture of the more idiosyncratic aspects at a certain level of the railway industry, or of a particular company or nation) the way towards standardisation will be far less simple: the examples of context variables given above may be suitable for a fixed, standardised coding only at the higher levels, but not at the important detailed levels (that is, not all companies are associated with all types of track, trains, or use all possible lines and stations). Thus we may expect a small set of fixed context variables (for example 'work experience' divided in a number of standardised classes), and within these, a large set of context variables which are specific to a certain organisational level (driver-conductor team; operations department; company) and /or to certain types of company (train operating company; maintance contractor, etc.).

Solutions for common safety problems piloted in other companies may thus be compared and possibly transfered efficiently. On top of that, at the (inter)national level a comprehensive overview of the current safety status of the entire railway industry may guide individual companies and regulators alike to identify, much earlier than before, new risks as well as monitor trends in the level of control of known risks. As laid out in the IOM report for the healthcare industry in the USA, this will require a strictly fixed set of standard failure and recovery causal taxonomies, combined with a partially flexible set of context variables.

Such standardisation will allow the steady build-up of perfectly comparable causal databases, between different departments in the same company as well as between entire companies, and even potentially between entire nations. This improved form of learning cycle (identifying new risks, and evaluating the success of interventions for known risks) may thus dramatically speed up, at less cost per company, the general safety and reliability level of the railway industry.

## References

Fahlbruch, B., and Wilpert, B. (1997), 'Event Analysis as Problem Solving Process', in A. Hale, B. Wilpert and M. Freitag (eds), *After the Event: From Accident to Organisational Learning*, Oxford: Pergamon.

Geurts, R.W.L. (1996), 'PRISMA in the Railway System' (in Dutch), MSc thesis, Eindhoven University of Technology.

Institute of Medicine (2003), *Patient Safety: A New Standard for Care*, Washington DC: National Academies Press.

Kirwan, B. (1992), 'Human Error Identification in Human Reliability Assessment. Part 2: A Detailed Comparison of Techniques', *Applied Ergonomics*, 23 (6): 371–8.

Rasmussen, J. (1986), *Information Processing and Human–Machine Interaction*, Amsterdam: North-Holland.

Reason, J. (1990), *Human Error*, Cambridge: Cambridge University Press.

Shorrock, S. (2002), 'Error Classification for Safety Management – Finding the Right Approach', paper presented at IRIA2002, University of Glasgow, 17–20 July.

Van der Schaaf, T.W. (1992), 'Near-miss Reporting in the Chemical Process Industry', PhD thesis, Eindhoven University of Technology.

Van der Schaaf, T.W. and Kanse, L. (2000), 'Errors and Error Recovery', in P.F. Elzer, R.H. Kluwe and B. Boussoffara (eds), *Human Error and System Design and Management*, Lecture Notes in Control and Information Sciences, 253, London: Springer Verlag: 27–38.

Wright, L.B. (2002), 'The Analysis of UK Railway Accidents and Incidents: A Comparison of their Causal Patterns', PhD thesis, University of Strathclyde, Glasgow.

# Chapter 38

# CIRAS – History and Issues Arising During Development

Maurice Wilsdon and Helen Muir

## Introduction

The rail industry's confidential incident reporting and analysis system (CIRAS) is one of the biggest confidential reporting systems in Europe. Currently it covers 80,000 staff and plans are in hand to double that with an extension to staff employed in the smaller contractors and subcontractors involved in maintenance and renewal work.

This chapter chronicles the main events in the development of the scheme since its inception in 1995 through the launch on a small scale in 1996 through the national rollout in 2000 to the current preparations for the next major extension. During that period, many lessons have been learnt and some of the major ones are detailed in this chapter.

## The Original Scottish Scheme

In 1995, a study by Vosper Thorneycroft, commissioned by ScotRail, identified requirements relating to human factors and human error. In particular, the study suggested that safety-related incidents were under-reported, as a result of the 'blame culture' perceived to prevail within ScotRail at that time. The blame culture appeared to shift responsibility for incidents until a discrete cause or culprit could be found. This made it difficult to identify the underlying causes of incidents, resulting in a lack of information to enable enhancement of safety management systems.

The study recommended the implementation of a no-blame third party reporting system, similar to CHIRP (confidential human factors incident reporting programme), which had been running in the aviation industry for over 15 years. ScotRail decided to take this recommendation forward in the form of the Confidential Incident Reporting and Analysis System (CIRAS), and approached the second author (Professor Helen Muir), because of her experience in human factors in the aviation industry.

She recommended an approach to the University of Strathclyde to develop the project. The reason for this was that the success of any confidential reporting programme is dependent on the establishment of trust between the employees and those responsible for operation of the system. It was agreed that Scottish rail employees were more likely to build up trust with professionals from a similar locality and culture than with a small university in southern England. Over time this decision has been demonstrated to be correct. ScotRail contracted the Centre for Applied Social Psychology at the University of Strathclyde to implement and operate the scheme.

A steering committee was established to oversee the general running of the project, to provide guidance and to monitor progress. The second author chaired the committee, which included representatives from ScotRail, Railtrack (now Network Rail), HMRI and the University of Strathclyde. The project received cross-industry support from ScotRail, Railtrack, the HSE, British Rail Safety Unit, ASLEF and RMT.

Prior to the development of CIRAS a review was undertaken of other confidential reporting programs (the CAA's CHIRP, British Airway's BASIS, ASRS in the USA, and the program in the nuclear industry). The review findings highlighted to the steering committee that a weakness in some aviation systems was a lack of feedback to close the loop; the incorporation of liaison groups and issue of quarterly journals in CIRAS has helped to address this issue. The system design took account of the nature and prevailing culture within the railway industry, and its recent privatisation. The system was available from September 1996.

CIRAS set out to encourage an open forum for the discussion of safety concerns and to raise awareness across staff and management through ongoing dialogue, quarterly journals and management reports. A mission statement was developed: 'To improve the understanding of what goes wrong when incidents and accidents occur and to develop methods that will assist staff and management to avoid situations where human error results in such incidents'.

Following this initial phase, GNER joined the scheme in December 1997, adding a further 1,000 safety critical staff who could submit reports to CIRAS. Subsequently, from late 1998, Railtrack Scotland, Virgin Trains, First Engineering, GTRM and a number of smaller contracting companies in Scotland also joined CIRAS.

Between September 1996 and June 2000, CIRAS had received 590 reports.

### The National Scheme

In June 1999, Railtrack's Safety and Standards Directorate (S&SD – a predecessor of the Rail Safety and Standards Board) called an industry conference to discuss the potential for implementing CIRAS on a national basis. There was not universal support as some organisations felt that such a scheme would not fit with their corporate culture.

However, the tragedy of the Ladbroke Grove accident on 5 October 1999 was to be a watershed for many aspects of safety management on the railways. There was a very high level of media and political interest in the days immediately after the accident. Although the linkage between the accident and confidential reporting is not immediately obvious, the decision to mandate the national rollout of CIRAS was taken in this period.

It was agreed that the first phase of the national rollout would cover safety-critical and safety-related staff[1] employed by Railway Group members. ScotRail generously agreed to allow CIRAS to be extended to become the national scheme and for the title to be used.

Railtrack S&SD agreed to provide accommodation and administrative support for project management of implementation of the system, and a cross-industry implementation group and steering committee were formed.

The steering group decided that there was a need for a core data facility to develop a national database; Railtrack S&SD offered to fund this element of the system. The steering committee decided that as the University of Strathclyde had already developed a procedure for data processing and the development of a database the industry would benefit from supporting them to be the core facility provider and to develop the historical database. They were invited to bid on a single tender basis.

The implementation group, aided by a project manager, developed the design for the regional data collection process. The initial view that rail employees would be more likely to trust and report to a local regional centre had been vindicated, so rather than having one national data collection centre, the country was divided into three regions. Using three service providers also offers more opportunities to develop and share good practice.

- Between October 1999 and January 2000 specifications were drawn up for the national system, and expressions of interest were sought via the official journal of the European Union (OJEU).
- In January 2000 tenders were invited for contracts to operate one of three regional facilities that would receive reports. Seventeen tenders were received, and eight tenderers were invited to make presentations to the cross-industry selection panel.
- Contract negotiations were completed and four separate contracts were let in June 2000:
    1   operation of the core facility was let to the University of Strathclyde;
    2   operation of region 1, which broadly covers the north of Britain, was also let to the University of Strathclyde and later transferred to a university spin-off company, Human Factors Analysts Ltd;
    3   operation of region 2, which broadly covers the middle of Britain, was let to WS Atkins Consultants Limited;
    4   operation of region 3, which broadly covers the south of Britain, was let to DERA (the defence evaluation and research agency – now called QinetiQ).

In July and August 2000 the University of Strathclyde provided training for staff from the successful tenderers for regions 2 and 3, to improve their understanding of CIRAS and the railway industry. From July 2000, meetings were held with management teams at Railway Group members to finalise arrangements, and briefing of line managers commenced. CIRAS uses cascade briefing to reach front line staff; CIRAS staff brief line managers within the client organisation, and those line managers then brief front line staff.

Members of the existing scheme in Scotland moved across to join region 1 of the national scheme as their contracts with the 'existing' scheme expired.

The first reports were received from newly briefed staff in September 2000. Briefing of front line staff involved in the first phase of CIRAS implementation was completed by 31 March 2001, at which point approximately 68,000 staff had access to the scheme.

Since then, further companies have joined on a voluntary basis. London Underground Ltd has been especially active and has enrolled staff from its three Infracos (infrastructure maintenance companies). By January 2002, total membership stood at approximately 80,000 with plans to extend membership later in the year to cover the staff of infrastructure contractors through a link to the Sentinel system of on-track competence registration.

## Important Issues

No confidential reporting scheme can succeed unless the users have faith in the scheme's ability to protect their anonymity. CIRAS has addressed that in a number of ways.

Trades unions have been closely involved in the development and running of the scheme. They have been very active in ensuring that the system is not only secure in fact but also is perceived to be wholly reliable. The perceptions of the reporting population are fundamental to the continued existence of the scheme. Having the unions playing an active role in the development went a long way to identifying and allaying potential staff concerns. It also gave the scheme a tacit seal of approval in the eyes of the staff.

These perceptions have to be underpinned by extensive measures to address both physical and process security. Physical measures include tight control of access to the CIRAS facilities, the availability and use of safes, shedders, etc. Audits of process compliance are undertaken annually and performance is measured quarterly.

Process security has been augmented by the establishment of a charitable trust to act as custodian of the data gathered by the system. That body ensures that these data, even though they are anonymised shortly after receipt, are not released to unauthorised bodies.

A recent survey by the polling organisation MORI found no staff concerns about the security of CIRAS. However, we have to recognise that this level of confidence could be destroyed by a single incident – or even by a rumour about an incident.

From the outset, the Health and Safety Executive has been an active supporter of CIRAS. However, the presence of the industry's safety regulator caused some unease amongst both staff and companies. In order to allay concerns, the HSE has published a protocol that explains that knowledge gained through CIRAS will not be used to initiate regulatory action unless the problem persists.

Staff are not the only group of stakeholders who need to be reassured about their security. The employing companies also have a right to protection. They can request, and receive analyses of data relating to themselves and to all other organisations in their market sector (for example train operating companies or infrastructure maintenance companies) but cannot have access to data relating to other individual companies within their sector.

It is interesting to note that there have been major changes in the availability and use of technology even in the lifetime of CIRAS. At the outset, mobile telephones were fairly rare amongst reporters. Many of those that did exist tended to be company issue and would be passed from one person to another at the end of a shift. Hence CIRAS recorders were loathe to return calls to mobile phones in case the user at that time was not the reporter.

The University of Strathclyde ran the Scottish scheme but the advent of the national system saw the introduction of two new suppliers from very different backgrounds. The diversity was an outcome of the competitive tendering process. It was expected that it might lead to competition in the development of best practice. This expectation has only partly been fulfilled.

The University of Strathclyde decided to create a spin-off company, Human Factors Analysts, to take on the role of the operator of region one. This transfer from academia to private enterprise has been achieved with very little apparent pain.

At the end of 2001, the contract to run the core – the national database and analysis service – expired and was retendered, with Atkins winning the work. The transfer exposed some issues over the ownership of the IPR of the database structure. These provided the impetus for a redesign with a new taxonomy which is now fully portable to future operators. The downside of this change was the requirement to recode all the existing records to fit the new structure.

Funding of the system is achieved by levying a subscription based on the number of staff each company declares to be enrolled in the scheme. This method provides an incentive for creative counting to minimise the cost to the less enthusiastic companies. Alternative funding mechanisms are being considered.

The fact that some companies are reluctant members can be seen in the quality of some of the responses. Enthusiastic companies tend to use the data constructively and they provide very positive responses that are published in the journals. However, some responses are terse and stick to a very brief statement of facts such as quoting the relevant entry in the rulebook. These do little to generate interest amongst the readers of the journals and may perhaps be viewed as being defensive statements from organisations that feel somehow threatened by the existence of CIRAS.

At the very outset, it was decided that a small local scheme would be best served by having a local provider who could establish a rapport with the users through common language and knowledge of the geography. That decision was correct for the initial scheme but was impossible to sustain as the scheme expanded and gained national coverage.

Regions were designed with a geographic basis derived from the boundaries of the Railtrack (now Network Rail) zones. However, the areas of operation of train companies and infrastructure maintenance companies rarely match these administrative areas. Indeed some companies operate on a totally national basis and have been allocated to a single CIRAS region. Hence we have a combination of local and national allocations. The picture is further complicated when companies win or lose contracts/ franchises and when mergers or acquisitions occur.

Furthermore, the diversity of the three regional service providers has tended to give different characteristics to the three regions despite having a common set of standards. This may not be aiding the creation of a single unified system.

All members of the Railway Group – holders of approved Railway Safety Cases – are obliged to belong to CIRAS. These have been joined by a number of voluntary members but there is a large population that is not currently involved. Beyond the large infrastructure maintenance companies, there are many companies supporting the maintenance and renewals activities through the provision of specialist services and staff. It is believed that there are in excess of 85,000 staff in this category who are outside CIRAS though it is widely felt that these individuals face issues which are potential reports for a confidential system.

Plans are well advanced for the extension of CIRAS to cover this new population though some changes have been necessary to make the system work without compromise. Many of the organisations have a relatively small workforce so there is an increased opportunity for the managers to guess the identities of reporters – either correctly or erroneously. Hence the process has been revisited to ensure that the anonymity of the reporters can be protected.

As the workforce of these contractors is highly itinerant, it has been necessary to adjust the arrangements for the feedback of information to the staff. Instead of journals there will be smaller newsletters which will address issues on a generic level rather than through individual transactions. These newsletters will be distributed electronically to the employers rather than being sent to home addresses. This change is a major component in the achievement of a substantial cost reduction for the new scheme. All reports will continue to be entered into the national database and can be analysed along with data from the existing scheme.

The industry is hungry for price reductions and work is in hand on designing the new data collection contracts with a view to giving the member companies better value for money. For the present, the funding remains through the imperfect per head subscription basis.

## Conclusion

Many lessons have been learned during the development and implementation of CIRAS. The challenges are many and various but it is vital to create a system that will guarantee the anonymity of reporters. The system must not only be leak proof but it must be perceived as being totally impervious. Managing the perceptions can be a lot harder than designing the system.

It is essential that all stakeholders are involved in the development and are content with the result. This will create ownership and minimise scepticism. Given a good foundation, running the system should be relatively easy!

Following a detailed consultation process, the plans for the new organisation outlined earlier in this chapter are about to be realised. The lessons learned during the development of CIRAS will be taken forward into the new centralised CIRAS unit, in operation from Spring 2005. The London-based unit will cover the entire UK rail network, almost doubling the number of staff able to use the system to 160,000. Building on the excellent work of those who set up the existing system, the new organisation will improve its communication with front line staff and industry management alike. By increasing the number and quality of reports received and working with industry to improve the quality of response. CIRAS is set to become a more dynamic and effective organisation.

The new system has been designed to continue to guarantee the anonymity of reporters, and will work to maintain the trust CIRAS has so far enjoyed amongst rail staff.

*Rail Human Factors*

**Note**

1  Safety critical staff are staff undertaking safety-critical work, as defined in the 1994 Railways (Safety Critical Work) Regulations. Safety related staff may be considered as staff involved in activities that have an indirect impact on the safety of the operational railway.

# Chapter 39

# Near Miss Versus Accident Causation in the UK Railway Industry

Linda B. Wright and Tjerk W. van der Schaaf

## Introduction

There is much controversy over the veracity of the common cause hypothesis: that near misses and actual accidents have the same causal patterns. This chapter describes the comparison of the causes of accidents and near misses collected from one UK railway company. The accidents were collected following the completion of the normal accident investigation process, while the near misses were collected from the CIRAS (Confidential Incident Reporting and Analysis System) database. The results suggest that for the railway industry the common cause hypothesis is valid. The implications of the hypothesis are discussed with reference to the railway industry in general.

The UK railway industry implemented CIRAS on a nationwide basis in June 2000, after a successful pilot period in a number of companies. CIRAS is designed to collect, analyse and feedback information on near misses and unsafe situations. The rationale behind introducing such a voluntary and confidential near miss reporting system was that it would remove the barriers to reporting near misses and therefore aid in the understanding of accident propagation and provide the ability to prevent more serious incidents on the basis of acting on the causes of near misses.

The assumptions for the predictive validity of near miss reporting systems are based upon the work of Heinrich (1931) who first proposed the common cause hypothesis. The common cause hypothesis suggests that near misses have the same causal factors as more serious incidents (those resulting in property damage or injuries and fatalities). If indeed near misses have the same distribution of failure causes as more serious incidents then they are a valuable resource for safety management systems and accident prevention.

The value and advantages of collecting near misses in addition to accidents are as follows (van der Schaaf *et al.*, 1991):

1   they occur in greater numbers than actual accidents, thus providing a larger data set on which to base prevention measures;
2   if near misses are precursors to later, possible accidents, then there is no need to wait until an actual accident has occurred to understand the causal process and begin prevention measures;

3   the successful error/failure recovery actions which prevented the near miss from escalating to a fully-fledged accident can be used to help encourage recovery behaviour.

However, if near misses and accidents have very different causal patterns then the learning from near misses may not be generalisable to accidents and may place an extra burden on safety management systems for a rather small return (in terms of safety benefits and reduced accidents).

## The Common Cause Hypothesis

The common cause hypothesis was first proposed by Heinrich (1931) who suggested that near misses had the same set of 'root' causes as actual accidents. The common cause hypothesis is not to be confused with the iceberg/triangle models of the frequency of occurrence and severity of accidents and near misses or the hidden costs of accidents (Heinrich, 1931; Bird, 1966), and see Wright (2002) for a full description of these problems. Since Heinrich proposed this hypothesis there has been little research adequately performed in testing the theory (mainly due to the confusion of the common cause hypothesis with ratio/iceberg models). However, one pilot study by Geurts (1996) compared the causes of no-consequence Signals Passed At Danger (SPADs) with the causes of SPADs resulting in damage or injury in the Dutch railway system. He used the PRISMA (Prevention and Recovery Information System for Monitoring and Analysis – van der Schaaf, 1992) incident analysis methodology and concluded that the causes of both types of incident were the same, at the level of technical, human and organisational causes.

The present study aims to test the validity of the common cause hypothesis in the UK railway industry and widens the focus from SPADs to all incidents involving train drivers.

## Rationale for the Study

This study was designed to determine if there were significant differences between the failure causes of actual accidents and near misses in the railway industry. If there are significant differences, then collecting near misses will shed no light on the development of incidents resulting in loss or injury. As such, near misses may be interesting for their own sake, but are not useful in preventing or mitigating more serious incidents.

## Data Collection

Three different investigation methods used by the UK railway industry and providing the data for this study are described below.

## Formal Inquiries

Following a major incident or an incident with the potential for a major loss an internal formal inquiry may be performed. In the case of a formal inquiry the company seeks to learn where both individuals and systems can be improved. Thus the investigation takes place at both the level of human error and organisational causes. Formal inquiries are performed by a panel of highly expert railway managers (usually four or five plus a union representative) who act as investigators and interviewers during the process. The investigations are intended to be comprehensive, with the aim of determining technical, human and organisational causes. Staff involved, including witnesses, are interviewed regarding their part in the incident and the interviews are transcribed verbatim and included in the final report, along with the conclusions of the panel. Further technical evidence such as speed calculations, brake tests, damage reports and re-enactments of the incident are also presented, as is scientific evidence relating to rail contamination.

## Signal Passed at Danger Investigations

SPAD investigations are performed following all incidents of Signals Passed at Danger (that is, red) without authority. They are often less comprehensive than Formal Inquiries and are investigated by fewer people. SPADs are usually detected automatically by the signaller, but on occasion are reported by the driver. Automatic detection is not the case in a minority of areas, which are not fitted with the appropriate technology (for example in depots). SPAD investigations are usually performed by the driver team manager, following an initial discussion between the driver who has passed the signal and the signaller who has detected it or has received the initial report from the driver. A SPAD investigation usually consists of a written report by the driver involved – no more than a dozen lines – and by any relevant member of staff who was present (for example conductor, guard or accompanying driver). Following the written report, the driver is interviewed by the team manager and the findings are written in a brief report. These investigations are not comprehensive and usually stop at the level of determining the human or technical causes. Organisational causes are often not discussed or investigated. SPAD investigations are rarely accompanied by technical reports such as brake tests or rail contamination tests, unless the driver has complained about rail or train characteristics. These reports have become much more comprehensive since the study was conducted.

## CIRAS Near Miss Report Investigations

CIRAS, the Confidential Incident Reporting and Analysis System, is the UK railway national system for the reporting and analysis of railway near misses (see Wilsdon and Muir in this book). CIRAS reports are made voluntarily by staff regarding near miss incidents and other safety issues. For this study only reports by drivers from the target company were selected and analysed. Reports are made on a form or by telephone initially detailing the incident or issue which the driver wishes to report. Following the initial report, CIRAS staff perform a critical incident interview (Flanagan, 1954) with each driver. These interviews include details about the 'what', 'when', 'where', 'why' and 'how' of the incident. The CIRAS reports used in this study all came from the same company and from approximately the same period as the Formal Inquiries and SPAD investigations.

For all three data sources, an incident is defined as a specific example of a failure of personnel or equipment to operate as planned or a specific example of general public behaviour that has safety implications. Table 39.1 shows the data source and level of severity of the incidents used.

**Table 39.1  Data source and level of severity**

| Severity level | Data source | | | Total |
|---|---|---|---|---|
| | **Formal inquiry** | **SPAD investigation** | **CIRAS report** | |
| Fatality/injury | 17 | 0 | 0 | 17 |
| Damage | 18 | 7 | 0 | 25 |
| Near miss | 11 | 81 | 106 | 198 |
| Total | 46 | 88 | 106 | 240 |

## CIRAS Analysis

The data from all three sources were thoroughly (re-)analysed according to the University of Strathclyde CIRAS human factors model which is hierarchical (see Davies *et al.*, 2000 for a full description of the system). According to this model individual causal codes are subsumed under one of four top-level categories: 'technical', 'proximal', 'intermediate' and 'distal' which are called the 'macro' codes. These macro codes each comprise an exclusive set of individual causal codes, which are termed 'micro' codes. Thus the common cause hypothesis can be tested on two levels: the more general level of macro codes, and the specific level of the individual micro codes. (Note: since December 2001, the University of Strathclyde CIRAS human factors model was no longer used to analyse CIRAS near misses.)

### Inter-rater Reliability

Inter-rater reliability is a vital (and often neglected) part of any analysis system. Data analysed via the CIRAS system are subject to periodic inter-rater reliability trials. The index of concordance was above 80 per cent for each trial. To ensure the data used in this study were also reliably coded two independent raters (other than the authors, and experienced in using the coding scheme) coded a total of 14 incidents from various classes of event used in this study. This resulted in an index of concordance of 78.4 per cent.

## Results and Discussion

Figure 39.1 shows how the four macro causal codes are distributed over the three levels of severity. A chi-square test for proportions showed non-significant differences at the 5 per cent level.

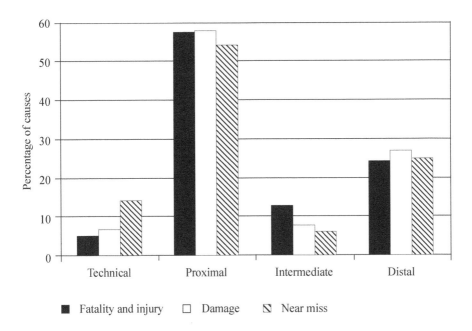

**Figure 39.1   Macro causal codes by level of severity**

Thus, at the level of the superordinate categories of technical, proximal, intermediate and distal no significant differences were found in the proportion of causal codes between the three severity outcomes (injury, damage and near miss). However, despite the fact that these results are supportive of the common cause hypothesis, this macro level analysis tests the common cause hypothesis only at a very general level, which would not give very precise guidance on how to prevent accidents.

At the more specific level of individual codes only three of 21 causal factors are significantly different, namely knowledge-based errors, training and procedures. In all three cases 'fatality and injury' have a greater proportion of these codes assigned than 'near miss' incidents. In terms of the differences found in knowledge based errors Embrey and Lucas (1988) suggest that such errors are more likely to be detected by someone other than the individual who made the error. It may therefore be the case that individuals reporting via the CIRAS system are unaware of knowledge based errors that they have committed and despite being interviewed these have not come to light.

There are also a number of possible explanations for the higher proportions of the causal codes training and procedures. Firstly, it may be the case that training and procedure causal factors are more prevalent in incidents with a more serious outcome. However, as training and procedures were high on the company agenda, it is more likely that such issues have been more frequently recognised by managers during the Formal Investigations than have been revealed by staff during interview after submitting CIRAS reports. Issues such as training and procedures are traditionally management-driven and factors that management expect to have an impact on adverse events and therefore these are more likely to be identified as causal factors in incidents that are investigated by managers. Therefore, overall, these findings provide qualified support for the common cause hypothesis within the railway domain.

## Limitations of the Study

There are a number of limitations associated with the present study. Firstly, the data were investigated and collected by different means. The near miss reports were all made voluntarily via CIRAS, while the reports from formal inquiries and SPAD investigations were mainly detected by the railway industry rather than being voluntarily reported. The depth, goal and scope of the incident investigation procedures were different for the three data collection methods – formal inquiries were investigated in greater depth than SPAD incidents. CIRAS investigations concerned only the individual making the report (due to confidentiality) while both formal inquiries and SPAD investigations were supplemented by evidence from witnesses. Finally, the data comprised a much smaller absolute number of serious accidents than of near misses. In an ideal situation for statistical analysis, the underlying causal factors would have come from a similar number of events for each class of severity.

Nevertheless, the data analysis attempts to take most of these limitations into account. The method of testing the common cause hypothesis is based upon the proportions of causal factors rather than the absolute frequencies (Fleiss, 1981). This therefore reduces the problems introduced by different investigation methods (formal inquiry, SPAD investigation, CIRAS interview) and also the problems associated with unequal numbers of incidents in the different severity levels.

A further limitation is that the specific human factors taxonomy used at the time of the study has since been replaced by a different one. This fact limits the direct comparison of these data (reported and analysed in the period 1990/6–2000) with more recent data or data from the same period from other companies. Therefore another, alternative causal taxonomy with a proven suitability for the railway domain (like the earlier mentioned 'PRISMA' method – van der Schaaf, 1992) would be necessary to test the common cause hypothesis for:

- the same company with a larger and more recent database;
- other companies both within and outside of the UK, with possibly other safety/reporting cultures;
- an even wider domain of railway incidents, other operating conditions (like maintenance), work situations, other types of staff, etc.

## Implications for the Railway Community

The findings suggest that near misses are a useful source of precursor data for understanding and therefore preventing actual accidents. However, given the limited nature of the data (only one company was used and incidents were confined to the driving function) it is suggested that a larger scale study be performed to determine if the common cause hypothesis is applicable to the industry as a whole.

Nevertheless, these first results are positive and have a number of implications for both the UK and European railway community.

Firstly, the railway industry should expend more effort not only in collecting near misses, but also in fully exploiting them. Currently, the CIRAS system in the UK is a good example of how to collect near misses. However, are these near misses being fully exploited? In order to be most effective, it is suggested that the analysis of near misses should feed the railway safety management system, allowing the most benefit to be gained by enabling a direct comparison with the consequential incidents currently held in safety management information systems (such as SMIS in the UK, or VIS and MISOS in the Netherlands). ProRail (rail infra provider in the Netherlands) are currently considering the use of PRISMA to causally analyse both accidents and near misses, thereby providing a direct comparison between the different classes of incident and allowing for more effective prevention measures to be generated (for example based on the full range of consequences and frequency of causes, rather than basing remedial measures simply on the seriousness of the consequence). In order for a direct comparison to be made at all levels of the severity continuum (that is, from near miss to fatality) all incidents should be investigated in the same manner, to the same depth and using the same taxonomy to assign root causes.

Secondly, an important but as yet untapped source of pro-active management information from the near miss reports is the system characteristics and conditions under which planned or unplanned 'recovery actions' are successful (van der Schaaf and Kanse, 2000; Kanse, 2004). How is it that most (indeed, possibly practically all) of the day-to-day problems with equipment, procedures etc usually are 'worked around' successfully? In which cases are such work-around practices adding new, unknown risks to the safe and efficient operation of the railway system? Near misses have not only information in them on the failure factors ('why things almost went wrong'), but also on recovery factors ('why things in the end were detected, understood, and corrected in time'). Just as insight into near-miss failure factors may lead to effective prevention, a parallel database of near miss recovery factors may suggest effective and efficient strategies to build 'recovery promotion' into the railway system, or strengthen and formalise it where it has already shown to successfully exist.

Thirdly, the actions (based on analysis of near miss causes) taken to reduce the recurrence of future accidents should be monitored for effectiveness by comparing the causation pattern and frequency of occurrence of similar accidents and near misses after implementation of remedial measures.

Fourthly, where near miss systems exist (such as CIRAS), it is important to audit and evaluate the system periodically to ensure that they are collecting actual near misses and not simply instances of possible dangerous situations (which contain less information than a real near miss and no information on recovery). Effort should be expended into promoting the near miss system and motivating individuals to provide reports of near misses.

Finally, as in any voluntary reporting system one should always be aware of possible reporting biases – they should therefore be regularly checked and monitored via confidential focus groups, personal incident diaries, etc. For a literature overview and empirical results on biases in reporting to a confidential near miss reporting system see van der Schaaf and Kanse (2004).

## Conclusion

These findings provide qualified support for the common cause hypothesis within the railway domain. However, as this study was limited to one group of railway employees in one company, using one type of causal taxonomy, it is recommended that further empirical tests of the common cause hypothesis be performed for a number of different employee groups and companies, and with other types of taxonomies (for example PRISMA). This would provide more robust evidence of the applicability of the theory. The safety and reliability performance of the railway industry would thus benefit from a more integrated approach towards the collecting and analysis of near misses as well as more serious incidents.

## References

Bird, F.E. (1966), *Damage Control*, Philadelphia: Insurance Company of North America.

Davies, J.B., Wright, L., Courtney, E. and Reid, H. (2000), 'Confidential Incident Reporting on the UK Railways: The CIRAS System', *Cognition, Technology and Work*, 2 (3): 117–25.

Embrey, D.E. and Lucas, D.A. (1988), 'The Nature of Recovery from Error', in L.H.J. Goossens (ed.), *Human Recovery: Proceedings of the COST A1 Working Group on Risk Analysis and Human Error*, Delft University of Technology, 13 October 1987.

Flanagan, J.C. (1954), 'The Critical Incident Technique', *Psychological Bulletin*, 51 (4): 327–58.

Fleiss, J.L. (1981), *Statistical Methods for Rates and Proportions*, New York: John Wiley and Sons.

Geurts, R.W.L. (1996), 'PRISMA in het railverkeerssysteem' ('PRISMA in the railway system'), MSc thesis, Eindhoven University of Technology.

Heinrich, H.W. (1931), *Industrial Accident Prevention*, New York: McGraw-Hill.

Kanse, L. (2004), 'Recovery Uncovered: How People Recover from Failures in the Chemical Process Industry', PhD thesis, Eindhoven University of Technology.

Van der Schaaf, T.W. (1992), 'Near Miss Reporting in the Chemical Process Industry', PhD thesis, Eindhoven University of Technology.

Van der Schaaf, T.W. and Kanse, L. (2000), 'Errors and Error Recovery', in P.F. Elzer, R.H. Kluwe and B. Boussoffara (eds), *Human Error and System Design and Management*, Lecture Notes in Control and Information Sciences, 253, London: Springer Verlag: 27–38.

Van der Schaaf, T.W., and Kanse, L. (2004), 'Biases in Incident Reporting: An Empirical Study in the Chemical Process Industry', *Safety Science*, 42 (1): 57–67.

Van der Schaaf, T.W., Lucas, D.A. and Hale, A.R. (1991) (eds), *Near Miss Reporting as a Safety Tool*, Oxford: Butterworth-Heinemann Ltd.
Wright, L.B. (2002), 'The Analysis of UK Railway Accidents and Incidents: A Comparison of their Causal Patterns', PhD thesis, University of Strathclyde, Glasgow.

# PART 12
# PASSENGERS AND STAFF

# Chapter 40

# Managing Violence

Claire Dickinson and Jeremy Bevan

## Introduction

This chapter describes the findings of a series of targeted inspections that considered the arrangements and measures for managing and preventing work-related violence in train operating companies. In general, train operating companies manage the risk of violence to their staff in the same way as other personal risks. Hence, the inspection considered risk assessments, reporting measures, investigation reports and evaluated the preventive, protective and response measures in place. A number of deficiencies were identified, the most serious of which concerned the arrangements and measures in place for contract staff employed to check tickets or in security activities. Action was taken by HMRI to ensure both the train operating company and the employers of the contract staff resolved this issue.

In recent years the number and severity of incidents of work-related violence has risen sharply across all employment sectors, including railways. This chapter describes the major findings of a targeted series of HM Railway Inspectorate (HMRI) inspections of train/station operating companies (TOCs), which reviewed the adequacy, against this rising trend, of current management arrangements for protecting railway employees and contract staff. The inspections were supported by interviews with key groups of workers. This work was undertaken in support of HSE's three-year programme on work-related violence.

## Legal Requirements

HMRI is part of the Health and Safety Executive, and is responsible for ensuring compliance with health and safety legislation on all UK railways, metros and light transport systems. The general duty to ensure the health and safety of employees so far as reasonably practicable is supplemented by a number of more detailed requirements to assess risks, implement control measures and have in place arrangements for ensuring risks continue to be prevented or managed properly. On the rail network, many of these duties are shared between different employers, and there are specific requirements on them to ensure cooperation and communication with each other to ensure health and safety. Railway employers also have legal duties to describe arrangements for managing work-related violence in the safety case (RSC) they are required to submit in order to be allowed to operate.

## Context

HSE has for many years regarded violence to staff as a health and safety risk which can and should be managed as any other workplace risk, and produced guidance to assist employers in this duty (for example IND(G)69 (HSE, 2002a) and HSG 229 (HSE, 2002b)). HSE has also played a major role in shaping standards drawn up by others, such as the National Occupational Standards for Managing Work-Related Violence produced by the Employment National Training Organisation (2002). In the rail sector, a combination of the Home Office's British Crime Survey (Budd, 1999), a 1999 conference on violence to staff in the industry, and the establishment of HSE's cross-sectoral Interdepartmental Committee on Violence to Staff provided the impetus for more concerted action on this topic than had previously been the case. In 2000, HSE (including HMRI) adopted a target of seeking to ensure a reduction of 10 per cent in the incidence of work-related violence nationally in workplaces. At the same time, the Department for Transport has been producing guidance for station operators (who are mainly the TOCs) on improving public security measures at stations, some of which are directly relevant to improving staff safety (Department for Transport, 1998). The British Transport Police is the arm of the police specifically responsible for preventing and detecting railway crime, and therefore also has a role in supplementing employers' responsibilities in managing the risk of violence to rail staff at work.

For some railway employees, experience of violence is an everyday occurrence. Many violent incidents stem from people being apprehended for fare evasion, and Revenue Protection Officers are identified as being at the greatest risk of violence of any group of railway industry employees. Station platform staff, guards and ticket gate staff form the front line response to members of the public frustrated with the delivery of operations and disruption to services. Other work-related violence incidents are linked to over-crowding or to assailants offering more challenging behaviours related to: alcohol, drugs, mental health or anti-social attitudes. Decisions to reduce staffing levels and, in particular, permitting access to the rail network from unstaffed stations have anecdotally featured strongly as factors contributing to incidents.

## Definition of Violence

HSE defines violence as any incident in which an employee is abused, threatened or assaulted by a member of the public in circumstances related to their work. This definition is consistent with that used by the TOCs, where violence is viewed as covering three categories: physical assaults, threats and verbal abuse.

## The Inspections – Method

The inspections of six TOCs considered what measures should be, and were in fact, in place to prevent or control work-related violence. This included arrangements and measures for preventing incidents, and measures that would protect the individual in the event of violence being offered, or would reduce the impact on the individual. The 'real-life' implementation of the arrangements which TOCs claimed were in place was explored with front-line staff, employee representatives and those in associated managerial functions. A question set was used to structure the discussions. Approximately 125 interviews were carried out to establish if the arrangements in place were used and sufficient. Findings of interviews were grouped to form case studies of situations where violence had been directed towards specific groups of workers or had occurred repeatedly at a particular location.

## Management of Violence

Arrangements for managing work-related violence are in principle the same as those needed to manage any other workplace health or safety risk, and have for many years been set out in HSE's publication 'Successful Health and Safety Management' (HSE, 1997). In outline, these are policy, organisational arrangements, planning and implementation (of control measures), measuring performance, audit and review (POPMAR for short).

## Policy

HMRI has published criteria in support of the Railways (Safety Case) Regulations 2000, defining what evidence is needed to make an adequate demonstration of the case for safety on particular topics. Criterion 7.5.9 states that a TOC should have a policy describing its commitment on managing violence to staff and setting out the management arrangements in place. The inspections found that some TOCs had no formal, written policy statement in place. However, there was *some* evidence of a company line or working position being operated and communicated through each company, albeit sometimes rather informal. There was no deficiency in senior management commitment to tackling the issue of violence towards staff, or neglect in implementing the stated policy arrangements to manage violence.

Policy statements typically recorded:

- zero tolerance approach towards assailants;
- the intention to give support and assistance in prosecuting assailants;
- the intention to provide appropriate support in the aftermath of an incident;
- the intention to reduce crime levels within stations or on trains by gathering improved, detailed information to allow preventive action to be taken by the British Transport Police (BTP).

The opinions of the front-line interviewees suggested there was insufficient publicity given to prosecutions arising out of violence towards staff. There was a frustration that establishing sufficient evidence to take a prosecution was very difficult and often not possible. Several TOCs assigned someone the role of supporting those involved in a prosecution because lengthy time demands led to a general reluctance to be involved. Some commented that resultant fines were disappointingly low. Nevertheless, all staff were aware that the BTP would take action, where possible, and that this was supported by their employer.

**Arrangements**

The frequency of violent incidents and the increasing level of violence being used were of concern to the rail companies involved in the current audits. Nevertheless, the true scale of violent incidents is not known. The reality is that verbal abuse is an everyday part of the job to many of the people interviewed and the reporting systems in place are not suitable for logging the significant number of such incidents that are experienced. Two TOCs had run local initiatives to count the incidence of verbal abuse at stations, and successfully used this as the basis for altering BTP deployment to ensure better targeting of the problem in the medium term.

Incidents of violence viewed as more serious, typically physical assaults, are generally reported using the same reporting and investigation systems in place for other personal injuries. Whilst generally this information system worked well, it was reliant on the quality of information being provided, and a level of commitment to recommend sound measures to prevent a recurrence. TOCs usually investigate incidents at local level, and periodically the quality of the incident investigations will be reviewed and audited by the TOC central safety department.

The range of measures recommended in incident investigation reports to prevent future violent incidents was found to be extremely limited, too often simply recommending further training to be provided. There was a serious lack of understanding by investigators as to what was effective and what could be done to prevent and respond to incidents. Recommended measures were likely to focus on the immediate errors leading to the incident and not include latent decisions or controls at all – for example decisions to reduce staffing levels at stations or have unstaffed stations where assailants can easily board trains. Alternative measures such as radio provision, greater use of CCTV monitoring, improved information on train services, improved levels of illumination or moving the position of a temporary barrier point were not apparent. Worse, example investigations were found leading to a report being placed on an individual's personal file and the person involved attending conflict avoidance training. Both actions served to give the staff member the message that they were to blame.

It was viewed as good practice to work in partnership with others. All TOCs inspected were running initiatives with the BTP and realising significant benefits from doing so. Successful liaison with local schools, community groups and local authorities were described. However, partnership with trade union safety representatives was less well developed.

Risk assessments to prevent work-related violence were of mixed quality and emerged as an area where improvements were needed. Risk assessments tended to be generic, and were not provided to local managers who could profitably have used them as a basis for site-specific assessments and to assure safety standards.

Issues associated with the delivery of operations are frequently a contributory cause of many reported incidents of violence. The provision of sufficient, relevant, accurate, timely information to ease frustrations in the event of delayed/cancelled trains or overcrowding was repeatedly cited as being a key element in eliminating the potential frustration and hence onset of violence. Hence, it was a positive finding of the inspections that TOCs were pursuing a number of initiatives to enhance communication systems. Increasing the number of information screens and improvements in information accuracy were viewed positively by front-line staff. The screens supplemented the individual's knowledge of the timetable, so aiding their competence in dealing with the public.

Arrangements for ensuring individuals' competence with dealing with the public began with recruitment and included sessions on avoiding confrontation both at induction and in training. It was one element considered by their supervisors and line management in individual performance appraisals.

Training in managing conflict or conflict avoidance aimed to reduce the incidence of physical or verbal abuse. Staff are taught to handle:

- difficult conflict situations with customers;
- complaints in a professional manner;
- antisocial customers, including people trying to avoid paying fares.

Staff are taught the skills of analysing a potentially difficult situation, reading and reacting to body language and listening and responding to customers. Role-play is used to support work books. The basis of course content was periodically refreshed by feedback from course evaluation and reviews of the company's incident data. For instance, further case study material was inserted to improve training in dealing with people under the influence of drugs and/or alcohol.

There is no blue-print for defining the arrangements for contracted staff, the extent of responsibilities of each party being dependent on the circumstances. Both employers and the TOC have legal duties under health and safety law to ensure the safety of contract staff. Further information is available in 'Use of Contractors – a Joint Responsibility' (HSE, 2002c).

Clearly, suitable contract staff should be recruited, and they should be given information and training to ensure their competence. Risk assessments should be prepared and mechanisms for incident reporting and investigations established. The inspections identified the management of contractors as the area of greatest deficiency in terms of providing arrangements for countering the risks from work-related violence.

After a series of incidents to its contract barrier staff, one TOC had started to take an active part in their recruitment and took on the role of providing their training. This was developed to drawing up a service level agreement to clarify the employer and TOCs responsibilities. It would be expected that if TOC employees are issued with a leaflet or pocket card, then such information should also be provided to the contract staff.

One contract ticket checker interviewed showed the reverse of his route timetable where he kept a note of the date, time and basic description of the incidents of violence he had experienced. He and his colleagues had not been told to whom they should report such incidents. His TOCs-employed supervisors did not appreciate they too should have initiated an incident report. Experience of violence, here, was said to be a reason for a very high turnover of staff and absenteeism. The TOC had embarked on the contract with the expectation of providing some 15 minutes of training in conflict avoidance. It was very quickly apparent that this had to be lengthened and substantially developed along the lines of their own staff training, given the extent of the staff turnover, which was proving to be costly. These instances, once apparent, were readily rectified by the TOC and the contract staff employers working together.

## Implementing Control or Response Measures

A number of points of good practice were evident in how TOCs ensure staff are protected from the risks of violence. In particular these were a zero tolerance stance towards assailants, the operation of business standards to support victims of violence, working in partnership with the BTP and utilising a wide range of physical and managerial measures.

TOCs are aware of the scope for successful recruitment and actively try to identify and de-select individuals who would be prone to being involved in incidents of violence. Individuals with a 'rules-bias' or with less well-developed interpersonal skills were perceived to be at greatest risk. Qualities looked for included an absence of apathy, suitably low level of aggression, genuine concern for quality of passenger journey and the ability to handle information correctly.

Techniques for avoiding or managing conflict were typically covered within the company induction period. The quality of this training, its evaluation and more importantly its application, were found to be mixed. In some TOCs interviewees described the content of their course and situations when they had applied it very positively. Others openly criticised inappropriate techniques that had been taught. The lack of relevance to working situations was a widespread criticism, especially for those involved in ticket office sales. Discussion focussed on the tactic of pulling down the blinds/screen to abusive members of the public and how this made matters worse. The instruction given to one group of barrier staff was to walk away from any confrontational situation and go to a place where they felt safe. Very often this was not realistic or possible and since they deliberately worked in pairs, would have left the sole remaining staff member at risk. Attendance at conflict avoidance training was not considered a core competency; many front line staff had never attended and saw their attendance necessitated only in response to their involvement in an incident. Refresher training was not undertaken, unless this need was (either justifiably or unjustifiably) identified following an incident.

Increased staff presence was widely voiced as an initiative that had impact in terms of preventing incidents of violence. Companies implemented this in a multitude of ways. The advent of Travelsafe Officers (TSOs) was particularly noteworthy and was a scheme being piloted by one TOC but receiving significant interest from others. The TSOs are employed by the TOC but strongly linked to the BTP, such that they booked on with the BTP, used BTP intelligence to determine their deployment, travelled on trains to maintain a staff presence and even wore a uniform which is a 'cross' between that of the TOC and the BTP. A parallel was drawn between their role and that of a Park Warden. They are not security guards but work with the BTP, bringing a resource of locally secured, front-line information. They do not have powers of arrest or detention, but are empowered to eject people from the railway and enforce the railway byelaws. Individuals were confident that initial indications point to TSOs having a beneficial impact for both the railway and the local station retail outlets.

Other companies arranged blitzes on a route or 'hotspot' location, deploying their Revenue Protection staff or additional security staff to travel to 'hotspot' locations or on 'troublesome' services. The installation of automatic ticket gates and manning of previously unmanned gates was seen as beneficial in reducing both fare evasion and the associated number of violent incidents. The uniformed presence of the BTP was strongly recognised by front-line staff as beneficial in making passengers/members of the public think twice about their actions.

The perception was that any action that lessened passenger frustration had the knock-on effect of reducing the potential for work-related violence. Replacing ticket machines that delivered multiple tickets too slowly or reorganising work to reduce queue length were noted as beneficial and had a bearing on the levels of violence to staff.

A range of measures are used by the TOCs to prevent, protect against or in response to work-related violence. These include:

*Prevention* – applying for Anti-Social Behaviour Orders (ASBOs), posters highlighting that a zero tolerance policy is in place, establishing assaults task force groups, policies on not working alone, installation of automatic ticket gates and intelligence-led deployment of staff or BTP uniform presence;

*Protection* – leaflets or pocket cards indicating what to do to avoid violence, selection at recruitment, training in both communication and conflict avoidance, provision of mobile telephones or radios;

*Response* – operating Chain of Care procedures, counselling, staff briefings to combat under-reporting of violence, support of staff during private prosecutions, presentation of CCTV coverage from both trains and in stations as prosecution evidence.

There was evidence of continuing investment in developing CCTV systems both on trains and in stations. Nevertheless, it was used to record and monitor events rather than provide a controlled response to developing incidents. There was no constant monitoring of CCTV monitors, despite this expectation from many front-line staff. Incidents captured on tape were of significant benefit to the BTP but the quality of systems was found to be far from perfect. For instance: snow blocking one camera view of the car park remained uncleared; poor illumination in one location well known for being a violent hotspot was persistently tolerated; and conversely, bright sunlight prevented any useful information being gathered along the train carriage since its installation some six months earlier. In one ticket hall the camera pointed to the ground because of anxieties over how the CCTV recording would be used, despite a reported incident and a need for information on the assailants.

Further attention is constantly needed to ensure that measures and arrangements are used, are effective and have impact because the inspections identified instances of physical assaults not being reported, or reported but then not investigated, of training not being undertaken or risk control measures not being known of at particular locations. All of these pointers signalled that despite commitment at the top of the company for tackling work-related violence as a health and safety issue, there was more general acceptance demonstrated at a local level that violence was part of the job and therefore to be tolerated.

## Measurement, Audit and Review

It was possible to track paperwork through each of the TOCs that described specific incidents of violence: from the initial incident report, subsequent investigation report and central recording on the industry-wide Safety Management Information System (SMIS), through to consolidated management information. The systems of gathering information were found to work well but it was less evident how this information was analysed for trends and used to improve the working conditions of staff at risk from violence. There was no evidence of specific performance indicators being set. Rather, the inspections found that general intelligence led to greater support for directing resources to specific action, usually in the form of a short-term local initiative to address violence.

## Safety Case Verification

The criteria in support of the Railways (Safety Case) Regulations (2000) referred to earlier require a policy and arrangements for addressing work-related violence. The current inspections took steps to verify that arrangements on the ground were as stated in the relevant RSC and were suitable and sufficient. Adherence to RSCs was broadly in place. The main gap identified was the absence of adequate arrangements for contract staff typically employed at the gateline to undertake ticket-checking duties or in security functions.

## Key Deficiencies

The key findings of the inspections are:

- improvements in the arrangements for contractors in, for example, ticket checking, Security is required. Both employer and the TOC need to agree who will be providing training and the nature of joint responsibility for example who to report an incident to, as well as ensuring that control measures adopted are compatible;
- more work was needed to ensure reporting of verbal abuse was carried out in a proportionate way;
- a concerted effort is needed to provide training on avoiding confrontation and instigate refresher training for all;
- there is scope for more joint initiatives with TU safety representatives;
- understanding on what works is emerging, but needs to be shared;
- there is a need for an improved understanding by investigators about what measures have impact, and how wider recommendations should feed in to securing broader 'whole-picture' improvements;
- there is scope for improved, site-specific risk assessments being made more accessible to/by local managers to ensure they are used, updated and accurately reflect the often widely-varying local picture;

- the quality of control measures often needs improving, for example CCTV quality, the resources given to Chain of Care procedures, the relevance of confrontation training.

## Conclusion

Whilst the audits concluded that TOCs were active in addressing work-related violence, there was scope for improvement and a need for greater sharing of practice in the industry. The signs are that this is beginning to happen with the setting-up of an industry Personal Security Working Group, to establish a website specifically for sharing information about what countermeasures are effective.

Two viewpoints were too commonly found amongst local and middle management of the TOCs. Firstly, of an acceptance that violence occurs for example the assailants bring the situation to the railway and therefore 'what can we do?' The second prevalent viewpoint was that it is the fault of individual staff members and therefore the attitude 'why should we spend time and effort on dealing with this?' was commonly encountered. The more active TOCs were working hard to counter such complacency that was seen as a hindrance to tackling work-related violence.

The inspections confirm that generally TOCs are becoming more committed to ensuring that standards for addressing work-related violence improve and that short-comings identified are addressed with outcomes that have impact.

As a result of the initiative, guidance for HMRI inspectors is in preparation, aimed at targeting areas of weakness for inspection and enforcement activity to secure the necessary improvements. The findings of the inspections will also be publicised to the industry, with the aim of fostering debate about how further improvements can be made. In addition, it is proposed to make the evidence criteria for assessment of those parts of the safety case describing arrangements for managing the violence risk more detailed and explicit.

## Acknowledgements

The authors would like to thank Mr Charles Twitchett and Mr Craig Richman (HMRI) who were involved in the inspections described in the current chapter.

## References

Budd, T. (1999) 'Violence at Work : Findings from the British Crime Survey', HSE and the Home Office, available via http//www.homeoffice.gov.uk/rds/bcs1.html.

Budd, T. (2001), Violence at Work: New Findings from the British Crime Survey 2000', HSE and the Home Office, www.homeoffice.gov.uk/rds/pdfs/occ-violence.pdf.

Department for Transport (1998), 'Personal Security On Public Transport. Guidelines for Operations', http://www.dft.gov.uk/stellent/groups/dft_control/documents/contentserver template/dft_index.hcst?n=10586&l=2.

Di Martino, V., Hoel, H. and Cooper, C.L. (2003), 'Preventing Violence and Harassment in the Workplace', http://www.eurofound.eu.int/publications/EF02109.htm.

Employment NTO (2002), National Occupational Standards in Managing Work-Related Violence, http://www.ento.co.uk/standards/index.php?catalogue=wrv.

Fox, B., Polkey, C. and Boatman, P. (eds) (2002), Managing Violence in the Workplace, Croydon: Butterworth Tolley.

HSE (1997), 'Successful Health and Safety Management', HSG65, Sudbury: HSE Books.

HSE (2002a), 'Violence at Work: A Guide for Employers', HSE IND(G)69(rev), http://www.hse.gov.uk/pubns/indg69.pdf.

HSE (2002b), 'Work-related Violence – Case Studies: Managing the Risk in Smaller Businesses', HSG229, Sudbury: HSE Books.

HSE (2002c), 'Use of Contractors – a Joint Responsibility', HSE IND(G)368, http://www.hse.gov.uk/pubns/indg368.pdf.

HSE (2003a), Railways (Safety Case) Regulations, 2000 including 2001 and 2003 amendments, L52, Sudbury: HSE Books.

HSE (2003b), 'Safety Case Assessment Manual', http://www.google.com/u/HSEC?q=Safety+Case+Assessment+Manual+2003&sa=Go&sitesearch=hse.gov.uk&domains=hse.gov.uk.

HSE (2005), 'Route Crime (Trespass, Vandalism and Assaults)', http://www.hse.gov.uk/railways/liveissues/vandal.htm.

Railway Group Standard (2003), 'Reporting of Safety Related Information', GE/RT 8047, Issue 2, February 2003.Railway Safety Research Programme/Risk Solutions (2002), Public Behaviour. Reducing Assaults on Railway Staff.

Chapter 41

# The Role of the Manager in Stress Management

Emma Lowe

## Introduction

Research into the effectiveness of stress management interventions generally demonstrates how they fall short of their intended purpose. One reason given for this is that little attempt has been made to find out what managers understand by stress and the extent to which they think they, or their organisation, has a responsibility to address stress related problems (Dewe and O'Driscoll, 2002). Indeed, very little research has been undertaken to examine the role of the manager as an intervention in their own right or in the role they play in facilitating other interventions. The purpose of this research, therefore, was to explore manager's perceptions of stress and their role in its management.

The results draw attention to a number of issues. Firstly, there is a role for the manager in identifying and managing stress, which is both direct (in terms of being able to identify stressors and initiate action to remove or reduce the impact of a stressor) and indirect, in terms of the culture engendered in order to encourage open discussion and reporting of the stress experience. Secondly, it is also clear that without a proper understanding of the stress process and the particular context within which it is to be managed, the manager's ability to identify and act can be limited.

Research over the last ten years demonstrates that job stress is a potentially serious problem with implications for both individuals, in terms of ill health, and for organisations, in terms of productivity and the costs associated with legal claims. This has led to a great deal of research about stress, in particular how it can be managed.

### A Stress Management Intervention (SMI) Framework

Researchers have over the years attempted to classify different types of stress management interventions (SMI),[1] which is useful in helping to understand the broad context of stress management and what the organisation is aiming to achieve by implementing SMIs. The classifications distinguish between the *focus* and the *aim* of the intervention. The focus tends to be either individual or organisational (De Frank and Cooper, 1987). Individually focused interventions typically include stress management training, counselling or health awareness initiatives. Organisational level interventions are generally concerned with workload, working hours/patterns, job design, reducing role ambiguity/conflict, improving the working environment and increasing social support. The aim of the intervention is usually either to prevent stress; react to stressful problems, such that the individual can cope with it; or treat the effects of stress (Cox, 1993; Reynolds, 1997).

Interventions are classified as:

- *primary interventions* which are concerned with managing the sources of stress in the work environment and will generally involve job redesign or workload reduction;
- *secondary interventions* which are geared toward helping employees to cope more effectively typically through stress management training programmes;
- *tertiary interventions* which are concerned with the treatment, rehabilitation and recovery of employees who have suffered, or are suffering from, serious ill health as a result of stress and typically involve provision of counselling and Employee Assistance Programmes (EAPs).[2]

*Evidence Supporting the Effectiveness of SMIs*

Recently evaluations of the effectiveness of such interventions have been more critical and not as clear-cut as the stress management business would like us to believe. For example, Parkes and Sparkes (1998) in their review of organisational interventions concluded that, taken as a whole, the research 'did not present a convincing picture of the value of organisational interventions'. Moreover, research suggests that while secondary interventions, such as stress management training, can improve employees' ratings of their psychological well-being, the effects are usually small and temporary (Sallis *et al.*, 1987; Murphy and Sorenson, 1988; Ganster *et al.*, 1982; Reynolds *et al.*, 1993). One reason given for this is that training may produce a non-specific 'feel good' factor associated with the process of participation rather than with specific content – in other words, a placebo effect (Bunce, 1997). Supporting this argument is the fact that many participants in these interventions are self-selected (Sutherland and Cooper, 1990). This raises a question about whether in fact these interventions cater more for the 'worried well' than the truly 'stressed'.

Tertiary interventions also have mixed success. There is evidence to suggest that tertiary interventions such as counselling have a positive impact on individual well-being (Briner, 2000; Cooper and Cartwright, 1994; Cooper and Sadri, 1991). However, there is little clear evidence to suggest it has a positive impact on objective work variables such as absence and performance (Briner, 2000). The main difficulty is the simplistic assumption that individual well-being, attitudes to work and work behaviours are linked and this relationship is the same for everyone. Evidence to support the more general concept of an EAP is described as 'embarrassingly thin, largely anecdotal and mainly American' (Carroll, 1996). The American bias is problematic in the sense that American EAPs are primarily concerned with programmes concentrated on drugs and alcohol abuse making it difficult to generalise to mental health counselling programmes in other countries (Arthur, 2000).

Researchers have offered a variety of reasons as to why interventions have been less than effective. However, the single most important reason why interventions appear to be limited is that they are not underpinned by a clear definition or explanatory model of stress.

What we mean by stress and how we explain the effect it has on the individual and organisation clearly has an impact on the identification of stress in the first instance and in the design, implementation and evaluation of an intervention in response (Dewe, 1994; Briner and Reynolds, 1993). Generally organisations are left to simply implement 'off the shelf' interventions such as EAPs or stress management training so it can be seen to be doing something but without really understanding what the problem is. In this way the organisation ends up with a 'non-specific management solution to an ill-defined problem' with the result that significant 'stress problems which do exist are likely to remain unexamined and undiagnosed' (Briner, 1996; Dewe and O'Driscoll, 2002).

## Investigating the Role of the Manager in Managing Stress

Dewe and O'Driscoll (2002) specifically saw the lack of a clear definition and explanatory model of stress as the main reason for organisations' tendencies to either do nothing or focus on treating the individual (McHugh and Bryson, 1992). Their study revealed that managers were able to identify signs of stress but listed 'response-based' behaviours which they argued indicated an over simplistic representation of stress. Further, they found issues regarding who should be responsible for stress management. While managers believed the individual had considerable responsibility, they also acknowledged that the organisation did.

Overall, the role of the manager is one area that appears to have received limited attention in the stress management literature. This is suprising given the potential impact they can have in identifying stress and in ensuring the implementation of various SMIs or indeed acting as an intervention themselves. Therefore the purpose of this research was to explore the role of the manager in stress management in more detail. Specifically, it aimed to understand:

- what managers perceive stress to be;
- how they identify it;
- what they perceive its impact on performance to be;
- what responsibilities they perceive they have for its management;
- what actions they take.

It was conducted within Network Rail, the company responsible for the safe operation of the rail network and for maintaining and improving the railway infrastructure. It employs 13,000 employees, 7,000 of whom are signallers who are responsible for the movement of 25,000 trains every day. Under the Railways (Safety Critical Work) Regulations 1994, signallers are defined as undertaking safety critical work, which is any work that affects the movement of trains. The company has specific legal responsibilities under these regulations for ensuring that the individuals are fit (physically and mentally) to work and that their performance is not impaired. To that end signallers and their managers are a relevant group to study.

## The Research

*Methodology*

Data were collected from signaller managers, asking a mix of open-ended and closed questions during face-to-face interviews. The questions were grouped into four parts:

*Part 1* included questions to obtain some basic biographical data about their age, experience on the railway and experience in signaller management.

*Part 2* was concerned with the nature of stress. Two types of question were asked. The first involved managers rating different stressors, as taken from the Railway Ergonomics Questionnaire (REQUEST),[3] on a five-point scale from 'no stress at all' to 'a great deal of stress'. They were asked to comment on any that they did not understand, or that they felt needed qualification. The second involved asking managers the following open-ended questions:

- for me personally stress is …;
- what do you think are the signs that signallers are under stress?
- in what ways do you think being stressed impacts on a signaller's performance?

Whereas the question about the signs of stress was aimed at eliciting information about how managers identified stress, the question about its impact on performance was designed to establish whether managers saw links between stress and the ability to perform the job safely, an area that seems to receive little attention in the literature on stress outcomes.

*Part 3* included questions relating to a manager's perceptions about responsibilities for stress management and one question relating specifically to manager's perceptions about who signallers would turn to for support for when stressed, in order to establish whether managers were part of a signaller's coping network.

*Part 4* was concerned with stress management interventions in terms of:

- frequency of use of stress management interventions by signallers;
- reasons for use/non-use of interventions;
- suggestions for improving signaller's use of interventions;
- actions managers take in the event of a report or observation of stress;
- effectiveness of the actions taken and why they may be ineffective;
- interventions managers would put in place if they were responsible for stress management.

*Findings and Discussion*

The primary aims of the research were to explore manager's perceptions of stress and their role in its management. Three main conclusions can be drawn from the results. Firstly, there is a clear need for a context specific theory of stress. The absence of one has implications for how stress is identified and managed. Secondly, in order to understand the management of stress, attention needs to be given to the issue of responsibility for stress management. Finally, there is an important role for the manager in facilitating a culture whereby stress is an openly discussed issue.

*The nature of stress and the need for stress theory*

In general the results suggested that signaller managers tended to have a simplified and in some cases incomplete understanding of the concept of stress. The majority defined it in simple stimulus-response terms, although some were at least able to recognise that it could depend on the individual and there may be some mediating variables. For example, not having enough to do is only a source of stress if it leads to feelings of boredom and an uncomfortable work environment is a source of stress because it leaves signallers feeling undervalued.

Interestingly, 9 per cent defined stress in terms of a fad. This suggests a sceptical view about stress that, while not significant, was a consistent theme throughout the study. It was mentioned again in relation to the difficulties in identifying stress because some use it as an excuse when a mistake has been made and others 'jump on the stress bandwagon'. Such comments could be an example of why having a broad definition of stress and no underpinning theoretical model is so limiting. Stress can be used to cover a range of ills making it difficult for the manager to identify genuine cases.

The traditional stimulus-response definition of stress, prevalent in this study, limits the way in which stress is assessed. Firstly this conceptualisation encourages managers to consider only one component of the stress process and not the process itself. Consequently, focus is given to the range of possible situations and responses rather than their inherent properties such as their frequency, duration, demand, intensity or severity for example (Cooper *et al.*, 2001). Indeed, only one manager recognised that the stress response may be dependent not just on the stimuli but the intensity and duration of the stimuli, distinguishing between chronic and acute stress.

Secondly, an incomplete understanding of stress has implications for the assessment of stress in terms of how it is recognised in the first place. Only 33 per cent of managers indicated that a sign of stress was ill health.

> I always assumed it came out when people crack up under pressure but I've just had a signaller go off with stress which has caused dermatological problems ...

They also failed to recognise that some of these responses may actually be legitimate coping strategies or that the type of outcome might be specific to the stimulus. For example, it has been suggested that there are different outcomes for stressors that are short-term or one-off events (episodic stressors) and those that are on-going (chronic stressors) (Bailey and Bhagat, 1987). There is also the issue of causality. Does stress cause these responses or are the responses adding to the levels of stress (Briner and Reynolds, 1993)? A model of stress that specifies how different stressors have an impact and includes consideration of the coping strategies used is needed.

*Stress management responsibilities*

The data presented in Table 40.1 reveal that managers perceive that they, the organisation and the individual all have responsibilities for managing stress. On the face of it this seems to be very positive. However, it does have some implications

*Rail Human Factors*

**Table 41.1 Individual, organisational and managerial responsibilities for stress management**

|  | None at all | A little | Some | Quite a lot | Totally responsible |
|---|---|---|---|---|---|
| If a signaller is showing signs of stress, to what extent is it their own responsibility to do something about this stress? | 0% | 0% | 33% | 67% | 0% |
| To what extent does the organisation have a responsibility to address problems of stress? | 0% | 0% | 19% | 67% | 14% |
| To what extent is it the line manager's responsibility to manage stress? | 0% | 0% | 29% | 71% | 0% |

for how managers discharge their responsibilities. The manager's comments as to the nature of the individual's responsibility suggest that they believe the individual's primary responsibility is to recognise that they may be stressed and bring it to their manager's attention, once again implying a passive role for the manager.

> They've got quite a lot of responsibility because they are the only ones who can bring it to a manager's attention. Most of the time its very difficult to spot whether someone is stressed or not, you think they're coping.

This supports the 'dual responsibility' findings of Dewe and O'Driscoll's (2002) study but presents something of a paradox. On the one hand managers recognise they have responsibilities, but on the other they are suggesting that it is up to the individual to come forward and report that they are experiencing stress. On this basis, it is not suprising that 'making signallers more aware' of the SMIs available to them and 'creating a culture which supports open reporting' are rated highly as interventions managers would like to see implemented.

However, given that the results also suggest that the manager is unlikely to be the first person the signaller turns to when stressed, discharging their managerial responsibilities could be quite difficult. To some extent this is also recognised by the manager, hence the frequency with which they suggest training and the need to develop the appropriate relationship with their employees as interventions.

Another interesting finding regarding responsibilities was that even where managers demonstrated some understanding of the need to remove the stressor, they felt they had little control over doing so. The frustration they felt at not being able to provide a comfortable working environment because of lack of resources and an overly bureaucratic process is a good example.

Of course within organisations there are usually layers of management and it is clear that further consideration needs to be given to the responsibilities of others in the organisation and not just the immediate manager. The importance of other key stakeholders is currently recognised in the literature on stress management to the extent that it is acknowledged that organisational context can have an impact on the effectiveness of SMIs, particularly organisational interventions. This is not suprising, organisational interventions by their nature are far more complex and usually require structural change that typically involves more than just the individual and their manager.

*Management action*

The rational for exploring what managers understood by stress, and what they rated as stressors for signallers, was that if there was no clear understanding about stress and its causes then managers would be limited in the actions they took. The nature of this study does not enable causality to be established. However, on the basis that the managers interviewed demonstrated an incomplete understanding of stress and then went on to focus on actions that primarily involved 'treating' the individual, it is not unreasonable to suppose the two are related. If stress was perceived as a transaction between individual and environment (Lazarus, 1991) rather than a stimulus and/or response, then more attention may have been given to interventions that focussed on coping skills or removal of the stressor altogether. Instead the most frequently cited actions managers said they would take were:

- 'having a chat' to identify what the problem might be;
- referring them to specialist support such as sending them for a medical or telling them about Care First, the organisation's EAP;
- Giving the individual experiencing stress time off or more flexible working arrangements.

> It's also good that they can go and get specialist help from somewhere. I mean I don't know what to say to a guy who's just lost his wife, I've never been through that but that's hopefully what Care First does.

When it came to suggesting what SMIs the organisation might implement, management training was mentioned a frequently mentioned intervention. This suggests that managers do not believe they currently have the appropriate knowledge and skills to discharge their responsibilities appropriately. This could also account for why they perceive the individual's responsibilities to be high and concerned with reporting stress.

There was a sense that managers rated some of the SMIs as 'should be implemented' simply because they were good things to do (that is, changed management style, career development programmes and performance evaluation), highlighting another potential problem with an inadequate understanding of stress. If managers do not understand the complexity of the stress process, there could be a tendency to choose those interventions that focus on employee well-being when in fact such interventions are essentially just good human resource management (Briner and Reynolds, 1993). However, if interventions are to reduce/remove stress they need to be designed with a specific stressor-strain relationship in mind and an understanding of why and how a stressor leads to strain, rather than being based on the assumption that a single intervention will have a universally positive effect as implied by the above (Briner and Reynolds, 1999).

*Implications for the Role of the Manager*

In accepting that individuals have a responsibility for stress management by reporting when they are experiencing it, managers take on a specific responsibility for ensuring a relationship with their employees that engenders a 'no blame' culture. Employees need to be able to feel that they can report stress without fear of negative consequences.

Developing a culture of this nature is not dissimilar to what Nytro *et al.* (2000) describe as 'cultural maturity'. While this term is used to describe an organisation's receptiveness to organisational interventions specifically it involves having the 'skills and willingness' to implement interventions. It is interesting to note that a fundamental step in ensuring cultural maturity is the need to 'educate managers and employers about the complex mechanisms behind the development of stress and illness at work'. The need for management training was evident from the frequency with which managers rated this as an intervention that the organisation should implement on the basis that a better understanding of what stress is and what interventions are available would enable them to support their employees more effectively. So, a manager's role can and should include promoting 'cultural maturity' but in order to fulfil this, training is essential.

## Limitations of the Study

There were a number of ways in which this research was limited either by its design or simply because of its scope. The most obvious omission from the study was that it did not include the signallers. This prevented any comparison of their perceptions about stress and its management with that of the managers and means only one half of the story is told. To have included the signallers would have been particularly useful for considering the answers to questions such as who the signaller would turn to if experiencing stress to establish whether, as perceived by the managers, they are more likely to turn to friends at work for support than their manager.

In addition, one of the consequences of the exploratory nature of the research, and its qualitative approach, was a small sample size. This meant it was not possible to assess if there were any statistically significant correlations between some of the issues being explored, such as the relationship between incomplete understanding of stress and the actions taken, or perceived responsibilities and actions taken. However, the study can be used as a basis for a much wider investigation making more use of the quantitative data gathered. With a bigger sample it would be possible to perform some statistical analysis of the data and investigate what some of the mediating variables might be which influence the extent to which the manager sees their responsibility for stress. For example:

• the length of time in a managerial role;

- manager's own experience of stress and well-being;
- manager's locus of control;
- number of staff to be managed;
- previous cases of stress encountered.

## Recommendations for Future Research

The main area for further research should be in understanding stressors and strains in the signalling context and particularly the impact stress could have on performance. While this study demonstrated that managers were able to identify a number of different ways in which stress could impact on performance, managers did not or could not provide any insights about why or how except to suggest that it could be a distraction and something that causes the individual to be preoccupied. This is not dissimilar to the limitations of the research to date in this area. While the dangers of stress are demonstrated by the association between accident risk and exposure to traumatic life events, such as bereavement and divorce (Selzer and Vinokur, 1975, Alkov *et al.*, 1985), existing work provides little information on the mechanisms that might link stress to increased accident risk (Sivak, 1981). The notable exception is the work of Matthews (2002) who recently outlined a transactional model of stress for car driving. In it he identifies specific stressors and suggests how they interact with personality factors to bias the appraisal and coping process which then have specific outcomes: subjective (for example, anxiety, anger and tiredness) or performance (for example, loss of attention, reduced risk-taking and impaired control). Having established that managers believe stress has an impact on safe performance, the application of Matthews' transactional model approach may provide further insights about why and how.

Furthermore, the importance of a model of stress that applies specifically to signalling is highlighted not only by the repeated references throughout this report that to be effective models of stress must be context specific, but also because at least one example was found of a stressor having particular meanings for the rail environment. 'Time pressure and deadlines' is a stressor because of the industry's performance regime.[4] It also prompted the suggestion for a stressor-specific intervention in that some managers saw they had a specific responsibility for acting as a 'buffer' to try and deflect some of the performance pressure.

## Conclusion

Overall, this study has demonstrated that there is a role for the manager in identifying and managing stress which is both direct, in terms of being able to identify stressors and initiate action to remove or reduce the impact of a stressor, and indirect in terms of the culture engendered in order to encourage open discussion and reporting of the stress experience. It is also clear that without a proper understanding of what the stress process is for the particular context within which it is to be managed, the managers ability to identify and act can be limited.

## Notes

1    An SMI is defined as 'any activity, programme or opportunity initiated by an organisation, which focuses on reducing the presence of work-related stressors or assisting individuals to minimise the negative outcomes of exposure to these stressors' (Ivancevich *et al.*, 1990).

2    EAPs are 'worksite focussed programmes to assist in the identification of resolution of employee concerns such as personal or work related matters, which affect or may affect performance' (UK EAPA, 1998).

3    The Railway Ergonomics Questionnaire is a tool that has been developed to provide a benchmark of current human factors issues and to evaluate the impact of a major route modernisation project (Wilson *et al.*, 2000)

4    The performance regime for the railways means that signallers must account for their delays. It can cause signallers to experience stress by creating an additional demand in terms of pressure to keep things moving and answering queries from performance staff about why delays occurred. It appears to promote a blame culture resulting in signallers reporting the feeling of being undervalued and inadequately supported by their company.

## References

Alkov, R.A., Gynor, J.A. and Borowsky, M.S. (1985), 'Pilot Error as a Symptom of Inadequate Stress Coping', *Aviation, Space and Environmental Medicine*, 56: 244–7.

Arthur, A.R. (2000), 'Employee's Assistance Programmes: The Emperor's New Clothes of Stress Management?', *British Journal of Guidance and Counselling*, 28 (4): 549–59.

Bailey, J. and Bhagat, R. (1987), 'Meaning and Measurement of Stressors in the Work Environment: An Evaluation', in S.V. Kasl and C.L. Cooper (eds), *Stress and Health: Issues in Research Methodology*, New York: John Wiley.

Briner, R.B. (1996), 'How Could "Improving" Psychological Work Conditions "Improve" Employee Well-being', paper presented at the conference *Work and Well-being: An Agenda for Europe*, Nottingham.

Briner, R.B. (2000), 'Do EAPs Work? A Complex Answer to a Simple Question', *Counselling at Work*, 29: 1–3.

Briner, R.B. and Reynolds, S. (1993), 'Bad Theory and Bad Practice in Occupational Stress', *Occupational Psychologist*, 19: 8–13.

Briner, R.B. and Reynolds, S. (1999), 'The Costs, Benefits and limitations of Organisational Level Stress Interventions', *Journal of Organizational Behaviour*, 20: 647–64.

Bunce, D. (1997), 'What Factors are Associated with the Outcome of Individual-focused Worksite Stress Management Interventions?', *Journal of Occupational and Organisational Psychology*, 70.

Carroll, M. (1996), *Workplace Counselling*, London: Sage.

Cooper, C.L. and Cartwright, S. (1994), 'Stress Management and Counselling: Stress Management Interventions in the Workplace: Stress Counselling and Stress Audits', *British Journal of Guidance and Counselling*, 22 (1): 65–73.

Cooper, C.L., Dewe, P.J. and O'Driscoll, M.P. (2001), 'Organisational Stress: A Review and Critique of Theory, Research and Applications', *Foundations of Organisational Science*, Sage Publications.

Cooper, C.L. and Sadri, G. (1991), 'The Impact of Stress Counselling at Work', in P.L. Perrewe (ed.), *Handbook of Job Stress*, Special Issue of *Journal of Social Behaviour and Personality*, 6: 7.

Cox, T. (1993), *Stress Research and Stress Management: Putting Theory to Work,* Sudbury: HSE Books.

De Frank, R.S. and Cooper, C.L. (1989), 'Worksite Stress Management Interventions: Their Effectiveness and Conceptualisation', *Managerial Psychology,* 2: 4–10.

Dewe, P. (1994), 'EAPs and Stress Management: From Theory to Practice to Comprehensiveness', *Personnel Review,* 23: 21–32.

Dewe, P. and O'Driscoll, M. (2002), 'Stress Management Interventions: What Do Managers Actually Do?', *Personnel Review,* 31 (2): 143–65.

Ganster, D.C., Mayes, B.T., Sime, W.E. and Tharp, G.D. (1982), 'Managing Organisational Stress: A Field Experiment', *Journal of Applied Psychology,* 67: 533–43.

Ivancevich, J.M., Matteson, M.T., Freedman, S.M. and Phillips, J.S. (1990), 'Worksite Stress Management Interventions', *American Psychologist,* 45: 252–61.

Lazurus, R.S. (1991), 'Psychological Stress in the Workplace', in P.L. Perrewe (ed.), *Handbook of Job Stress,* Special Issue of *Journal of Social Behaviour and Personality,* 6: 1–13.

Matthews, G. (2002), 'Towards a Transactional Ergonomics for Driver Stress and Fatigue', *Theoretical Issues in Ergonomics Science,* 3 (2): 195–211.

McHugh, M. and Bryson, I. (1992), 'Will the 21st Century Bring Healthier Profits Through Healthier People?', paper presented to the British Academy of Management Conference, Bradford, September.

Murphy, L.R. and Sorenson, S. (1988), 'Employee Behaviours Before and After Stress Management', *Journal of Organizational Behaviour,* 9 (2): 173–82.

Nytro, K., Saksvik, P.O., Mikkelsen, A., Bohle, P. and Quinlan, M. (2000), 'An Appraisal of Key Factors in the Implementation of Occupational Stress Interventions', *Work and Stress,* 14 (3): 213–25.

Parkes, K.R. and Sparkes, T.J. (1998), *Organisational Interventions to Reduce Work Stress: Are They Effective? A Review of the Literature,* Sudbury: HSE Books.

Reynolds, S. (1997), 'Psychological Well-being at Work: Is Prevention Better Than Cure?', *Journal of Psychosomatic Research,* 43 (1): 93–102.

Reynolds, S., Taylor, E. and Shapiro, D.A. (1993), 'Session Impact in Stress Management Training', *Journal of Occupational and Organisational Psychology,* 66: 93–113.

Sallis, J.F., Trevorrow, T.R., Johnson, C.C., Hovell, M.F. and Kaplan, R.M. (1987), 'Worksite Stress Management: A Comparison of Programmes', *Psychology and Health,* 1: 237–55.

Selzer, M.L. and Vinokur, A. (1975), 'Role of Life Events in Accident Causation', *Mental Health and Society,* 2: 36–54.

Sivak, M. (1981), 'Human Factors and Highway-accident Causation: Some Theoretical Considerations', *Accident Analysis and Preventions,* 13: 61–4.

Sutherland, V.J. and Cooper C.L. (1990), *Understanding Stress,* London: Chapman and Hall

UK EAPA (1998), *UK Guidelines for Audit and Evaluation for Employee Assistance Programmes,* London: UK EAPA.

Wilson, J., Nicholls, S. and Cordiner, L. (2000), 'Railway Ergonomics Assessment Package', in D. Harris (ed.), *Engineering Psychology and Cognitive Ergonomics,* Vol. 6, Ashgate Publishing Ltd.

# Chapter 42

# Getting Passengers Out – Evacuation Behaviours

Louis C. Boer

## Introduction

When disaster strikes, mass transportation means mass evacuation. The issue is especially urgent if, despite precautions, a train comes to a stop in a tunnel and there is a fire. Appropriate behaviour of passengers is a major success factor for an evacuation. Passengers should replace their original (travel) goal with the evacuation goal and – once the new goal has been established – know what to do, what to take with them and where to go. The information the passengers need (by PA, instruction by personnel, illumination, signage, etc.) should be designed so as to reduce any behaviour problems. The study reported in this chapter, based on (unannounced) evacuation exercises in a road tunnel, examined possible behaviour problems, together with remedies. The psychological issues which may prevent train personnel from taking decisive action are also discussed.

## Getting Passengers Out – Evacuation Behaviours

Mass transportation means that incidents and accidents can lead to many fatalities. Fires in tunnels are notoriously dangerous. Hot and toxic fumes cannot escape in the open air but are trapped in the same tube as the passengers. Many remember the fires in 1999 in the Gothard, Mont Blanc and Tauern tunnels (see Voeltzel, 2002). The main conclusions are that people will stay in their cars as long as they do not recognise the danger from the threat of fire.

Mass transportation is on the increase, bringing a greater number of tunnels. In The Netherlands, the high-speed rail link Amsterdam–Brussel–Paris will start with a 4.4 mile bored tunnel under the Dutch wetlands. The light-rail link between Rotterdam and The Hague starts with a 2.5 mile tunnel in Rotterdam. Amsterdam Underground will soon have its North–South line sometimes 40m below street level.

All kinds of precautions are made to prevent fire in a tunnel – such as use of inflammable materials – and there are measures to prevent the train from stopping in a tunnel. In the unlikely case that disaster strikes, passengers should still have a chance of surviving. The usual solution is to construct escape ways that lead to safety. Tunnels of the 'cut-and-cover' type may have a small middle channel which can be used as an escape footpath. The footpath runs parallel to the tubes and ends in the open air, together with the tubes. Most bored tunnels are without a separate escape footpath (an exception is the El-Azhar road tunnel in Cairo where the escape footpath is the cable channel *under* the road). Bored tunnels have cross corridors between the two tubes. People can leave the 'disaster tube' through a cross corridor, and find refuge in the other 'safe' tube. It is obviously essential to stop all traffic in the safe tube before people run into it. Self-closing fire-doors and pressure differences are used to keep escape ways free from smoke and heat.

Will people actually find and use the escape ways in emergencies? There are reports of emergency exits that were ignored; of people remaining where they were or returning where they came from. Why is this so? Why is emergency behaviour sometimes so woefully inadequate?

A popular belief is that *panic* is to blame. Panic would be the magic wand that turns people into cattle. In the 1950s in the UK, the prevailing government view was that 'Panic is when an assembly audience results in a crowd jamming the exits and causing injuries quite apart from injury by fire'. This is a *stampede* definition of panic. Sime (1995) discusses the conception of panic (see also Proulx and Sime, 1991). The evidence of panic is often said to be a sense of self-preservation at all costs, characterised by asocial or non-social behaviour in which even family ties break down. Sime shows that the crowd, far from being a homogeneous mass of individuals who necessarily ignore each other, maintains close ties to the group prior to and during flight to exits. In a situation of potential entrapment these ties increase in strength. He concludes that panic is mostly a rational reaction to the circumstances at the time and therefore that improving the circumstances would avoid both panic and the problems that result. Summarising 50 years of research, Quarantelli (1999) comes to similar conclusions.

Panic can be used to describe people's desperation in a situation that is desperate. If the situation is life-threatening, and if the possibilities for escape are perceived as non-existent or poor, that situation is desperate or hopeless. The best way to avoid panic is to avoid desperate situations occurring, rather than providing a *don't panic* sticker. Panic should be avoided by (a) warning the people early, before the fire is out of control; (b) offering them escape ways; and (c) showing them where the escape ways are.

Early warning, the first factor in evacuation safety, is not always possible because some accidents develop like explosions. The fire in the Volendam pub in The Netherlands in the early morning of 1 January 2001 heated the air up to 900 Celsius in 50 seconds. Almost all of over 300 occupants suffered burns, 70 sufferred stampede injuries and 13 died. The other two factors in evacuation safety – construction and marking of escape ways – are *always* possible. They are the responsibility of the authorities.

Ordinary goals and behaviours can make warnings ineffective. People have goals of their own, and are used to keep focussed on their business. They will protect their goals and behaviours from interruptions and distractions – the homeostasis principle. They will discard the early signs of danger as 'another rehearsal exercise' or similar, and continue life as normal. Moreover, early signs of danger are often ambiguous, especially for lay people who, for example, cannot imagine the exponential escalation of fire. In Figure 42.1, the behaviour resulting from the first down arrow 'disaster', is not always effective.

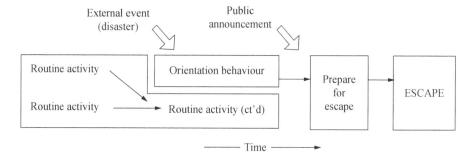

**Figure 42.1   The genesis of evacuation behaviour**

Note: Signs of the disaster lead to orientation but not necessarily to giving up daily routines.
      People give up daily routines and decide to find refuge after a threshold of perceived
      danger is exceeded. Public announcements – or other official warnings – help to bring
      this decision about.

It follows that the authorities must help the people by providing usable and accurate
information and direction on where to find safety – the second down arrow in Figure
42.1. Austrian Railways recommends that the train driver makes the following
announcement: *Achtung, achtung! Eine Durchsage Ihres Zugchefs! Wir haben
ein Brand im Zug! Lassen Sie Ihr Gepäck zurück! Wir müssen den Zug verlassen.
Steigen Sie zum Randweg aus! Helfen Sie sich gegenseitig! Ich wiederhole! Lassen
Sie Ihr Gepäck zurück! Helfen Sie sich gegenseitig!* [Attention, attention, this is your
train driver speaking. We have a fire in the train. Leave your luggage behind. All
passengers have to leave the train. Get out at the platform. Help those who need
assistance. I repeat: Leave your luggage behind; help those who need assistance.]
Austrian Railways gives no indication what to do after alighting. Dutch Railways
recommends that the train personnel instruct the passengers to alight at the side of
the platform, and to follow the signs 'emergency exit'. The recommendation does not
mention the word *fire*. It also relies on the signage along the tunnel walls, assuming
that this offers sufficient guidance to the passengers.

**What Will People Do?**

If you want to guide people, it helps to know their sensitivities, concerns, and state
of mind. Wrong or inadequate knowledge of these makes guidance less effective. If
the train driver imagines that the word *fire* will create panic among the passengers,
he will avoid the word in his announcement (he will not inform the passengers) or,
worse still, he may postpone the moment of announcement.

Often, it is *assumed* that people will react in some particular way. Some form of
*analogical reasoning* can be applied (Meister and Enderwick, 2002: 123): a specialist
identifies with the public, and substitutes his knowledge of or feeling for how the
public would react for the missing or ambiguous data. Such assumptions can be right
or wrong – witness the panic myth. The thesis of this chapter is that *observation* of

people's behaviour in (almost) real emergency situations is what is required to gauge how people react, and what they will do in their evacuation behaviour.

A study of people's behaviour when unexpectedly faced with a burning HGV in a tunnel is presented here. This was in a road tunnel rather than a train tunnel, but the study makes clear what a behavioural test is, and what information comes from such a test.

### Behavioural Test: 'Tunnel on Fire!'

The essence of a behavioural test is that participants have the same sensitivities, concerns, and states of mind as the target population would have in a real situation. Hence, the participants were not told that they would be stopped by a fire and would have to find refuge in the tunnel. The aim was to create a situation as unexpected and unannounced as in real life.

Hence, participants were invited for a test on '(driving) behaviour in a tunnel'. They were told that they were to drive through the tunnel a couple of times; the first time for familiarisation. It was added 'things can happen during a ride; there is no real danger, but behave as you would do under normal circumstances'. They entered the tunnel in groups of 45–50 cars. Then, on the very first ride (their only ride), the simulated fire accident happened. Seven such tests were organised.

Two hundred metres down the tunnel they encountered a HGV driving slowly in the fast lane, its alarm lights flashing. After 300m, and in the middle of the tunnel, the HGV came to a stop across the road, blocking both lanes, smoke billowing from its inside. No instructions were given for five and a half minutes. What motorists did was recorded on video. Figure 42.2 shows the layout of a typical test and Figure 42.3 shows the camera views. After five and a half minutes, 'explosion danger' was announced (repeated once). Two minutes later, 'leave your car' was announced (repeated once).

### Results

The main results of the tests are presented below. A fuller description can be found in Boer and Wittington (2004).

Motorists remained in their cars and took no action. In six out of seven behavioural tests, they just waited for the traffic congestion to clear while an increasing amount

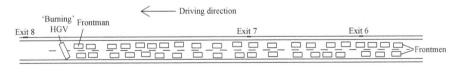

**Figure 42.2   Cars are stopped behind a smoking HGV – an example of a behavioural test**

Note: Motorists can escape through Exits 5, 6, and 7; Exit 8 is past the HGV.

**Figure 42.3   Images of the video cameras**

Note: Date and time at top of images: '8 Jan 2002; 20:46 h'.

of smoke spread throughout the tunnel. After all, 'no-one else did anything'. This agrees with the conclusions of Voeltzel (2002) about behaviour during actual fires in the Mont Blanc and Tauern tunnels.

Another observation was that motorists leaving the car was not enough for an evacuation. In one out of seven tests, motorists started evacuating immediately. The first ones reached the emergency exits while the cars at the back of the congestion were still closing in. But they did not continue evacuating, and motorists returned from the escape footpath into the disaster tube again; some went all the way back to their car. Others used the elevated doorstep in front of the emergency exits as observation posts and five minutes into the fire accident the evacuation had come to a standstill. The public announcement 'explosion danger' (given after five and a half minutes) got people out of the disaster tube.

A short official announcement ('explosion danger', repeated once) was sufficient to get people out in all seven tests. Copying the behaviour of others, or a herd instinct, helped to make the evacuation complete.

Dense smoke impeded evacuation behaviour, probably because it took the herding or group feelings away – people were unable to see their companions. In one particular test, smoke completely enveloped the cars in front (see Figure 42.4; in the other six tests, smoke drifted slowly away from the cars). The message 'explosion danger' was not effective for these motorists; they stayed in their cars while the others (not in the

**Figure 42.4  Smoke enveloping front cars**

Note:  Within five minutes smoke envelops the first cars completely, but drivers remain in their cars despite announcements of explosion danger

smoke) left. Two minutes later, there was a second announcement 'please leave the tunnel' (repeated once). This announcement finally got these people out of their cars.

Finding the emergency exits was no problem for participants. That was also a matter of the herd instinct: motorists saw others exiting and copied this behaviour. The role of the herd or group became apparent when compared with a behavioural test performed with lone individuals. In this test the motorist was completely alone in the tunnel. He was instructed to imagine a fire ahead and instructed to 'find safety'. Of the 29 participants, seven (24 per cent) ran away *via the main roadway*; the others used the emergency exits. In the group tests, no-one went via the main roadway. It seems that the exits by themselves are not attractive enough; their use needs the behavioural model of others. Hence, the marking of the escape route should be improved.

The contrast of this study with the conclusions of an earlier study is remarkable. In the latter, we asked 115 regular tunnel drivers what they would do when trapped behind a tanker on fire. A high proportion, 60 per cent, reported that they would get out via the main roadway. In the more realistic behavioural tests, no-one went via the roadway and, when alone, only 24 per cent went via the roadway. This illustrates the limited value of questionnaire data, at least in reference to conditions that people are not familiar with.

Another finding was that no-one used the emergency exit beyond the fire. This is relevant for at least 49 people; that is, for each of seven tests, this exit was nearest for at least seven motorists. The conclusion is that motorists will not use the exits beyond the fire. This result was at variance with the original expectations.

*Conclusions*

The behavioural tests reveal where evacuation problems exist and where they do not. Passivity is a major problem. Motorists continue their routine behaviour, and wait for the congestion to clear; they do not heed the signs of danger. Announcements by the operator can be used to start the evacuation. A word of warning is required, though

– getting motorists out of their car is important but it is not enough! We observed an evacuation that started spontaneously but came, also quite spontaneously, to a complete standstill. So announcing 'leave your car' by PA or illuminating 'leave your car' signs is not enough. The announcement should point out the threat in the tunnel, by using a term like 'explosion hazard'.

The behavioural tests also revealed that the isolation of people, caused here by smoke, deprives them of seeing what others do and of copying their successful behaviours. Such conditions required improved marking of the emergency exits. Inadequate marking was also apparent in related behavioural studies with smoke, including studies with sound beacons over the exits (Boer and Wittington, 2004). Too many test participants walked along without using the emergency exits.

The behavioural tests also reveal where the problems are not, a point that is less obvious than revealing where the problems are. Nevertheless, exposing a problem as imaginary saves valuable time and money as well as preventing implementation of inappropriate evacuation strategies. Consider the belief that people would go beyond the location of the fire in order to reach the nearest exit which would bring them right into the danger area. The current behavioural tests exposed this as an imaginary problem, and we can save the cost of measures to persuade people against this behaviour. As another example, consider the concern that people would not know how to operate the handle of the heavy sliding fire-doors (see Figure 42.5). The behavioural tests revealed that no-one had any problem opening the door. This information saves the cost of measures to clarify the opening procedure or revise the design of the door.

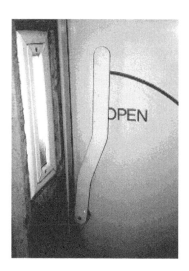

**Figure 42.5  Opening fire doors**

Note: Opening doors like these was no problem (all participants opened it without hesitation).

## Behavioural Tests Are Not Rescue Worker Drills

Behavioural tests give important information on how people react to emergencies. Participant's behaviour – rather than their opinions – is the focus of the study. Rescue workers' drills (RWDs) look similar to behavioural tests. In a rescue workers' drill, a disaster is simulated, complete with 'the public', 'victims' and sometimes even 'hooligans'. An example is a RWD in a train tunnel in the Dutch city of Best, six miles North of Eindhoven. The drill involved the police, fire brigade and hospital services. There were 1000 passengers, divided between two trains, one at the platform, the other halfway down the tunnel.

The focus of the RWD was these rescue workers, their communications and efficiency. The public was invited to add reality to the operations of the rescue workers. The public knew that it was an evacuation exercise, starting at about 10 in the morning. They boarded the trains without a real destination and without any wish for timely arrival – the RWD was on a Sunday morning. Then they waited (with anticipation) for the go signal. The train was full of firemen in their bright safety vests. They supervised the exercise. During the evacuation, they stood on the platforms, maintaining discipline. Fifty passengers had been instructed to play hooligans. For want of further instruction, they pulled the emergency brake. The train stopped while still halfway along the platform. Knowing that this was an evacuation, the hooligans used the door's emergency handle, and were out of the train in no time at all (on the wrong side, to be sure). The other passengers took this as the signal to start evacuating themselves, and copied the exit behaviour (that is, quite a few exited on the wrong side). The main differences between a behavioural test (BT) and a RWD are summarised in Table 42.1.

**Table 42.1  Key differences between a behavioural test and a rescue workers' drill**

| Behavioural test | Rescue workers' drill |
|---|---|
| Public has real everyday goals | Public has evacuation goals only |
| Evacuation unannounced | Evacuation announced in advance |
| Focus: behaviour of the public | Focus: rescue workers' performance |

Ideally, behavioural tests and RWD should go hand-in-hand. The behavioural tests provide the information on how passengers behave during an emergency, and the RWD uses that information to instruct the public more adequately about how to behave. The reality is less ideal. A behavioural test *in train tunnel evacuation* (comparable to the level of the behavioural road tunnel tests) has not been done. The consequence is that RWDs instruct the public inadequately, thus reducing the value of the RWD.

We recommend that *behavioural tests on train tunnel evacuation* be carried out. Such tests require careful information management to prevent leakage of the real purposes. Consider, for example, what a genial remark of the train guard like 'you're all for the evacuation, I see' could do. Or, in the context of passenger evacuation from a ship, the recommendation of the shop assistants in the supermarket ''don't buy now, wait until after the evacuation' (MEPdesign (Boer and Skjong, 2001)). Or consider the newspaper article 'large evacuation exercise upcoming'. The behavioural tests described in this chapter were hidden from the media. When we were approached for further information, we referred to the RWD under the supervision of Rotterdam fire brigade which was planned for one month after the behavioural test.

## Personnel Problems

The behavioural tests reveal that the public needs guidance such as messages on the public address system. For train evacuation, this would be a task for the train driver or the guard. Both of them are human beings and have their own goals like the passengers. The scheme of Figure 42.1 applies to them as well.

There are at least three problems for train personnel: lack of routine and guidelines, reluctance to raise the alarm, and fear of panicking the passengers. Lack of routine, Problem 1, has to do with emergency training. Fortunately, disasters are rare events. In addition, train personnel have many other tasks to do. Questioning 15 train drivers, five of them for some hours, revealed that train personnel of Dutch Railways are not very well prepared to cope with a fire in a tunnel (Boer and Weitenberg, 2002). There are guidelines (such as the one mentioned in the introduction) but these guidelines are not readily familiar to personnel. Personnel are mostly aware that there are guidelines but they would have to look them up. As a consequence, they rely on related procedures such as train evacuation in the open-air situation, and professional wisdom and intuition. They are uncertain who to call and what their evacuation tasks are. An additional problem is that the existing guidelines are not complete. They don't specify the role of the train personnel for passengers who have alighted. They seem to rely on the existing signage, which is not always adequate (Figure 42.6 shows that there can be maintenance or housekeeping problems). Moreover, the problem of where to go when there is smoke is not addressed at all.

Problem 2 is reluctance to raise the alarm. Usually, the early stages of a disaster do not look so bad, and the alarm will disrupt all normal services for quite some time. It is possible that the company will blame the personnel for a false alarm. Is the incident really as bad as that? This leads to orientation behaviour, waiting for more and more definitive information (see Figure 42.1); in short, it leads to a postponement of the alarm.

Problem 3 is fear of panicking the passengers. The popular belief that the word *fire* will result in a stampede is still at the back of the mind of staff, with the result that train personnel are unsure what to announce. This risks postponing the announcement.

All these problems can be addressed in a special training course (assuming that the guidelines themselves are complete).

## Conclusion

Behavioural tests are well suited to learn how the public reacts to unexpected emergencies like fire in a tunnel. Such tests have been carried out for road tunnels,

**Figure 42.6   Pollution obscures signs**

Note: The sign on the left, here illuminated by a pocket-torch, can be illuminated from the inside but pollution makes the text NOODUITGANG (emergency exit) difficult to see. The folded sign on the right is also covered with dirt; moreover, it points to two directions at once.

and have inspired several remedial measures that are cost-effective (Boer and Varkevisser, 2002). In addition, the tests can be used subsequently to instruct the participants of a rescue workers' drill with regard to 'realistic behaviour'. No such tests have been carried out to date for rail tunnel evacuation but based on the results from road tunnels, we expect much from a rail tunnel behavioural test.

## References

Boer, L.C. (2002), 'Behaviour of Motorists on Evacuation of a Tunnel', Report TM–02–C034, Soesterberg, NL: TNO Human Factors.

Boer, L.C. (2003), 'Geluidsbakens in tunnels' [Sound Beacons in Tunnels], Report TM–03–C060, Soesterberg, NL: TNO Human Factors.

Boer, L.C. and Skjong, R. (2001), 'Human Behaviour in Ship Evacuation: Contributions of MEPdesign', TIESMS 2001 Conference, Oslo, 19–22 June.

Boer, L.C. and van Lijngaarden, S.J. (2005), 'Self-explaining Sound Beacons in Evacuation', submitted to *Ergonomics*.

Boer, L.C. and Varkevisser, J. (2002), 'Herkenbaarheid vluchtwegen tunnels – Een ontwerpschets' [Espying Escape Ways in Tunnels – A Design Sketch], Report TM–02–C055, Soesterberg NL: TNO Human Factors.

Boer, L.C. and Weitenberg, A.I.M. (2002), 'Verloop Evacuatie Spoortunnels' [Events during Evacuation of a Rail Tunnel], Report TM–02–C058. Soesterberg, NL: TNO Human Factors.

Boer, L.C. and Wittington, D. (2004), 'Auditory Guidance in a Smoke Filled Tunnel', *Ergonomics*, 47: 1131–40.

Meister, D. and Enderwick, T.P. (2002), *Human Factors in System Design, Development, and Testing*, Mahwah, NJ/London: Erlbaum.

Proulx, G. and Sime, D.J. (1991), 'To Prevent Panic in an Underground Emergency: Why Not Tell People the Truth?', in C. Cox and B. Langford (eds), *Fire Safety Science, Proceedings of the Third International Symposium*, London: Elsevier Applied Science: 843–52.

Quarantelli, E.L. (1999), *Disaster Related Social Behavior: Summary of 50 Years of Research Findings*, Newark, DE: Disaster Research Center.

Sime, D.J. (1995), 'Crowd Psychology and Engineering', *Safety Science*, 21: 1–14.

Voeltzel, A. (2002), 'Compared Analysis of the Mont Blanc Tunnel and the Tauern Tunnel Fires', PIARC WG6.

Chapter 43

# Rail Passenger Perceptions of Risk and Safety and Priorities for Improvement

Lauren J. Thomas, Daniel J.A. Rhind and Katie J. Robinson

## Background

Recommendation 60 of Part 1 of Lord Cullen's Ladbroke Grove public inquiry stated that 'comprehensive market research in regard to safety related measures should be carried out in order to take account of the views of informed passengers'. In response, the Rail Safety and Standards Board Limited (RSSB) commissioned the Human Factors Group at Cranfield University to conduct a survey of railway passengers. The aim was to establish passenger perceptions and awareness of risk and safety at stations, on platforms and on trains, and to investigate passenger preferences for implementing safety interventions. Based on qualitative interviews conducted with both frequent and infrequent rail travellers, two separate questionnaires were developed and piloted. The questionnaires covered topics such as the perceived risk of injury or fatality while travelling by rail; the relative risk of travelling by rail compared to other modes of transport; passenger behaviour in incidents and accidents; and passenger preferences with regard to potential safety interventions. Both questionnaires were administered at 15 hub railway stations nationwide. Stations were selected on a regional basis to include a range of train operating companies and different types of rolling stock. The quota sample for each survey was approximately 900 people, including commuters, business travellers, those who used the train for leisure/pleasure purposes, and passengers with mobility limitations. Participants were approached face to face, and if they met the quota requirements and agreed to take part, they were given a questionnaire for self-completion and postal return. The results provided a valuable indicator of passenger perceptions of risk, especially when compared to quantitative assessments of actual risk. The findings suggested that the relationship between perceived and actual risk is not particularly strong, and reasons why this might be the case are discussed. The data from the passenger preferences questionnaire were analysed using conjoint techniques, and the results provided valuable information on passenger preferences for improvement. This information is likely to be of value to the railway industry in prioritising the implementation of safety interventions according to passenger preferences, and in understanding rail passenger behaviour.

**Introduction**

Part 1 of the Ladbroke Grove Rail Inquiry reported that passenger perceptions may not always reflect the real risks on the rail network, and may even conflict with proposed safety initiatives. For example, many passengers may believe that it is advisable to leave the train in the event of an incident. The reality is that, in most cases, passengers are likely to be safer if they stay on the train, and simply move to an adjacent carriage. Similarly, while passengers may prefer to travel in carriages with spacious tables and seating arrangements, such features may not always be optimal in crashworthiness terms. Hence, Recommendation 60 stated that: 'comprehensive market research in regard to safety related measures should be carried out in order to take account of the views of informed passengers' (Cullen, 2000).

This recommendation was made with a view to understanding the extent to which passenger preferences and perceptions might influence safety interventions. For example, passengers may wish to increase the number of potential escape routes on a train by having additional removable or breakable windows in train carriages. However, in the event of an accident, an increase in the number of escape routes could also increase the risk of fuel entering the carriage. The risk of passengers being ejected from the carriage on impact, rather than being contained within the vehicle, would also increase. Hence, in this case, passenger preferences may actually conflict with initiatives designed to improve safety, and minimise the risk of injury or fatality.

Risk on the United Kingdom rail network is assessed and quantified via the Safety Risk Model (SRM). This model is a mathematical representation of the causes and consequences of potential accidents arising from rail maintenance and operations on the Network Rail Controlled Infrastructure. The SRM is used to quantify current levels of risk, and to determine priorities for risk reduction and safety improvement. Risk is quantified in terms of equivalent fatalities, so that the total risk – excluding suicides – is currently predicted to be 138 'equivalent fatalities' per year. Equivalent fatalities allow injuries to also be taken into account: one equivalent fatality is equal to ten major injuries, or two hundred minor injuries (Railway Safety, 2001).

While major accidents such as Ladbroke Grove are very serious, they are in fact extremely rare. It is perhaps unfortunate that this type of event is given a high level of media attention, since this naturally influences public opinion. Therefore, public perceptions of the level of risk and safety on the rail network may not necessarily be accurate. It is reasonable to suppose that passenger perceptions may be skewed towards the more serious accidents, those which tend to involve a relatively high number of injuries and/or fatalities. This effect has been found within aviation, where passengers were found to believe that their chances of being in an aircraft accident were greater than they actually were. They also perceived that their chances of surviving an aircraft accident were far smaller than they would actually be (Fennell and Muir, 1992).

Apart from the actual and perceived absolute level of risk when travelling on the rail network, there is also the issue of perceived relative risk between different forms of transport. For example, motorcycles are a notoriously risky form of transport, whereas it is known that people tend to feel safer in cars. Actual levels of risk across several different forms of transport are provided in Table 43.1. Again, it should be noted that there may be a tendency for passengers to perceive the relative risks inaccurately, and therefore be unable to rank order the different risks across transportation modes correctly.

**Table 43.1  Passenger fatalities per billion passenger journeys, across transport modes**

| Mode of transport | Fatalities per billion passenger journeys |
|---|:---:|
| Bus/coach travel | 3 |
| Rail travel | 12 |
| Water travel (boat and/or ship) | 14 |
| Car travel | 38 |
| Air travel | 39 |
| Motor cycle travel | 1940 |

*Source*: DLTR figures, from Nelson, 2002: 32.

It is unfortunate that passenger perceptions of risks and consequence, although they may be inaccurate, are likely to influence their behaviour. For example, passengers who believe that they will be unlikely to survive an aviation accident are less likely to pay attention to the safety information. It could be argued that passenger misperceptions of the relative 'riskiness' of air travel almost primes them to expect the worst. In such cases, passengers might misinterpret a routine – but still infrequent – situation as a threat to safety, and take emergency action when in fact there is no need to do so.

Within the United Kingdom railway industry, a number of safety initiatives have been researched with a view to improving passenger safety, and the outcomes of safety critical situations. However, little research has been conducted regarding passenger safety preferences. When looking at passenger preferences, one has to rely on what passengers say that they would prefer. Stated preference techniques allow researchers to investigate passenger preferences for a range of service or product attributes, even when that service or product is still in the development or planning stages (Pearmain *et al.*, 1991).

In relation to the rail industry, such techniques can assist in understanding passenger preferences and priorities. By allowing passengers to express their preferences through such research, it should be possible to take their opinions into account when making decisions which have an impact on passenger services. This can assist in making more informed decisions about upgrading or improving the rail network, in order to maximise customer comfort and satisfaction. Asking passengers to prioritise interventions in this way can assist in allocating scarce resources to those areas in which investment would have the strongest or most positive impact. However, it must be noted that passenger preferences will not necessarily always be translated into passenger spending behaviour (Abley, 2000).

Nevertheless, London Underground Limited use market research methods such as these to inform investment decisions on upgrading passenger facilities. In this way, their spending can be targeted at those areas where improvement will most enrich or maximise passenger comfort and satisfaction (Wilson and Gutmann, 1998). It could be argued that including passengers in the investment decision process is vital in ensuring that safety and comfort interventions truly address passenger needs. For example, previous research has indicated that transport managers' perceptions of passengers needs do not accurately reflect actual passenger needs. These circumstances could result in a misallocation of resources, and a growing passenger dissatisfaction with transport services (Koushki, Al-Saleh and Al-Lumaia, 2003).

**Method**

To establish the scope and range of the project, and ensure that the survey was likely to answer the research questions of relevance to the rail industry, consultation was undertaken with key stakeholders. In addition, a series of qualitative interviews was conducted with passengers to assist in questionnaire development. Focus groups were conducted with a range of passengers, including commuters, business travellers, and those who used the train for leisure and pleasure purposes. The 22 participants in the qualitative phase were volunteers, and each was paid £10 for taking part.

All focus groups and interviews were transcribed verbatim as soon as possible after the interview had been conducted. Transcription was always undertaken by a researcher who had been present at the interview. Completed transcripts were then content analysed to identify the main themes and concepts. Because of the large volume of data, it was decided to develop two separate questionnaires. This approach would ensure that each questionnaire would be a reasonable length, an important consideration in obtaining reasonable response rates. However, the use of two questionnaires would mean that comprehensive data on a wider range of themes could be obtained. It was decided that one questionnaire would address passenger safety and risk awareness on the rail network, and the second would examine and evaluate passenger preferences for a series of interventions which might be implemented to improve passenger safety and comfort on the rail network. The draft questionnaires were trialled and piloted via an iterative process of test, review and redesign.

It was decided to sample rail passengers nationwide to ensure coverage of a range of train operating companies, using different types of rolling stock on different routes. In addition, it was necessary to ensure adequate representation of sub-groups, to enable the results to be generalised with some degree of confidence to the wider travelling public. To achieve this, it was decided to sample passengers proportionate to the number and types of journeys undertaken within each rail region. For example, more passengers would be invited to take part in busier rail regions, and invitations to take part would also reflect the proportion of commuter, business and leisure/ pleasure journeys within each region.

It was decided to offer an incentive to passengers to complete the questionnaires, and thereby ensure a higher response rate. For each survey, a charity donation scheme was organised. Participants could choose which, of three pre-determined charities, they would like to support. Cranfield University undertook to donate £1 to the chosen charity, provided that the completed questionnaire was returned by the cut-off date. In addition, a Cranfield University pen was provided with every questionnaire. Each questionnaire was administered to 905 people over a four week period, according to the quota sample.

Of the 905 administered safety and risk awareness questionnaires, useful data were provided on 266 questionnaires returned to Cranfield University. Of the 905 administered safety preferences questionnaires, useful data were provided on 286 returned questionnaires. The overall response rate was 30.5 per cent. All data were entered into a software package for statistical analysis. The results are based on the total number of people deemed to have provided an answer to the question.

## Results

### Perceived and Actual Risk

Passengers were asked to estimate the percentage of injuries/fatalities that would be caused by each type of incident or accident, and the mean percentage risk estimates are shown in Table 43.2. The actual risk, as derived from the Risk Profile Bulletin using equivalent fatalities, is also shown (Railway Safety, 2001).

**Table 43.2  The mean perceived risk of injury/fatality in a range of incidents and accidents, compared with actual risk (equivalent fatalities)**

| Incidents or accidents which cause injuries or fatalities | Perceived risk | Actual risk |
|---|---|---|
| A passenger train colliding with another passenger or freight train | 16.7% | 5.2% |
| A passenger train colliding with an object on the track | 11.0% | 5.7% |
| People being struck whilst on level crossings or footpaths | 5.8% | 5.2% |
| Passengers injured whilst getting on or off train | 9.1% | 4.9% |
| Overcrowding on the train or platform | 5.6% | 0.1% |
| People trespassing on the track | 10.1% | 44.9% |
| Spillage and release of flammable or toxic load | 0.1% | 0.1% |
| Staff or employee risks | 9.1% | 10.0% |
| Slips, trips and falls | 11.4% | 10.4% |
| Train derailment | 9.2% | 3.1% |
| Fire | 3.8% | 0.4% |
| Any other incidents | 6.5% | 10.0% |

### Safety of Rail Travel Relative to Other Transport Modes

The safety and risk questionnaire also asked participants to estimate the relative risk of fatality when travelling by rail, compared to other modes of transport. The results for perceived risk were analysed according to the number of passenger journeys for each single fatality. The results for actual and perceived risk are provided in Table 43.3.

**Table 43.3  Actual and perceived number of passenger journeys per single fatality, across transport modes**

| Transport mode | Actual number of passenger journeys per fatality | Perceived number of passenger journeys per fatality (mean) |
|---|---|---|
| Bus or coach | 333.33 million | 11.11 million |
| Rail | 83.00 million | 14.86 million |
| Boat or ship | 71.43 million | 42.47 million |
| Car | 26.32 million | 0.22 million |
| Aeroplane | 25.64 million | 44.84 million |
| Motorbike | 0.52 million | 10.39 million |

*Source*: Actual fatality statistics from Nelson, 2002: 32.

*Preferences for the Car Park and the Way into the Station*

One section on the preferences questionnaire related to four initiatives to improve safety and security in station car parks and at the entrance to railway stations. These were: having a member of staff in the car park, having good lighting, having monitored CCTV, and having secure fencing around the car park. The limited number of options meant that each could be matched to every other using six pair-wise trade-off grids, and therefore a conjoint approach to analysis was undertaken. This allowed the utility values of each option, relative to every other, to be evaluated. The results are shown in Table 43.4, and indicate the utility that passengers on different journey types attach to each initiative.

**Table 43.4  Mean utility values for the four initiatives relating to the car park and the way into the station, by journey type**

| | Commuting passengers | Business passengers | Leisure/pleasure passengers |
|---|---|---|---|
| A member of staff in a booth in the car park | 21.9 | 20.4 | 26.5 |
| Good lighting in and around the car park | 28.5 | 28.3 | 26.9 |
| CCTV cameras monitored by a member of staff | 31.8 | 32.5 | 27.9 |
| Strong and secure fencing around the car park | 17.8 | 18.8 | 18.7 |

A repeated measures ANOVA on the utility values from all passengers (regardless of the purpose of the journey) showed that having strong and secure fencing around the car park had a significantly lower utility value than the other three options. Having a member of staff in a booth in the car park had significantly less utility than CCTV monitored by a member of staff. CCTV cameras monitored by a member of staff and good lighting were not significantly different in utility, indicating that both would be preferred by passengers.

*Preferences for Railway Stations and Platforms*

With regard to the railway stations and platforms, a larger number of options were identified as suitable for inclusion on the questionnaire. It was decided to group individual options into overall categories, and to ask passengers to conduct a simulated spending exercise to indicate their priorities for each category. The categories were: the type of information provided for passengers, how information for passengers should be provided, the way that stations and platforms are monitored, the environment within the station and on the platforms, the facilities available at the station and on the platform, and the waiting areas available on the platforms. Passengers were asked to allocate £100 between these options to indicate relative priorities. The results are shown in Table 43.5.

**Table 43.5  Mean simulated spend for railway station and platform categories, by journey type**

|  | Commuting passengers (N=120) | Business passengers (N=38) | Leisure/ pleasure passengers (N=123) |
|---|---|---|---|
| The type of information provided for passengers | 15 | 15 | 12 |
| How information for passengers should be provided | 16 | 17 | 14 |
| The way that stations and platforms are monitored | 27 | 21 | 28 |
| The environment within the station and on the platforms | 12 | 15 | 13 |
| The facilities available at the station and on the platform | 12 | 11 | 13 |
| The waiting areas available on the platforms | 12 | 13 | 16 |

The data in Table 43.5 were entered into a repeated measures ANOVA. The results indicated that the way in which stations and platforms were monitored was the most important priority for passengers, since the spend on this option was significantly higher than it was for all other options. There were also some significant differences between some of the remaining least preferred options.

*The Trains and Carriages*

A repeated measures ANOVA indicated that the overall priority was the safety of the whole rail network, since this was allocated a significantly higher proportion of funding than each of the other options. The second highest priority was the performance of the train service, followed by the way that trains and carriages were monitored.

Although passengers indicated that overall, safety was the highest priority for the trains and the carriages, with the performance of the train service being the second

**Table 43.6 Mean simulated spend for train and carriage categories, by journey type**

| | Commuting passengers (N=122) | Business passengers (N=36) | Leisure/ pleasure passengers (N=122) |
|---|---|---|---|
| How safety information for passengers is provided | 6 | 5 | 5 |
| The way that trains and carriages are monitored | 13 | 11 | 15 |
| The environment on the train and in the carriages | 10 | 12 | 9 |
| The seating and standing facilities in the carriages | 15 | 10 | 11 |
| The performance of the train service | 18 | 18 | 16 |
| The role of the train driver | 7 | 7 | 7 |
| The safety equipment on the train and in the carriages | 9 | 9 | 8 |
| The safety of the rail network overall | 20 | 26 | 27 |

highest priority, there was a trade off with regards to service. Passengers were asked to split £1 between three train service parameters to indicate their priorities. The results are shown in Table 43.7.

**Table 43.7 Mean simulated spend on the performance of the train service, by journey type**

| | Commuting passengers (N=123) | Business passengers (N=38) | Leisure/ pleasure passengers (N=123) |
|---|---|---|---|
| Having a reliable train service, which departs and arrives on time | 41 | 41 | 37 |
| Having a frequent service, with trains running regularly | 32 | 35 | 26 |
| Having a safer service, although this may mean more delays | 22 | 20 | 31 |

With regard to the performance of the train service, significantly more of the simulated spending budget was spent by passengers on having a reliable train service. This option was a priority over both frequency and safety of service. Frequency and safety received similar proportions of the budget, with the difference between the spend on these two options being too small to reflect a genuine difference in preference.

## Conclusion

Unfortunately, there can be no guarantee that the returned questionnaires accurately reflect all those participants who accepted a questionnaire, or indeed, that those passengers who accepted a questionnaire were truly representative or typical passengers. It is also possible that administrators only approached people who seemed most approachable, and not necessarily those who were most representative of rail travellers. However, the range of people included in the final sample, and the fact that a quota sample was used to obtain it, has certainly reduced the biases that might have been evident if people had been approached to take part in the study purely on an opportunity basis. It is simply a function of human nature that not everybody will agree to accept a questionnaire, and that those who do may not return it.

The first major finding was that passengers appeared to have a distorted view of the actual sources of risk on the rail network. As an example, a very large proportion of actual risk on the rail network is a result of trespass. Trespass incidents are not only very frequent, but they can evidently have extremely serious, if not fatal, consequences. However, passengers did not appear to rate this type of incident as contributing to a large number of injuries and fatalities. Clearly, passenger opinions of risk and safety do not reflect the actual situation. This is perhaps to be expected given the media attention which has been focussed on recent incidents and accidents within the railway industry.

With regards to the relative safety of rail travel, statistics issued by DLTR (Nelson, 2002: 32), shows that travelling by bus/coach carries the lowest risk of fatality, in that a passenger would have to travel 333.33 million journeys before being killed. Second safest is rail, with a journey to fatality ratio of 83.00 million to one. Passenger perceptions of relative risk did not appear to reflect reality: responses given by passengers indicated that they thought cars to be the riskiest form of transport, followed by motorbikes, bus/coach, rail, boat/ship, and aeroplane. It may be the case that passenger perceptions of relative risk were influenced primarily by the frequency of such events, or the frequency of journeys using each mode of transport. Certainly, the rank positions of each form of transport do not correspond to the actual level of risk.

The second questionnaire was related to safety preferences, and aimed to establish what preferences, overall, passengers had for improving safety on the rail network. With regards to station car parks and entrances, passengers gave significantly higher utility values to having CCTV cameras monitored by a member of staff and having good lighting in and around the car park. In terms of overall preferences for the stations and platforms, passengers again spent a significantly higher proportion on the way that stations and platforms were monitored than they did on the remaining options. Having staff available to monitor, either in person or via CCTV, appears to be a key priority for passengers.

Passengers were also asked about their priorities and preferences for the trains and the carriages. Perhaps predictably, the simulated spending patterns showed that passengers placed a significantly higher value on the safety of the rail network overall than they did on any other option. The second highest priority was the performance of the train service, followed by the way that trains and carriages were monitored. Overall, safety may be regarded as the number one passenger priority, when the train service is assessed on an overarching level. Interestingly, passengers were also offered a choice between several specific performance parameters regarding the train service: having a reliable service which departs and arrives on time; having a frequent service, with trains running more regularly; or having a safer service, although this may mean more delays. When the features of the train service were examined in this manner, on a component by component basis, reliability was the first priority among all passengers, with a significantly higher simulated spend. In reality, making repairs to infrastructure may cause delays to train services, notwithstanding that such work is generally planned to minimise disruption. These findings suggest that where maintenance work is taking place, passengers will be more likely to tolerate fewer trains running, and will be less likely to tolerate delays on the services that do run.

## References

Abley, J. (2000), 'Stated Preference Techniques and Consumer Decision Making: New Challenges to Old Assumptions', Cranfield School of Management Working Paper 2/00, Cranfield University.

Fennell, P.J. and Muir, H.C. (1992), 'Passenger Attitudes Towards Airline Safety Information and Comprehension of Safety Briefings and Cards', CAA Paper 92015, London: Civil Aviation Authority.

Health and Safety Executive (2001b), *The Ladbroke Grove Inquiry: Part 1* ('The Cullen Report'), Sudbury: HSE Books.

Koushki, P.A., Al-Saleh, O.I. and Al-Lumaia, M. (2003), 'On Management's Awareness of Transit Passenger Needs', *Transport Policy*, 10: 17–26.

Nelson, A. (2002), *Annual Safety Performance Report 2001/2*, London: Railway Safety.

Pearmain, D., Swanson, J., Kroes, E. and Bradley, M. (1991), *Stated Preference Techniques: A Guide to Practice*, 2nd edn, London: Steer Davies Gleeve and Hague Consulting Group.

Railway Safety (2001), *Profile of Safety Risk on Railtrack Controlled Infrastructure: Risk Profile Bulletin*, Issue 2, July, London: Railway Safety.

Wilson, A. and Gutmann, J. (1998), 'Public Transport: The Role of Mystery Shopping in Investment Decisions', *Journal of the Market Research Society*, 40 (4): 285–93.

Chapter 44

# A Common System of Passenger Safety Signage

Gary Davis and Ann Mills

## Background

In May of 2002, Davis Associates Ltd and Interfleet Technology Ltd commenced a 12-month safety signage research project for the Rail Safety and Standards Board (RSSB). The work had been commissioned on behalf of the Association of Train Operating Companies (ATOC) and was one of several projects commissioned in response to the recommendations of the Cullen Report into the Ladbroke Grove rail accident.

The Cullen recommendations most directly addressed by this project are shown in Table 44.1.

**Table 44.1 The Cullen recommendations most directly addressed**

| Cullen recommendation | Key objectives |
|---|---|
| 71   The requirement for emergency signs to be luminous should be made retrospective | Luminous signs |
| 72   So far as is feasible, emergency signs on all trains should be capable of being understood by passengers without the need to read text | Graphical symbols |
| 73   There should be research with the aim of arriving at a system of signage which is common to all trains in Great Britain | Common signage system |

The key challenges for the project were therefore:

(i)    Luminous materials: what graphical rules are appropriate to ensure symbol and text legibility in darkness, low light and smoke-filled conditions?

(ii)   Graphical symbols: what graphical symbols are required? What methods would ensure comprehension of the symbols?

(iii) Common system: what graphics rules should be applied to ensure a consistent style and quality of signs? What process should be followed to develop new symbols?

The scope for the project covered the following safety sign categories:

- safe condition;
- fire equipment;
- passenger alarm/ommunication devices.

The following sign categories were excluded from the scope by RSSB: hazard warnings, prohibition signs, mandatory signs and all non-safety signs.


## Research Programme

Davis Associates proposed an outline research programme but it was clear that some initial research would be necessary in order to determine what relevant information and research material was already available before the programme could be refined and finalised. Therefore, Stage 1 of the programme was commissioned to define current best practice.

*Stage 1 – Defining Best Practice*

The research programme commenced with a literature review covering: graphics and typographics guidelines, the characteristics of luminous materials and their application to signs, the psychological and physiological aspects of legibility and comprehension particularly in low light and smoke-filled conditions, relevant standards, guidance and best practice in the railway industry and other industries. The review included a search for the most appropriate and practical testing methods to evaluate the relevant factors.

As a result of the information obtained during Stage 1, the outline research programme was reviewed and the following programme stages were scoped and agreed:

Stage 1  defining best practice (previously completed);
Stage 2  information needs analysis;
Stage 3  benchmark legibility tests;
Stage 4  symbol development and comprehension testing;
Stage 5  instructional comprehension tests in context;
Stage 6  legibility and discriminability verification testing;
Stage 7  standardisation (defining guidelines and procedures).

The scope was confirmed, including the development of graphical symbols for at least 40 referents and an initial batch of 100 safety signs. Through a parallel project, all of the deliverables would be made available via the ATOC website.

*Stage 2  – Information Needs Analysis*

The primary objective of this stage was to define the list of referents for which graphical symbols were required. A referent is a written description of a single element of information such as an object (for example 'Fire extinguisher') or action (for example 'Pull handle').

An industry stakeholder workshop was facilitated to discuss:

- all possible safety information that might need to be communicated via fixed signage in trains;
- the characteristics of the user group (including staff and emergency services as well as passengers);
- the scenarios and circumstances under which the information might be needed;
- the associated emergency equipment and the actions required to operate it.

An existing 'catalogue' of UK safety signs was reviewed, although it was not an exhaustive collection and featured many examples of poor practice. In addition, Interfleet had previously conducted a signage survey on a wide sample of UK and overseas trains and had engineering knowledge of the associated emergency equipment and how it should be operated. Each existing sign was broken down into its constituent information elements and referents were defined for each. Certain elements had to remain text only (for example 'Penalty for improper use'), and others were so specific to a particular device and vehicle that those too had to remain text only.

Stage 2 culminated in a list of 48 referents. Of these, eight were to be based on existing British Standard symbols (for example fire extinguisher) and a further five did not require comprehension testing (for example direction arrows). There remained a list of 35 referents requiring comprehension testing.

*Stage 3 – Benchmark Legibility Tests*

Before any symbols or signs could be developed it was first necessary to define the graphics guidelines that would ensure their legibility when printed on luminous material and when viewed under low-light and smoke-filled conditions. This was necessary because the literature review had revealed that no such guidelines existed.

A custom-designed viewing chamber was constructed to allow the controlled introduction of smoke and the representation of various lighting conditions (see Figure 44.1). The chamber was three metres long with a shielded viewing port at one end facing the visual target area at the other. The chamber was sealed to contain the smoke and to enable the smoke density to be maintained at a consistent level throughout each trial session. The smoke was a non-toxic, oil-based aerosol with small droplet size providing a longer persistence time.

Test graphics were inserted into the viewing area on boards, through a sealed slot in the side of the chamber. There was a separate compartment within the chamber containing ultraviolet lamps in order to rapidly energise the luminous material. All tests were conducted with the material at or near maximum saturation. The smoke/light conditions set in the viewing chamber are shown in Table 44.2.

Test graphics were designed and printed onto the luminous material. The graphics were of two types:

(i) Snellen charts – to test different text sizes and stroke widths, in red/green and positive/negative permutations;

**Figure 44.1  The purpose-built viewing chamber (access panels shown open)**

**Table 44.2  Smoke/light conditions in the viewing chamber**

| Condition | Representing | Illumination level | Colour temperature |
|---|---|---|---|
| 'Daylight' | Carriage illumination during daylight hours | 1700 lux | 4,750K |
| Carriage lighting | Carriage lighting at night or in tunnels | 410 lux | 3,450K |
| Emergency lighting | Conditions under the emergency lighting system | 20 lux | 2,650K |
| Total darkness | Failure of the emergency lighting system at night or in a tunnel | 0 lux | Not applicable |
| Snap wands | The glow from yellow snap-wands (in otherwise dark conditions) | 1.5 lux | Not measured |

(ii) Landolt rings – of different sizes to determine the minimum legible graphical elements, also in red/green and positive/negative permutations.

The series of legibility tests was performed with 33 people representing a range of visual acuity levels.

The following key findings of the legibility tests enabled the graphics design of the symbols and signs to commence.

•   Helvetica medium font (as specified in BS5499–1) proved adequate for use on luminous signs;

- signs were more legible if printed in positive format (that is, light text on dark background);
- upper case characters should be 7.5mm high, and lower case characters should be 5.5mm high, for every one metre of viewing distance;
- symbols should feature a minimum line thickness of 0.86mm per viewing distance;
- critical details within symbols should be a minimum of 2.6mm per metre of viewing distance;
- the maximum ratio of critical detail to symbol size should be 1:16.5.

*Stage 4 – Symbol Development and Comprehension Testing*

The objectives of this stage were to:

- generate alternative graphical concepts for each of the referents defined in Stage 2;
- refine the design style for all the graphical symbols;
- conduct comprehensibility judgement tests on up to ten symbol variants for each referent;
- conduct comprehension tests on up to three of the best symbol variants for each referent;
- conduct mutual confusion tests on the set of most comprehensible graphical symbols.

*Generation of graphical concepts*

Graphical symbol concepts to represent each referent were conceived during a brainstorm session amongst the project team, involving ergonomists, designers and engineers. The graphical symbol concepts were then drawn-up, in line with the minimum sizes of line thickness and critical detail defined in Stage 3. Up to ten of the most promising graphical symbols for each referent were selected and taken forward to comprehensibility judgement testing.

*Comprehensibility judgement testing*

The test procedure used was based on that in ISO 9186 (2001). Test sheets were produced for each referent with up to ten graphical symbol variants on each sheet (see Figure 44.2). Members of the public were approached on board trains and asked to complete booklets of several test sheets. The participant was informed of the symbol's intended meaning (that is, the referent) and then asked for their estimation of the percentage of the general population that would correctly understand the meaning of the symbol. Over 300 train passengers were consulted for this test. Analysis of the data enabled the shortlisting of graphical symbol variants for comprehension testing. In some cases, none of the variants for a particular referent achieved the required 66 per cent pass mark, and so the symbols were redesigned (or new symbols created) followed by further testing.

*Comprehension testing*

This test procedure was also based on that in ISO 9186 (ISO, 2001). There were up to three graphical symbol variants for each referent. Each test sheet showed one symbol variant together with information about the context of use of the symbol – but with no indication as to the meaning (see Figure 44.3). Participants were asked to write on each sheet what they thought the meaning of the symbol was. Over 300 train passengers were consulted for this test. The responses were then assessed by a panel of three independent judges and assigned one of seven possible categories. A weighted scoring system was then used to calculate the overall level of comprehension for each symbol tested. Most symbols achieved the ISO pass mark of 66 per cent. Some instruction symbols were redesigned and re-tested within the context of an appropriate instruction sign.

**Figure 44.2 Example test sheet for the comprehensibility judgement test**

**Figure 44.3 Example test sheet for comprehension test**

*Mutual confusion test*

A mutual confusion test was then performed on the most successful graphical symbol for each referent in order to identify potential confusions between the meanings of any pair of symbols. A number of test sheets containing a matrix of symbols were compiled (see Figure 44.4). Symbols for similar referents (such as 'Strike window in corner' and 'Strike window on green dot') were placed on separate sheets because they would never appear together on the same train. Testing took place at major rail stations around the UK involving a total of 330 members of the public (see Figure 44.5).

At the end of Stage 4 the results of the symbol development process showed that:

**Figure 44.4 A typical symbol matrix being used for the mutual confusion test**

**Figure 44.5 Mutual confusion test applied at a main railway station**

- 23 referents achieved final comprehension scores of 66 per cent or higher;
- eight referents achieved final comprehension scores of between 50 and 65 per cent;
- four referents achieved final comprehension scores of less than 50 per cent;
- all symbols passed the mutual confusion test.

Any symbols that achieved less than 50 per cent comprehension score were redesigned and subjected to further testing. In a few cases, it was concluded that the referent would have to remain text-only (for example 'This hammer is alarmed') or use an existing symbol supported by text (for example 'Emergency door release' in conjunction with the running–man symbol).

*Stage 5 – Instruction Comprehension Tests in Context*

The objectives of this stage were:

- to verify the comprehensibility of the graphical symbols when presented within the context of a rail vehicle by means of user testing with a sample of the general public under stressed conditions;
- to evaluate the usability of the draft graphics guidelines through the production of signs for a specific vehicle.

The tests were conducted over three days at Connex South Eastern's Slade Green Maintenance Depot with 31 participants on board a BREL 465/0/ vehicle and a METRO-CAM 465–2 vehicle (see Figure 44.6). The devices to be tested were: passenger alarm, emergency door release, and emergency hammer.

The appropriate instruction symbols for each device were compiled into signs following the draft 'Graphics Guidelines' which were being developed in parallel

**Figure 44.6   User trials underway with the passenger alarm device in the rail vehicle**

with Stages 4, 5 and 6. The signs were test-fitted to the vehicle and various alternative concepts were also compiled for testing.

One participant at a time was briefed on a particular emergency scenario and then asked to execute the appropriate task, that is, sound the alarm, open the doors, or break the window to escape (the latter task stopped short of actually breaking the window). Participants completed all three tasks in a balanced presentation order and were then interviewed. The tests were video recorded for detailed analysis.

The results of the instruction comprehension test in context showed that:

- 21 out of 31 participants (68 per cent) claimed not to have studied the signs, but instead acted on instinct, relying on their preconceptions and using whatever visual cues they could see from the devices themselves;
- of those participants who did study the signs, most did not read past the header text or the first couple of instructions;
- participants tended to return to the signs only when they got stuck at a particular stage of the task;
- design attributes of the rail vehicles used for the tests significantly affected the results, making it difficult to draw conclusions about the signs themselves. For example:
  - the emergency devices in the doorway were mounted in the ceiling—out of normal line of sight for most people;
  - the upper door and ceiling was poorly illuminated—approx. 7–13 lux compared with 220 lux of the surrounding panels;
  - for operational reasons, two variants of the 465 vehicle had to be used during the tests. The design variations between the two vehicles also affected the results, making it difficult to compare different sign set-ups;

- ○ inconsistency in the breaking strength of the device covers meant that some participants had great difficulty in breaking the covers;
- ○ on two occasions, the emergency door release device was partially operated without the cover being completely removed. The doors partially opened and then slammed closed again when the participant let go of the handle;
- ○ despite the practical difficulties described above, the user tests provided invaluable information about the effectiveness of the emergency devices and the effect of the signage.

The user tests lead to the following conclusions relating to the draft 'Graphics Guidelines' for safety signs:

- instruction symbols should not be reduced below standard size (33mm square) if at all possible, particularly if they are located in areas which are poorly illuminated;
- some of the symbols were too complex and detailed for their meaning to be assimilated in stressed conditions. These symbols required redesign and re-testing;
- the words 'Break glass' should be replaced with 'Break cover';
- the size of the header words 'Emergency door release' should be enlarged;
- user tests of this type provide invaluable information about the usability of the safety devices and associated signage. Such tests should be performed whenever a new installation or refurbishment takes place.

*Stage 6 – Legibility and Discriminability Verification Testing*

This stage revisited the viewing chamber in order to:

- verify the text size, line thickness and critical detail size specified earlier in Stage 3;
- evaluate each of the new symbols in terms of legibility and discriminability under adverse viewing conditions and identify opportunities for further improvements;
- investigate any subjective differences in legibility between three different types of luminous material;
- subjectively review the print quality limitations of the screen-printing process on luminous material.

To verify the legibility of the text and graphical symbols, the full set of symbols, and some example signs, were printed on to 'Jalite AAA' luminous material. A sub-set of symbols were also printed on Class B and Class C materials. Thirty-six participants then viewed the symbols and signs in the viewing chamber under various light conditions, in both clear and smoke filled conditions.

The results of the legibility tests showed that:

- the upper case text size recommended in Stage 3 was validated (that is, 7.5mm);

- the recommended symbol size of 33mm × 33mm at one metre viewing distance was validated;
- the absolute minimum recommended symbol size of 23.1mm × 23.1mm at one metre distance was validated;
- symbols printed smaller than 23.1mm × 23.1mm were not reproduced reliably enough by the screen-printing process and the legibility was unsatisfactory;
- five symbols required redesign in order to improve legibility and discriminability under adverse viewing conditions;
- in clear conditions, subjective ratings showed that Class C material was perceived to have the sharpest detail, compared to Class B and 'Jalite AAA'. In smoke viewing conditions, there was little difference between the three materials.

*Stage 7 – Standardisation*

In this final stage of the research programme, the following deliverables were produced with the aim of standardising a new Common System of Passenger Safety Signs:

- an initial set of 138 graphical symbols representing 43 different referents;
- a detailed set of graphics guidelines for the design of safety signs;
- an initial set of over 100 passenger safety signs for five different types of rail vehicle, developed according to the new graphics guidelines and incorporating the new graphical symbols;
- design guidelines and testing procedures for the development of new symbols.

Through a parallel project, Interfleet produced the new ATOC website and all of the above deliverables are now available on a special section of the site. This website enables any authorised individuals and organisations to download all the necessary resources to design signs which are fully compliant with the new Common System of Passenger Safety Signs.

**Recommendations**

*Standardisation*

It is recommended that the Common Passenger Safety Sign System should be mandated for all UK rolling stock new-build and refurbishment projects, and that a date be set by which time all rail vehicles in the UK shall have to comply.

*Other Safety Sign Categories*

The rail industry should undertake to extend the work of this research project to encompass those safety sign categories not currently covered (that is, hazard warning signs, prohibition signs and mandatory signs).

*Guidance for Defining Signage Requirements*

Guidelines should be developed to ensure that an appropriate and consistent process is adopted throughout the rail industry for the definition of passenger safety signage requirements in each vehicle type. The requirements should include: information content, location within the vehicle and required viewing distances.

*Standardise the Design of Devices*

The design of safety devices (for example emergency door release devices) should be standardised as far as practicable. A study should be performed to identify the optimum designs, including user trial evaluations with a representative sample of train passengers.

*Integrated Signage Design*

It is recommended that safety signs should be considered as an integral part of the vehicle design process – the information content, sign locations and required viewing distances should be defined and suitable mounting spaces provided in the vehicles.

*User Trials*

As part of the design process of new-build vehicles, and for major refurbishments, the vehicle safety system (including all safety devices and associated safety signs) should be evaluated by means of user trials involving a representative sample of train users.

## Conclusion

The deliverables from this project have satisfied those specific recommendations from the Cullen Report listed in the first section. The project has demonstrated that rigorous human factors research can be responsive to the needs of the rail industry and can deliver practical benefits which the industry can implement. It also demonstrates the benefits of effective cross-industry collaboration – in this case between RSSB, ATOC, the rolling stock leasing companies, and the consultants.

## References

British Standards Institute (2002), 'Safety Signs, Including Fire Safety Signs. Part 1: Specification for geometric Shapes, Colours and Layouts', BS 5499–1.
International Organization for Standardization (2001), 'Graphical Symbols – Test Methods for Judged Comprehensibility and for Comprehension', ISO 9186, Geneva: ISO.

# PART 13
# HUMAN FACTORS
# INTEGRATION

# Chapter 45

# Human Factors within LUL – History, Progress and Future

Simon Pledger, Caroline Horbury and Andy Bourne

## Introduction

Human factors (HF) is an area where London Underground Limited (LUL) continues to learn lessons. In projects, this was recently seen with the extension of the Jubilee line where the outstanding HF issues in control rooms, operational philosophy, workload, and alarm design ended up on the critical path for project completion. In an attempt to avoid the pitfalls from the past an Human Factors Integration (HFI) Standard was developed.

Similarly LUL is learning how to consider HF more formally from an operational safety perspective, for example in the management of change process, formal incident investigations and the reliance on procedural 'fixes'. An HF strategy was developed to attempt to address these key issues and ensure buy-in from the company's Health and Safety Executive Committee (HSEC) to the improvement of HF across the company. Key elements of the strategy include raising awareness, incorporating HF into the training of incident investigators and change assessors, as well as ensuring that lessons are learned from incidents and recommendations appropriately applied.

The chapter outlines the company's actions and progress against these initiatives, as well as the actions being taken to ensure that the application of HF within LUL and its infrastructure company partners continues to improve. Areas discussed are:

- Human Factors Forum – working in partnership to progress the application of the HFI Standard, and define aims and objectives for the continued improvement of its application;
- supporting the use of the HFI Standard, in terms of: guiding application; defining competence of User Acceptance Managers, and HF acceptance criteria;
- research to support new standards, for example HF in SPADs, and evaluating shift patterns to identify good practice for train operators;
- the future, for example incorporating maintenance work and customers into the HFI standard; development of performance indicators to track improvements on HF.

LUL helps over 3 million people to travel around the capital each weekday and about the same number each weekend, resulting in over 950 million customer journeys last year on the LUL network. LUL is now owned by Transport for London (TfL), and has implemented a new organisation, effective as of July 2003.

The LUL organisation currently employs some 10,000 people. The great majority of these can be considered to be operators. In the course of their work they 'operate' station facilities (253 owned/275 served), trains (507) and control systems in order to safely and effectively deliver London's tube service. Inevitably, operation of the network involves a variety of hazards that give rise to safety risks to members of the public and those who work on the Underground which need to be suitably controlled.

Railway operations place significant reliance upon the effective combination of equipment, personnel and operational procedures. This can pose a particular challenge in an environment of rapid technological development and changing roles for operational staff. Achieving safe and reliable performance, therefore, requires explicit attention to be given to the human requirements within system design.

During the year 2002/2003 a major derailment at Chancery Lane resulted in the closure of the Central Line until the trains had undergone safety modifications. Although this was the worst rail accident LUL have experienced in over 25 years, there were fortunately no serious injuries sustained. The accident has been subjected to a number of internal and external investigations. The findings and recommendations from all of these investigations have been, and will continue to be, responded to appropriately by LUL to ensure that the issues raised are addressed.

In addition, experiences with recent projects have highlighted the problems that can arise in integrating human factors (HF) into system design. In particular, the Jubilee Line Extension Project introduced a number of new and advanced control systems, which posed particular challenges in terms of HF and resultant threats to effective project delivery. This clearly highlighted to LUL the need to address the approach taken to human factors integration (HFI) in future projects.

As a first step in a response to such problems, a standard has been developed that addresses the integration of HF into future developments of railway control systems. The standard focuses on the management of all elements relevant to HF performance, including the specification and training of the people who will use the system, derivation of operational processes, the detailed design of equipment, and the working environment.

This chapter begins by summarising the problems that have been encountered in addressing HF within LUL. It then describes the progress of LUL in addressing HF, and provides some insight into the perceived challenges of the future.

## History

*Lessons Learned from the Jubilee Line Extension Project (Bourne and Carey, 2001)*

The JLE Project introduced new trains, a new service control centre built at Neasden, and station control rooms provided at each new station. Significant effort was put into the ergonomic analysis and design for the Neasden station control rooms. In the course of the project, and for a variety of reasons, changes were made to the delivery strategy, which were not subject to the same level of HF analysis as had been the case for the original design. Furthermore, these changes had implications for safety assessments, staffing procedures and training development, which proved difficult to progress in an integrated manner across the various contractual and organisational interfaces. As a result, a number of key operability, training and system reliability concerns were raised by HMRI at a late stage in the project.

These issues posed not only a threat to the safe and efficient operation of the railway, but also a major project delivery threat. To resolve the concerns a joint group made up of operators, project team members, Chief Engineer representatives and other stakeholders was formed, called the Human Factors Operability Group (HFOG). This was supplemented by a team of HF experts. This joint approach was essential to understanding the issues and quickly resolving them, and the expert perspective gave the group credibility with regard to the HMRI. This strategy was successful in addressing many of the issues raised and demonstrating sufficient evidence for safe operation of the system by the operators. However, it was also expensive (involving late changes to systems, significant management and technical effort at an already critical phase in the project). In 1999, LUL opened the Jubilee Line extension (JLE), allowing passenger services between Stratford in East London and Stanmore in the northwest via several major interchanges, London's Docklands and of course the Millennium Dome at North Greenwich.

In the process much has been learned about HF and how LUL's own processes can be improved to address safety issues such as operator workload and human error to the benefit of both operators and customers. A number of key lessons have been learned on the JLE Project, namely:

* the need for HF analysis early and throughout the project lifecycle;
* the need for an overall systems integration role;
* the need to formally incorporate consideration of HF into the engineering decision making and project lifecycle;
* to recognise the impact of change on the human machine interface;
* to improve LUL standards for HF.

*HF Awareness*

Major incidents around the world have demonstrated the importance of HF in determining safe and effective operations, especially in environments such as London's tube network where a substantial part of its operations take place underground. Some of these incidents (for example Ladbroke Grove and Chancery Lane) have been specific to rail, and have therefore resulted in a growth of HF awareness in the Rail Industry. These incidents, and the influence of the Cullen Report, have highlighted the need to understand and develop LUL's approach to HF.

HF awareness in LUL has very much been driven as a result of the JLE project, and as projects have been carried out as required within directorates, HF support and research work has been carried out on an *ad hoc* basis for the specific directorates.

*Human Factors and the Private Public Partnership (PPP)*

LUL is embarking on a new era of investment through a public private partnership (PPP). Under this arrangement LUL, as a publicly owned company, remains the infrastructure controller and safety authority, and operates the trains, stations and signalling systems. Three infrastructure companies, Tube Lines, Metronet Rail BCV and Metronet Rail SSL (known collectively as Infracos) are responsible for renewal and maintenance of the assets under contract. These Infracos operate under Safety Cases accepted by LUL. The Infracos, along with other suppliers (including three major private finance initiative (PFI) suppliers) are supplying LUL with engineering services and assets to enable LUL to operate and improve the tube for its customers. These high level performance based contracts require significant transfer of risk to the supplier and hence proscription is not always possible with regard to standards.

This is significant from a HF perspective, since the operators of the equipment will be on the LUL side of the contract and the assets they are required to operate on the other. The performance of the assets when in use will be of importance not only to LUL (as user) but to the Infracos, given that returns on investment in new equipment will be determined by resulting service performance.

## Progress

*The Regulators' Top HF Issues*

The following list provides an insight into the regulators' (Her Majesty's Railway Inspectorate (HMRI)) top HF issues. London Underground is focusing significant effort in addressing some of these issues, which will be described in later Sections:

1   organisational change and transition management;
2   demanning, staffing levels, shiftwork and overtime;
3   training and competence;
4   alarm handling;
5   driver/signaller communications;
6   vigilance devices and reminders;
7   human reliability (COREDATA, Railway HEART);
8   measuring workload and devising strategies;
9   human factors integration;
10  compliance with safety critical procedures;
11  safety culture/blame culture;
12  ergonomic design of interfaces and cab design;
13  maintenance error.

*Human Factors within the LUL Group*

Figure 45.1 provides an overview of the current key internal and external stakeholder roles in the provision of HF within the London Underground group (LUL and Infracos), and their relationships (key communication lines).

## Progress – HF within LUL Group (2003)

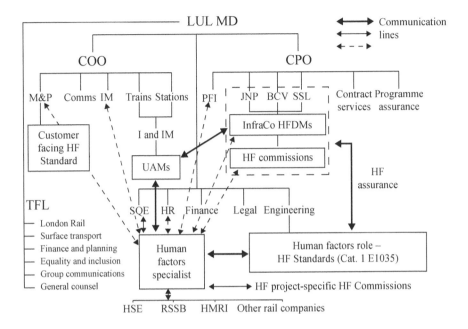

**Figure 45.1   Human factors roles in the LUL group**

*The London Underground Human Factors Standards*

*Background research activities*

The JLE Project and the PPP highlighted the need for improved HF standards and guidance within the London Underground group of companies. Subsequently, LUL commissioned a project to capture lessons learned from previous LUL projects and identify HF best practice in engineering projects (across sectors), to propose an approach to HF for future projects and to develop standards as necessary.

Input to the analysis phase of the project was obtained from a variety of sources including: HSE/HMRI; other rail contexts (notably the West Coast route modernisation project); air traffic control, including UK NATS; nuclear industry; and military systems, focusing on UK MOD, but also examining US military approaches. These were recognised as areas where specific effort has been placed into HFI into system design processes. A literature review and a review of standards was carried out. The analysts' experience in other industrial areas (for example process control) was also applied in reaching conclusions.

*Specification of the standards*

On the basis of the research phase, recommendations were made to drive the production of new standards and guidance, aimed at non-HF specialists. These were presented, discussed and endorsed at a workshop attended by representatives from key engineering and operations personnel from across the LUL group of companies.

The HFI standard E1035 has been developed primarily for control applications areas such as signalling, station control and train cabs. It also focuses just on the operation of equipment by LUL staff. Under the PPP arrangements, requirements related to equipment maintenance are not regulated by LUL standards, except in those areas where poor maintenance may impact upon operability. The Infracos are expected to implement their own standards to cover those aspects of design which affect their own staff.

The standard was developed to target management of the HF process and does not dictate the use of any specific methods or criteria. Advice on how to interpret and apply the standard is provided in an accompanying Manual of Good Practice.

*Benefits realised to date*

The HFI standard has followed a well-trodden path in aiming its requirements at an organisational level within projects. It is not a highly detailed or technical document. However, initial evidence suggests that it is already succeeding in changing awareness, culture and practice in current projects. There is every reason to believe, from reviews to date, that the results will be better end-products to the benefit of the LUL, its suppliers and ultimately its customers.

Having created the awareness, the next step is to develop further supporting materials and capabilities within LUL and the Infracos to address HF in a competent and mature manner. There is still a long way to go, but initial steps have been encouraging.

*Human Factors Forum*

A requirement was identified for cross-company grouping that exists to coordinate the integration of HF considerations into the operations and assets of the company. The use of champions and a HF forum to assist in this process was endorsed by the Board Health, Safety and Environment Committee in July 2000.

The HF Forum aims to:

- facilitate an integrated approach to managing HF by bringing together key players from different areas of the business to increase cross-directorate and cross-business understanding and cooperation;
- provide a focus in the business for HF issues, including raising awareness and spreading good practice;
- enable interaction with outside bodies with similar interests;
- ensure that LUL and its partners can demonstrate competence in HF issues.

*What is LUL doing?*

In 2002, the SQE Strategy and Planning team produced a memorandum to the LUL Health, Safety and Environment Committee providing a strategic view of the application of HF in London Underground. The following issues were highlighted:

1   there is evidence of an awareness within LUL of the need to consider HF in the design of jobs and equipment. However, with this awareness comes a generally poor understanding within the business of what this means in practice;
2   job designs, and procedures for carrying out those jobs, are generally not informed by good practice in HF;
3   the HF standard requires our suppliers to provide HF expertise, but LUL still has obligations which require an understanding of HF in establishing its operating philosophy. The standard does not set requirements to ensure that HF is considered from either the customer or maintainers' perspectives;
4   a poor understanding of HF root causes from either incident reporting or incident investigation.

The focus of the remainder of this section, and the 'Future' section, is to describe the additional activities of LUL to: address the internal HF issues identified by the HF strategy; address the regulators' concerns; and, incorporate HF into PPP, PFI and in-house projects

*Human factors in trains (cabs) (ongoing)*

The purpose of this commisioned work is to provide ergonomics support for the development of the cabs of new rolling stock and conduct an analysis of the existing LUL rolling stock operator facilities, to identify the HF Issues during normal modes of operation that can be: reactively removed from existing stock through feasible mitigating actions, and; proactively designed out of future LUL stock through the generation of an LUL Operators Facilities HF Standard.

*Human factors in stations (ongoing)*

There are currently many ongoing projects whereby stations are being modified or refurbished. This is being done on an individual office basis (for example the Fulham Broadway ticket office), a station site basis (for example the Kings Cross redevelopment) and on a global basis (for example the introduction of the Connect digital radio). Some of the projects are also aiming to deliver generic good practice designs (for example ticket offices). The introduction of digital CCTV, Help Points, and the influence of legacy design/systems/equipment, are each introducing significant challenges for HFI.

*New/existing projects under PPP (ongoing)*

There are a vast number of significant new projects, or projects that have been 'handed over', as a result of the PPP contract. The following are just some of the major projects requiring substantial HF effort, being provided through the Infracos: Kings Cross redevelopment; CONNECT digital radio integration; TIMIS (train identification and management information system); FOCUS (control room project incorporating CCTV monitoring and customer service improvement facilities).

*Human factors in quantified risk assessment (QRA) (ongoing)*

The original LUL risk model incorporates some human reliability assessments. Periodic updates to these models includes reviews of these HF precursor analyses and human reliability considerations in our QRA models, and further development work is already planned.

*Customer facing human factors (ongoing)*

The HFI Standard E1035 focuses on the operational angle (ie not maintainers and customers). Therefore, guidance has been produced on HFI in relation to customer service delivery standards. The aim of this guidance is to ensure that the equipment and assets design, as customer-facing applications, consider the human factors requirements of their users. The link between this guidance and the HFI standard is still required.

*Signals Passed at Danger ( SPADs) (ongoing)*

LUL have commissioned a feasbility study to investigate the application of a human factors checklist to LUL SPAD incident reporting (based on Network Rail version). HF awareness has also been incorporated in duty manager SPAD training.

*Shift work and work-life balance (ongoing)*

London Underground has commissioned the University of Surrey to monitor any changes to individual train operators' work-life balance during the trial introduction of fixed-link shift-working. As part of the monitoring project all train operators currently on, or soon to be on fixed or mixed link schedules, are being asked to volunteer to take part in the study.

**Future**

*Challenges*

A number of challenges are apparent for the long-term future of HF in LUL:

1   to ensure that HF is effectively integrated into project planning;
2   to ensure that adequate and appropriate consultation takes place to increase uptake of the findings, and how to implement those findings;
3   how to continue to provide support to the development and interpretation of the HFI standard and people using it that is, it is not a 'fire and forget' process;

4   to provide assurance, and agree, a necessary level of competence in all roles;
5   to provide metrics for measuring the effectiveness of HFI;
6   to be assured that there is an holistic supply chain awareness;
7   to ensure system integration and consistency;
8   management of end user wishes vs HF best practice;
9   use on large projects (organisation, specialists);
10  aligning with contractual issues (cannot ignore the contract).

*Future – Next Steps*

*Gaining more experience*

Human factors have emerged as a major influence on safety performance, and LUL is dedicated to achieving a greater understanding and control of these risks. We will continue to work both within the LUL Group and externally (RSSB/HMRI/HSE/The Ergonomics Society, etc.) to seek opportunities to increase our understanding and knowledge leading to increased options for control.

*Addressing a Culture of Blame*

The organisational culture which makes our staff reluctant to admit to errors or to freely challenge procedures that they have difficulty following is deeply rooted. LUL has embarked on a programme of breaking out from old behaviours and attitudes, based on values such as trust and respect. This presents the best opportunity for changing this culture and underpins specific safety improvement actions.

*Support (templates)*

There is a need to continue to provide guidance and support on: the implementation of the HFI standard E1035; the incorporation of HFI for customers and maintenance (with guidance linked to PPP contract); and roles and responsibilities for human factors competence assurance.

*Incident reporting and investigation*

We have an existing safety improvement programme to improve the quality of our incident investigations by providing a cadre of investigators trained to identify root causes and improving the quality of actions arising from the recommendations made. Some progress is being made but the improvement in this area needs to continue. The SQE HF specialist will be involved in Formal Investigation (FI) teams, or review recommendations from the FI reports to analyse any wider HF implications.

*HF awareness (LUL group and TfL)*

There is a need to continue to spread awareness of human factors throughout the business, not only in terms of what the study of human factors is, but also what this means in practice, and the benefits of appropriate and effective integration. There may also be a future need in spreading this awareness to Transport for London (TfL).

*Competence definition and assurance*

Significant changes to organisational arrangements, job designs, shift patterns etc. fall under our safety review and change control requirements. As part of our existing safety improvement programme there is an action to provide more skilled safety assessors to assist those accountable for the change to plan and carry out the safety assessment. The training for these safety assessors includes the application of HF assessment techniques to such changes.

The standard for HF integration requires a plan for HF integration. The key here is the quality of the plan; the assessment of what we expect of our people, the HF addressed in the plan, and the implementation of the plan. Assurance crosses traditional asset and operational assurance areas, and more clarity is required on who receives assurance and the resources required to provide competent HF advice.

*Develop human factors metrics*

The LUL HF specialist has the responsibility to determine proactive indicators that can be monitored to identify the effective application of human factors.

*System integration and consistency*

Perhaps the most significant challenge ahead for the London Underground group of companies is ensuring that human factors are integrated within systems appropriately, and consistently. It is expected that the HF Forum will be the key facilitator in ensuring the necessary stategies, standards, processes and policies to ensure appropriate integration and consistency across the LUL Network.

**Conclusion**

LUL are currently taking a positive approach in incorporating human factors. The organisation and Infraco partners carry out a lot of HF work, and are supported by a business awareness and commitment to HFI. Furthermore, the companies now have an HFI Standard and an HF Forum, as mechanisms for ensuring good HFI practice.

However, much of the HF work is still carried out on an as needed basis by the business, is done in a micro versus macro manner, and relies heavily on competence and self-assurance. In addition, there are some pockets of reinventing wheels (that is paying for same thing twice) and it is hard to identify performance and progress.

The future offers many challenges for HFI within LUL, but the commitment and enthusiam of all stakeholders, aligned with the groundwork already in place, suggests that the London Underground Group is well placed to meet these challenges.

## Acknowledgements

The authors wish to acknowledge the role that colleagues in LUL, its suppliers, and consultants have played in many of the areas discussed in this chapter. Permission of LUL to write and present this chapter is also acknowledged with thanks.

## Reference

Bourne, A. and Carey, M. (Amey Vectra) (2001), 'Integrating Human Factors into the Development of Railway Systems', IEE Conference *People In Control 2001*, Manchester, June.

# Chapter 46

# Human Factors Integration – A Company Case Study

Sarah D. Garner

## Introduction

There is sometimes a reluctance, perceived or otherwise, in engineering communities to accept the validity and importance of human factors. The intention of this chapter is to discuss the experience of Lloyd's Register MHA Limited (LRMHA) in addressing some of the integration and interface issues between the engineering and human factors activities in the UK railway industry. The objective of this study is not to present the 'right way' or the 'only way', but to share the author's experience and approach.

The author's experience is that a human factors specialist must be able to communicate in the technical language and demonstrate the in-depth domain knowledge that is important within the rail engineering fraternity. Conversely, there is a need for engineers to understand the potentially critical role human factors can play in overall system concept, design and operation.

At the inception of LRMHA's human factors function, a human factors specialist was co-located within the safety engineering team, thus enabling an exchange of information in both directions. Human factors knowledge was informally imparted to the engineers by real application to safety engineering projects in a pragmatic and appropriate manner. Conversely, the close involvement of the human factors specialist allowed the engineers' railway domain knowledge to be transferred. More formally, the human factors specialist conducted an interactive workshop for the Safety Engineering and Rolling Stock Teams which deliberately stayed away from the traditional railway ergonomics areas, but instead demonstrated the breadth of the subject, and to indicate the scientific basis behind human factors work.

This mutual exchange of knowledge has led to a much better understanding of the impact of physical systems on the operator by both functions, which has enabled engineers to see when and where human factors is needed, and the human factors specialist to understand safety and engineering processes. This has meant that both parties can work more closely to a common aim such that the complex interaction between the engineering solutions and the human interfaces are understood and the impact on overall system performance is determined. This approach allows sensible and practical solutions to be developed.

The pragmatic, mutual knowledge transfer approach has led to a truly integrated team and the use of a human factors specialist as a junior safety engineer has opened many doors. There is still a lot of work to be done but the important battle to convince colleagues of the value of human factors is well on its way to being won.

## Lloyd's Register

Lloyd's Register is an independent non-profit distributing organisation, founded in 1760 and now operating in over 100 countries worldwide. It provides safety assurance services to the marine, offshore, land-based industrial and transport sectors, and contributes to international standards making in these fields. Lloyd's Register employs over 6000 professional staff worldwide with a wealth of experience within the transport sector.

While the form of these services varies considerably between sectors, the common element is that all Lloyd's Register's principal services involve some form of independent verification and validation, and Lloyd's Register has an international name and reputation as a provider of robust and authoritative independent appraisal. The services cover the design and through-life integrity of plant and equipment, and the management of safety throughout the lifecycle, including the assessment and certification of safety management systems.

Lloyd's Register MHA Limited (LRMHA) is a member of the Lloyd's Register group of companies whose headquarters are in Fenchurch Street, London. LRMHA is an established international consultancy offering a wide range of services to the transport sector. LRMHA's head office is in Reading, and has further offices in York, Leatherhead, Glasgow, Derby, Crewe, Bristol, Preston, Birmingham, Singapore and Hong Kong.

LRMHA was established in 1993 and registered as a limited company in July 1996. LRMHA employs over 120 professional staff with expertise including signalling, telecomm, infrastructure engineering, civil engineering, rolling stock, energy, safety and risk engineering, RAM, system integration, independent verification and validation, and management consultancy services.

## Background

As many will know, human factors has been a rapidly expanding area in the rail industry, and regulations and standards are evolving to incorporate a more complete understanding of the issues, particularly in the safety engineering and safety justification area. It was this requirement which led to the need for in-house human factors experience at LRMHA.

In mid-2000, LRMHA advertised for a human factors specialist to join its consultancy. As is often the case there were not enough seasoned HF consultants to go round. Instead of waiting LRMHA chose to employ a recently graduated ergonomist with some experience until a senior consultant could be found to lead the group. As there was no established human factors team within LRMHA, the ergonomist was placed with the Safety Engineering Team. These two factors, the junior ergonomist and their placement within the Safety Engineering Team produced a mutually beneficial method of working which we believe has led to a better interaction and understanding of HF within the company. The ergonomist had no preconceived ideas as to how human factors should work in a consultancy environment which enabled the group to develop their own ideas and strategies without being constrained to previous regimes. Both parties were learning as they went along.

The fact that the engineers were in the same office enabled the ergonomist to learn to work with them and gain a better understanding of the railway system. This was particularly important in the initial stages of the relationship. As with any discipline it takes time to build up a body of work, and during this time the ergonomist was able to train and work as a junior safety engineer, again gaining a hands-on learning experience of the railway and the safety engineering process.

## Problems

The engineers had some understanding of human factors and had worked with human factors consultants in the past but had also experienced problems (whether perceived or otherwise). Obviously working closely with the engineers meant the ergonomist was only too aware of the issues. Whilst there was the usual 'it costs too much', or 'adds no value', there seemed to be a lack of understanding between the two groups, engineers and ergonomists, and no real way for the two to fit together despite very obvious overlaps in basic principles.

It became apparent that one of the most prominent problems was the perceived lack of applicability of human factors reports to safety engineering work or, perhaps more correctly, a lack of understanding by both parties. This was a major concern as any value or importance the human factors reports contained was at risk of being lost.

Surely as human factors professionals, if the engineers we produced reports for did not understand those reports then we, as a profession, were not doing our jobs properly? It is the same as designing a control panel that the user cannot possibly use. We are all trained to focus on the user, it is the whole point of us being here, but we need to recognise who our users are. It seemed that we had forgotten about the engineers and only concentrated on the end user.

In any mix of disciplines it is accepted that there are going to be disagreements and there is often compromise, but it became obvious that to fully integrate into the engineering team these problems of communication and understanding needed addressing by both parties. It is unlikely that any of these issues were unique to LRMHA as the problems seemed to be apparent at project level rather than just at company level.

## Laying the Foundations

As already mentioned there was the opportunity for the ergonomist to train alongside the safety engineers. The ergonomist was encouraged to ask questions and seek information on the railway system. In the same vein the ergonomist was also able to advise, and answer questions about their own field. The ergonomist found that the more they understood about the railway system the better able they were to interpret external human factors reports so the engineers could understand why a recommendation had been made.

Some six months after joining the company there was an opportunity to undertake a presentation to LRMHA's Derby Office (the centre for the Safety Engineering and Rolling Stock Team) about human factors, as part of an away day. This was seen as an opportunity to show how wide the human factors discipline is and to explain some of the science behind the 'common sense engineering' that human factors professionals seem to be accused of.

Previous experiences had shown the ergonomist that most engineers have a limited understanding of what human factors actually is and they tend to be limited to either the one module they studied at university or the exposure they have had in the workplace. It seemed that the only thing the ergonomist was ever asked about was workload or human error. Both valid and important areas, but only a small part of the human factors portfolio.

The aim of the presentation was to demonstrate:

- that there are close links between human factors and in particular safety engineering;
- that it is not 'commonsense engineering'. There is real science behind it!;
- the breadth of the subject and how it can be applied.

It was also important for the presentation to be remembered (no point doing it otherwise), so an interactive workshop format was adopted.

A task analysis session was used to show how close engineering and human factors can be. Many of the engineers used fault trees on a regular basis and so the presentation of information in this format was nothing new. Past work which had been completed prior to joining the company was discussed to show how those skills could be transferred and the similarity between basic tasks. The training to gain an ergonomics degree was explained to highlight the importance of the scientific approach, and explain how having an understanding of, for example, anatomy can help with design. This helped to illustrate the science behind the common sense.

Perhaps the most successful part of the presentation was the empirical workshop, developed from one the ergonomist had taken as part of their degree. The ergonomist wanted to show the engineers how users can be limited by the system. In a safety critical industry these limitations are so minute it is difficult to demonstrate, for example so what if the signaller gets a bit hot/is a bit busier? It was decided that by replicating a physical condition it could be more easily demonstrated by experience how we can be restricted and limited by our environment. It was a simple adaptation of the original idea involving the simulation of arthritis in the hands. The engineers taped buttons to their knuckles to restrict the movement of the knuckles. They were then asked to complete simple tasks, such as writing their name, or unwrapping sweets, and were asked to start thinking about how they would be affected whilst completing a typical journey on a bus, train or even just to the shops. It had the right effect, everybody remembered that bit of the day.

When this workshop was held at a later date with some graduate trainees, one of the questions asked was what comes to mind when you think about ergonomics. One of the most rewarding comments was that 'ergonomists think differently to engineers'; not that it is just common sense but simply that we think differently. If both parties understand this then they can start to build on that understanding.

## Building Blocks

It seemed to the ergonomist that the profession had missed the point by forgetting that the engineers and designers are our users as well as the signallers and drivers. Whilst it is all well and good completing the textbook study (assuming there is time and money for it), if the engineers do not understand the results there is little point, and worthwhile recommendations will be given no merit.

We would not talk to signallers, maintainers or drivers in the language of human factors so why do we write reports to engineers in it – indeed why do we expect engineers to take the time to explain the meaning of specific railway terms if we cannot be bothered to start finding out for ourselves?

It was important that the work produced by the ergonomist could be integrated by the engineers into their projects. The ergonomist also needed the engineers to see the need for human factors so that they could sell the service to their established customers. To do this the ergonomist needed to be able to talk to the engineers in their language and to convince them of the value by showing them examples which they could see the point of. The ergonomist also recognised that, particularly in the rail industry, there was a certain reluctance to change, that is, an 'it is always been done like that' mentality, and that to gain any respect you have to learn about the railway and at least be able to understand the basic technical language and terminology. The inclusion of the Ergonomist within LRMHA's Safety Engineering Team meant that both parties were in an ideal position to pick up this language and system knowledge.

## Working Together

The more the engineers understood about human factors and how it can be used the more able they were to see opportunities within their own field. That is not to say they were not continually reminded, but the idea of integrated projects was starting to work.

At the same time the ergonomist realised that whilst the ideal of a full study on a given project was nice, in many instances it was not practical or appropriate. Nonetheless the possibility of offering a little bit of advice or help was still worthwhile. At the same time the Ergonomist was trying to educate the customers that there was an easier way to address the HF issues. At least by trying to help and by being flexible it was a 'foot in the door'. The feeling was that even if a human factors study resulted in only one change, that would make it better than nothing at all. This was an approach that seemed to work. The pragmatic viewpoint and understanding of the system was enough to gain repeat business, and each time the ergonomist was asked to undertake a little bit more.

## An Example

The ergonomist was involved in a project as the Hazard Log Manager, 'it will just be half a day a week, an easy job', for a fairly small scale resignalling scheme. However, events conspired to complicate matters and whilst the original job was still a small scale resignalling scheme it had a much higher profile. At one point our engineer asked the customer, 'Who's doing the human factors work?' The customer responded 'Nobody'! A job of this size previously would have had no human factors involvement and to some extent it is understandable, but this time there were issues which our engineers recognised would need addressing. So the half day a week 'easy job' turned into something slightly larger and more problematic. Yes, the ergonomist was brought in late in the day but we were still able to make changes. The signallers felt more involved in the project as the ergonomist visited the signal box to talk to them. Most importantly the customer listened too, and was involved in the recommendations that were made. Where possible, virtually all the recommendations made were enacted.

There were still compromises and things the project would have liked to change that they were not able to, but on the whole, considering no HF work was planned, it was generally agreed that a lot had been achieved. The second phase of work is now being discussed. This time, however, instead of being an add-on, the human factors issues have been considered from the outset.

By working with our engineers and the customer so that both parties understand the difficulties and sometimes apparent vagueness of human factors work, it has meant that the recommendations have been better understood and more favourably received.

## Way Forward

This study is an example of how, through some direction and a degree of serendipity, human factors was integrated into the Safety Engineering Team. It seems to work in our Derby office but we know we still have a long way to go before human factors is fully integrated in the company, but that is true for many specialist disciplines in many companies.

We have made a start and are committed to trying to resolve the issues raised and to understand all users. Working as we do in multidisciplinary teams allows for a greater exchange of learning and information between parties and more support, that is, I am not the sole ergonomist sat at the table because the engineers understand why I say what I say and back me up. Yes, there are still compromises but we believe that giving in a little is better than being alienated and ignored.

This may be seen by some as a dilution of human factors. We should not give up the 'holy grail' so to speak. As a profession, however, we need to recognise that a formal Human Factors Integration Plan does not mean that our work will be fully integrated. The work described was integrated because the team were, not because there was a plan laid out. The arguments were won because the whole team believed in them not just the ergonomist.

We need to recognise that it is not always possible to complete things by the book. Sometimes partial compromise is the only way forward and we are reliant on adapting techniques as best we can. At least if we understand the system and the engineers understand us, we can work together to try and fix the problem. We have to learn to adapt what we know to fit the situation. Human Factors is important but it is not the only cog in the system and unless we can be flexible it is felt we will be left behind with cries of 'it costs too much' and 'you don't understand.' From the experience we have, it is seen that true human factors integration will only ever happen if it is pursued at company or project level. To do that we must come back to one of our first principles: understanding the user.

## Disclaimer

This chapter is not intended to propose a right or wrong way to approach human factors integration. Nor is it intended to be anything more than a general discussion; a download of ideas. The work is an account as to the author's experiences and why they feel that so far the process seems to be working. There was no formal plan for integration, indeed to some extent various happy accidents have contributed to what was really a unique situation to learn from.

It should be added that this is purely an account of the author's experiences and does not necessarily represent the view of the whole company.

Chapter 47

# Ergonomics Standards in the UK Rail Industry

John Wood and Theresa Clarke

## Introduction

Less than three years ago a trawl though UK rail standards using the terms 'ergonomics' or 'human factors' produced almost no returns. For an industry so heavily dependent on human intervention – whether for driving trains, signalling or maintenance – this was clearly a matter to be addressed. An early initiative of the newly appointed Head of Ergonomics at Network Rail was to put in place an ergonomics standards writing programme. As a result of this initiative the available standards now include ergonomics 'policy' for Network Rail, human factors integration in major programmes, alerts and alarms, control rooms, human–computer interfaces and signal box design.

This chapter outlines the scope of the new standards showing how they relate to each other and their target audiences. The specific challenges in writing a wide-ranging 'human factors integration standard' is discussed.

## A Framework of Standards

The UK rail industry generates two classes of industry-related standard – 'Group' and 'Company'. These are not only issued by different organisations but also have differing objectives (see Figure 47.1).

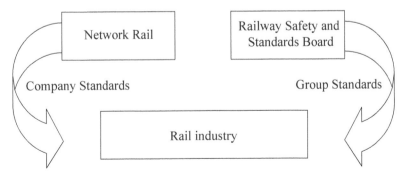

**Figure 47.1 Delivery of rail industry standards**

515

Group Standards are prepared and circulated by Railway Safety and Standards Board and set out requirements for safety and safe interworking. Their scope covers the control of risks to passengers, railway workers and members of the public relating to the rail infrastructure. As the Railway Group Standards Change Procedure outlines, they mandate 'what' is to be achieved rather than 'how' objectives are to be achieved. As far as ergonomics is concerned they are setting performance requirements.

Network Rail 'company' standards are issued in two forms – Company Specifications presenting specific provisions that should be met and 'Company Procedure' in which the overall process to be adopted is spelt out. The boundary between specifications and procedures is not always a hard one – with the control room specification discussed here, for example, guidance is also provided on processes. Company standards provide basis for achieving the requirements presented in Group standards – thus addressing some of the 'how' issues.

**Policy Standard**

Not many industries can boast of having issued a standard on 'ergonomics policy' as recently issued by Network Rail (NR). The document sets a framework for the incorporation of ergonomics in the rail industry and provides the cornerstone for the set of interlinking standards now been issued by Network Rail (2002). The document provides managers and suppliers with an indication of the company's broad intentions and expectations in the area of ergonomics.

Ergonomics policy requirements have been developed for the main areas where the subject should make a contribution. These include the following:

*   general application of ergonomics;
*   control facilities;
*   human machine interfaces (HCI);
*   job and task design;
*   work organisation;
*   systems integration;
*   communications;
*   station and related facilities;
*   interaction with the public;
*   personnel support systems.

For each of the topic areas high level policy statements are presented together with the associated rationale. For example, on control rooms, many of the specific NR recommendations cite the need to recognize the ergonomic requirements presented in the emerging ISO standard on this subject (ISO, 2005).

The policy standard draws on, and distils, the extensive experience of the application of ergonomics both within the rail industry and elsewhere. Its writing involved a careful balance between aspiration – necessary for things to improve – and what was practical within an industry coping with simultaneous demands on many fronts. It was not an easy document to prepare – something acknowledged by a number of the highly experienced ergonomists who contributed.

**Figure 47.2 'York IECC' – Rail Network Control Room**

**Control Room Design Specification**

Even in the early days of running a rail service it had been accepted that rail control rooms play a major role in the safe and efficient running of the network. What has taken longer to recognize is the role that human factors must have in order to achieve these goals. Many innovations in signalling have arisen from accidents, and those steeped in the history of the railways, can always rattle off the accidents which led to a particular change. Change has often been based on technical innovation rather than a deep consideration of any underlying human factor limitations which may have caused the problem. The fact that human factor problems exist has been recently exposed through the Cullen Report (HSE, 2001) where human factors was mentioned in relation to task analysis, alarm presentation and training and supervision.

The underlying driver for this specification (Network Rail, 2003a) is the goal of improving the ergonomic quality of NR signal boxes. The overall structure of the specification is based around that of the emerging International Standard on the Ergonomic Design of Control Rooms (ISO, 2005). The rail standard has been geared to the industry's special needs and does not slavishly follow that of the ISO, though where specific requirements are presented in the ISO, these are also presented in the rail specification.

The specification is intended as a practical document providing both guidance on the control room design 'process' and well as specific comment on human factors related features that need to be addressed. These features are themselves subdivided into 'requirements' that need to be met and 'recommendations' which represent good practice and the document structured along these lines.

Following the process presented in the document will not necessarily guarantee that weaknesses will not be incorporated into the design though the risk of doing so will be substantially reduced – this would be true for any domain where the ISO standard is used. In one area the document addresses issues not covered by the ISO – though commonly encountered in practice – that of refurbishment of control rooms. Control room refurbishment raised particular problems for project managers in that achieving a minimum level of ergonomic acceptance may be both technically and financially impossible. The risks involved in continuing with the existing working environment needs to be weighed up against the benefits of implementing change. The NR specification provides guidance to project managers on how to address this difficult problem.

Unlike the 'greenfield site' situation, where the ergonomic ground rules can be drawn up from scratch, the challenge faced during refurbishment is very different. Decisions need to be made on whether features which just cannot be changed are so constraining that they are likely to totally undermine any attempt to achieve a satisfactory ergonomic solution.

The specification presents a series of tables in which some of the conditions that have been previously found to lead to an unsatisfactory solution – 'ergonomic showstoppers'– are highlighted. The judgment as to whether the risk can be taken, once having established an 'unacceptable' constraint can only be taken by a domain experienced ergonomics engineer. The guiding rules will vary from case to case.

The decision process centres around those elements which would be difficult to change – referred to in the standard as 'fixities' and these vary depending on the stage of the project. A check on budgets at the start of the programme, which results in a conclusion that no resources have been allocated, will obviously stop any ergonomics input in its tracks.

The criteria listed are defined as follows:

'Go'      No foreseeable problems whatsoever
'Critical' Likely to be a problem – constraint takes solution just outside usually acceptable ergonomic standards
'Abort'   Fixity would result in solution well outside acceptable ergonomic solution

The control room specification covers the following topics:

* the ergonomic design process;
* space selection and control room layout;
* access and circulation in control rooms;
* environmental ergonomics;

- finishes;
- 'control suite' layout;
- general requirements associated with workstations;
- interface design.

The specification offers general requirements for workstations but not in detail – this is covered in the workstation design specification (Network Rail, under preparation) discussed later in this chapter.

The standard also includes a 'compliance checklist'. It can be used at any stage of a control room planning and design process to identify ergonomic tasks that may need to be undertaken. Requirements in the checklist have been cross-referenced to the main text. In each case these requirements have been classified as 'Requirements' (RQ) which are mandatory or 'Recommendations' (RC). Mandatory requirements reflect those to be found in International/European Standards.

## Human Factors Integration Procedure

The procedure covers the replacement, upgrade and new design of systems which are electronic rather than mechanical (Network Rail, 2003b).

The satisfactory integration of human factors in major development programmes has been a challenge for ergonomists since the early days. MANPRINT (DERA, 2000) was a framework developed by the Ministry of Defence and adopted by them to get human factors introduced at the appropriate stages of weapon systems design. The documentation associated with this was voluminous and one of the criticisms was that the application could be overkill on small projects. The difficulty project managers had was in deciding which programmes fell below the threshold where the wholesale application of the MANPRINT programme would cease to be cost-effective.

The development of a human factors integration standard for Network Rail benefited from the work undertaken by London Underground Limited (LUL) who had prepared a similar standard for their needs (LUL, 2001). Meetings were held with LUL users of their own standard and discussions were also held with MoD on their experience of applying MANPRINT.

An early decision was made to prepare the procedure in two parts – a mandatory section encapsulating the essence of the required process supported by a series of Guidance Notes.

The mandatory section defines roles and responsibilities within a human factors integration programme – how projects should be organised and the main tasks and 'deliverables' for each project stage. The procedure relates the framework found in the ergonomics standard on Human-Centred Lifecycle Process Descriptions (ISO, 2000) with that of the 'engineering lifecycle' familiar to rail industry engineers. The 14 Guidance Notes provide the user with advice on such topics as 'preparing an ergonomic brief', 'writing an ergonomics integration plan', 'contacts with HMRI and approval bodies' and 'human reliability and hazard assessment'. In the Guidance Note on Project Organisation, for example, the supporting guidance provides a simple flowchart for determining the level of ergonomics input required (Figure 47.3).

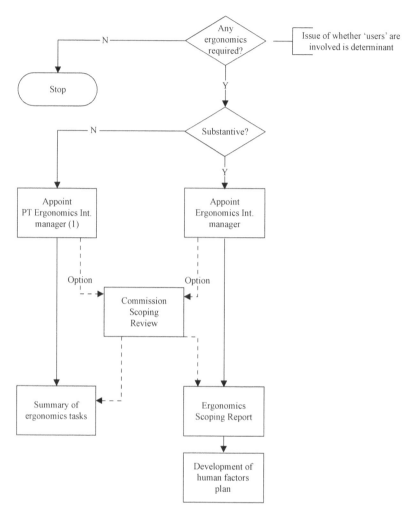

**Figure 47.3   Framework for deciding ergonomics support**

The standard has been written with the lay reader in mind. It was recognized during the drafting, however, that the application of some of the methods and approaches would require expertise in ergonomics.

## Workstation Design Specification

There are over 1,000 signal boxes controlling the UK network. They range from lever frames – where signals and points are controlled directly by actuating levers pulled by the signaller – to computer-based systems where control is exercised via VDUs. NR has a rolling programme to upgrade and replace signal boxes as they become obsolete and the specification provides a framework for achieving a satisfactory ergonomics baseline.

This specification (Network Rail, under preparation) presents requirements and guidance to cover the design of VDU workstations in railway control centres. The specification covers the design process for workstations from requirements definition though manufacture and installation to evaluation. The purpose of the specification is to ensure that user needs are met in the design of any workstation.

The structure of the specification encourages standardisation. This is based on best practice and the process presented leads the design team to select from a set of 'standard' workstation elements presented within the document.

The scope at present in restricted to all workstations that accommodate VDUs within a railway control room and this covers zone controls, electrical control, station controls and CCTV level crossing control. Though the scope is restricted to environments where display screens are used discussions are already underway to apply the same approach to environments where 'conventional' equipment is used – hard-wired control panels and 'leverframes'.

The document covers 'process' as well as 'solutions'. The overall process involved in the specification and design of workstations is described in detail and advice given on the steps to be undertaken – often via flowcharts as presented here:

A substantive portion of the document is devoted to providing practical advice on achieving satisfactory designs. Topics covered include:

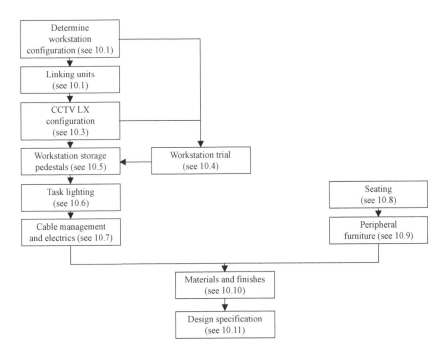

**Figure 47.4  Sample workstation design process chart**

- defining workstation configurations;
- conducting workstation trials;
- design of storage pedestals;
- design of task lighting;
- cable management;
- seating;
- materials and finishes.

Throughout the specification illustrations have been used to guide the user – an example of workstation ergonomics is presented in Figure 47.5.

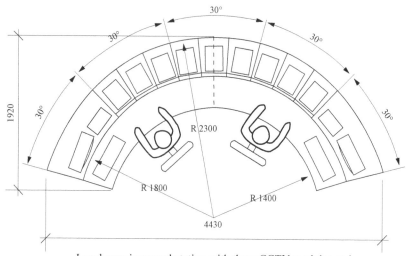

Level crossings workstation with three CCTV modules and
two voice comms/systems modules

**Figure 47.5  Workstation ergonomics**

The workstation specification also assists the project manager in the selection of the appropriate workstation arrangements to achieve the ergonomic layouts generated, Figure 47.6.

The standard also offers an exemplar programme to assist project managers in the planning of workstation design programmes and the procurement, manufacture and installation of these items.

**Ongoing Initiatives**

Apart from the Company Standards reported on there are currently a number of other initiatives underway upgrading existing standards on human–computer interface design and the specification of alerts and alarms. Work being undertaken on the West

Tight arc workstations                    Open arc workstations

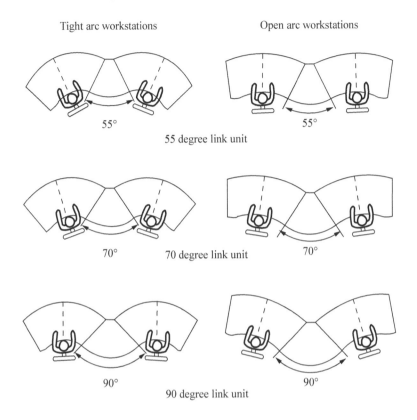

55 degree link unit

70 degree link unit

90 degree link unit

**Figure 47.6  Examples of alternative workstation arrangements**

Coast upgrade programme, under the Network Management Centre project, has also revisited icon design for VDU-based systems and it is hoped that this work will also feed into Company Standards.

The RSSB has recently commissioned a review of all group standards as far as coverage of human factors is concerned. The overall scope of over 380 Group Standards is being examined and existing coverage of human factor issues compared to what might reasonably be expected.

## References

Defence Research Agency (DERA) (2000), 'A Guide for Integration Project Teams and Capability Working Groups: A Practical Guide to Human Factors Integration', DERA-CHS.

Health and Safety Executive (2001), *The Ladbroke Grove Inquiry: Part 1* ('The Cullen Report'), Sudbury: HSE Books.

International Organization for Standardization (2000), 'Ergonomics of Human-centred Lifecycle Process Descriptions', ISO TR/18529.

International Organization for Standardization (2005), 'Ergonomic Design of Control Centres', ISO 11064, http://www.iso.org/iso/en/CatalogueDetailPage.CatalogueDetail?CSNUMBE R=19042&COMMID=&scopelist=.

London Underground Limited (2001), 'Engineering Standard. Integration of Human Factors into Systems Design', E1035 A1, January.

Network Rail (2002), 'Ergonomics Policy', RT/ENGP/11, Issue 1, September.

Network Rail (2003a), 'Control Room Design, Specification, Process and Guidance, Network Rail Company Specification', RT/E/S/24017, Issue 1, April.

Network Rail (2003b), 'Incorporating Ergonomics Within Engineering Design Projects: Requirements and Guidance', Network Rail Company Procedure, RT/x/P/yyy, Issue A, March.

Network Rail (under preparation), 'Workstation Design Specification', Company Guidance Note.

# Chapter 48

# Task Specification in Designing for Human Factors in Railway Operations

Andrew Shepherd and Edward Marshall

## Introduction

Human error is often related to design flaws which might have been avoided had greater attention been paid to human factors knowledge and methods throughout various stages of system development. The chapter emphasises the importance of adopting an *operational* perspective in understanding how tasks are carried out, since this ensures that subsequent human factors advice is focused upon real needs. It also considers the issue of *timeliness* of human factors input. Failure to consider these issues at a time when other engineering decisions are being taken often means that rectification of problems identified later is resisted by managers and engineers for reasons of cost and delay in meeting project milestones.

Human factors has gained prominence in the railway industry through the identification and investigation of incidents involving human error. Railway companies acknowledge the relevance of human factors through the commitment of senior management and by employing safety specialists who use inspection and audit processes to investigate actual incidents, near misses and potential problems. Managers and engineers usually appear willing to acknowledge the use of human factors *retrospectively* as part of audit or accident investigation especially in accounting for an incident. However, there is often reluctance to appreciate human factors as something that could *contribute* to making design and management decisions. This response is understandable when taken in the wider context of managerial responsibilities, where remaining within budget and meeting development and operational targets may be paramount. Resolving a design problem identified after completion of any phase of system development is costly. It may require substantial rework and investment in additional equipment. It will incur delays in meeting project targets and will cause the system to be unavailable when it should be making a profit. Human factors issues raised *during* system development or the normal phases of operation are rarely greeted with enthusiasm by managers who have other, more obvious, demands for their attention In any case, concerns about human factors do not mean that a system failure will necessarily occur. It may even be assumed that conscientious and experienced staff will avoid severe problems or take them in their stride and resolve them before serious consequences arise. Managers and engineers may prefer to take the risk and rely on human resources and management to deal with remaining problems through personnel selection, training and conscientious supervision and operation.

Such risk-taking, which can be part of the culture of some areas of management or some professions, can frustrate the human factors specialist trying to engage constructively in system development. If wider organisational pressures mean that managers and engineers are reluctant to acknowledge this contribution, it becomes an issue for human factors to consider how best to support managers and engineers in taking human factors into account at a time in system development where problems can be identified and solved with minimum expense and inconvenience.

This chapter will argue that human factors should be promoted, not only to enable its use retrospectively, but also by demonstrating ways in which it can be accommodated sensibly and economically as an *input* to provide genuine support for people in industry responsible for managing system operation and design. To this end, it is argued that human factors should be undertaken from an *operational* perspective so that (i) managers and engineers appreciate how such issues relate to the operational goals for which they are responsible and (ii) human factors specialists properly understand how the task context affects these issues. Furthermore, both industrial management and human factors specialists engaged in the development and management of systems should also be aware of the *timeliness* of dealing with human factors issues. This view is not incompatible with psychological or biological perspectives that may need to be taken to resolve some issues, since even these perspectives should only be adopted when the operational context is clear.

Making best use of human factors advice during system development will be repaid by reducing the costs and consequences of subsequent incidents and need not incur significant additional cost. To achieve this, however, attention must be paid to engaging the right expertise at the right time and providing appropriate access and support.

## The Case for Human Factors

Human factors is important because the reliability of people employed to work in a system directly affects system performance. It is also directly concerned with the health, safety and wellbeing of workers for which managers have a duty of care. Human beings are important components of complex systems because, within limits, they are physically and mentally versatile, so engineering does not need to solve all functional problems. People are skilled and can do difficult physical tasks where automation has not been devised or is uneconomical. They can also carry out tasks entailing judgement and decision making, often with greater flexibility than machines, although this may be accompanied by greater unreliability. One of the most potent benefits to be gained from employing human beings is their versatility. This often enables them to overcome flaws in system design and deal with unforeseen circumstances.

People have limitations in the manner in which they process information in order to make judgments about system status and to control actions. They are limited by their anatomy, physiology and psychology in the actions they can carry out, in terms of the control of movements that may be required, the forces they can exert and the physical and mental effort they must sustain. The design of work resources will affect efficiency and reliability of human performance. The efficiency and effectiveness with which workers are able to use these resources is affected by the competency of personnel, how they cooperate with each other, and their working environment.

Human factors knowledge has been established through psychological and biological research. Such research is often based on experimental studies, carried out under controlled conditions enabling researchers to attribute experimental effects to psychological or biological causes. This approach enables hypotheses about behaviour to be formally tested. Another strand of human factors research has focused on the investigation of different occupational contexts, for example radar watch-keeping, air traffic control and process control, usually by examining performance in real contexts or through simulation. Such research often relies on making inferences from observation, analysing verbal utterances, collecting questionnaire data, carrying out task analysis, or analysing response patterns to different types of event. Such research is open to criticism from a strictly scientific perspective, since it is difficult to control and may be subject to bias as the researcher must place an interpretation on these findings. Despite this, such studies substantially add to how we understand applied situations and have greater context validity than most experimental studies.

## Human Factors in Management and Design

People are recruited, trained, equipped and managed in order to complement engineering assets. *Recruitment* and *training* are generally seen to be the concern of *human resource management* although this is, arguably, an aspect of human factors anyway. *Equipping* people is an engineering responsibility, both in terms of designing the means by which the worker is able to control equipment and in terms of providing tools that will enable the worker to be more effective, by amplifying power, assisting decision making, or enabling remote operation, for example. *Management* involves providing adequate resources to operate and maintain systems – including specifying the size and roles of work teams to match system demands – set goals, targets and operating standards, organise work and supervise work to ensure it is carried out appropriately.

Different demands placed on the system place different requirements on what workers have to do. This may cause human error through excessive workload, inadequate information and facilities, and task difficulty. Work is designed to help achieve business goals. However, inappropriate design of work – including, the workplace provided, the range of tasks, workload, information displays and controls – can contribute to system failure. Managers and designers must balance their aspirations for system performance with reality concerning human limitations. Better still, they should take responsibility for better management of human factors in order to limit the extent to which their aspirations are compromised.

Human factors is not an optional extra in system development. It does not add extra features to a system which one could, arguably, do without, like interior decorating. All work features that would be the concern of human factors will be present in some form. In the absence of human factors guidance there will still be controls in trains and signal boxes; information will still be provided through windows, instruments and via communications devices; there will still be a thermal, visual and mechanical environment in which people must work. Design engineers may have always paid attention to these issues, but may have regarded them as commonsense, but this is not always the case as many human errors and accidents have demonstrated. Human factors expertise is required to provide methods to evaluate requirements properly, design how people use information and controls, how the workspace should be set out, how the working environmental should be designed, and supply ergonomics standards.

Designers work by specifying how materials or components should be used to fulfil a functional part of the design. They must observe, for example: constraints on costs; constraints on materials, in order to withstand thermal, mechanical and chemical stresses; compliance with electrical regulations. They must also fit in with other objects being designed. A feature that is often missing is a specification of how people are supposed to use the equipment. If this is not known, the design may not provide enough space, for example, to carry out the necessary tasks. If tasks are anticipated properly, then it may transpire that facilities should be provided for maintenance and normal operations. The tasks involved may require equipment to be replaced, covers to be removed, components cleaned or replaced, or controls to be manipulated. Without sufficient knowledge of these tasks, the designer cannot be clear what space is warranted for access and control, and what information and control facilities will be required for actions to be carried out. It is surprisingly rare to find designers of plant and equipment being provided, early on, with task descriptions of sufficient detail to enable them properly to take into account what workers must do to fulfil their roles. Many designers work with functional specifications and a stereotypical notion of the sorts of things that workers are likely to do; this is rarely sufficient. Often *task analysis* is not carried out until it is required to inform training design. At this time much of the task analysis will be concerned with working out how best to fit human operation in with how equipment has been constructed. Sometimes this means being satisfied with compromising on the original system requirements.

Further examples can be seen in system control tasks. Software engineers often observe human computer interface design guidelines to create screens that conform to standards of character height, colour consistency, glare and screen flicker rate. Yet failure to understand the tasks that have to be carried out can mean that the information required to support operating decisions is not sufficient or provided in a satisfactory form. Thus, the screens that are designed do not support the tasks. Where engineers focus solely on the functional requirements of design and fail to attend to operating requirements, they can create problems that will emerge later on as a result of a safety review, an operating problem or an accident. Rectifying such problems incurs greater cost the longer it is left. Even where the engineer is aware of the potential problems for human operation, unfamiliarity with human factors knowledge or methods can compromise design. Many human factors problems encountered can be attributed to design decisions being taken without full consideration for the implications for human performance.

In order to make judgements about all of these features it is necessary first to understand the tasks that workers and operating teams must undertake. Task analysis must be suited to examining these issues, but it must also enable those people responsible for the successful operation of the system to be fully confident that their operational requirements are being addressed.

## The Operational Perspective for Understanding Tasks

An operational perspective focuses on what personnel are required to do in terms of the system's operating goals. Human factors specialists gain from adopting an operational perspective since it takes better account of the context in which the task is carried out, including, the environment in which people work and the time pressures they experience. It takes account of wider system goals and activities and the other duties that a worker must fulfil. It takes account of the system values, including the costs and consequences of error. By adopting an operational perspective, human factors specialists and their clients can collaborate on a common agenda.

This operational perspective also gives designers and managers more freedom to consider different approaches to choose solutions that are most beneficial for the system. If a task entails dealing with faults or incidents, for example, it might be supported by: referring the problem to highly experienced and qualified staff; by training problem-solving skills; by developing decision aids; by specifying case-by-case procedures; by addressing the cause of faults and reengineering the system in some way; by providing better information and access for repair; or by using a combination of these things. Design choices are, therefore, far less constrained if an operational perspective is taken.

### *Examining Tasks from an Operational Perspective*

Adopting an operational perspective is a long-standing approach in the application of human factors. It has been used to develop task analysis methods which focus on the real and agreed requirements of an organisation without imposing inappropriate solutions. Hierarchical task analysis (HTA) (Annett *et al.*, 1971) is an important example of this. Duncan (1972, 1974) details the merits of focusing on operational requirements to provide a basis for task analysis that takes proper account of a client's needs, whilst acknowledging that further investigation of skills from a psychological or biological perspective may also be warranted in order to refine design suggestions. Detailed analysis from, say, a psychological perspective may often be helpful, but this still benefits from understanding operational aspects. To understand the cognitive processes involved in training train-operator skills, for example, it is important to understand the range of tasks that the system requires of the train operator and, thus, the context in which cognitive performance must be considered.

HTA examines tasks by considering the broad operational goal that the client – the person responsible for the system – judges to be appropriate. In the railway context, this could be concerned with track maintenance, signalling, train operation or station supervision, for example. Usually, an analyst works with the client, or a task expert responsible to the client, to agree an appropriate operational description of the task at this broad level. This general task description, the starting point for HTA, is expressed as an *imperative* – in the form of *verb–object*. For example, tasks could be expressed as 'maintain track', 'carry out signalling duties', 'supervise station'. At this point, none of these three examples implies any detail concerning how these tasks will be performed.

HTA progresses by examining the overall goal through a process of *redescription*. Redescription sets out 'subordinate goals' and a 'plan'. The plan describes the conditions under which each subordinate goal is carried out in order to achieve the overall goal. Figure 48.1 shows a redescription of the operational goal of 'parking' a passenger train in a depot – this is actually part of a much broader analysis. All of the *goal* statements in boxes are expressed in the verb-action form. The *plan* refers to each of the subordinates, indicating when it is appropriate to carry them out. Shepherd (2001) gives extensive examples of HTA from a number of industries, demonstrating how plan hierarchies account for task complexity.

Each subordinate goal is considered in turn to decide whether it warrants further attention. This judgment is made by considering the *likelihood of inadequate performance* and the *costs or consequences of inadequate performance*. Together, these estimates indicate the *risk* attached to carrying out the goal. If the risk is judged unacceptable then, either a solution is sought to overcome the problem or further redescription of the subordinate goal in question is carried out. This process continues until the task is sufficiently well understood to satisfy the analyst and the informant that all of the risks in the task have been considered and that solutions have been proposed to deal with issues of concern. This systematic process results in a hierarchical description of the task in terms of goals, subordinate goals and plans.

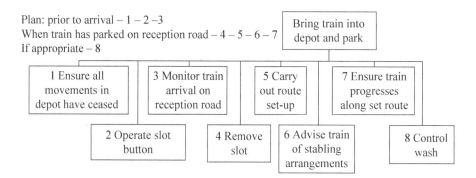

**Figure 48.1   Illustration of a task redescription**

*Using Several Experts*

While HTA is often carried out by an analyst working with a single informant, the reality is many times that the effort has to be more than this. 'Maintain trackside equipment' is a complex task, where many different facets must be taken into account, especially because it intrudes on service delivery and has safety concerns. *Management* concerns include judging urgency, setting priorities and specifying when it is suitable to carry out this maintenance. *Safety* concerns include providing appropriate isolation of the track, issuing permits to work, guarding the track and supervising the system to ensure continued protection. *Operational* and *signalling* concerns take account of the consequences of the maintenance on the wider system. There are *technical* aspects concerned with dealing most efficiently with the diagnosis, repair or replacement of equipment. *Communications* are concerned with providing information on system status to various other parties whose own work will be affected. This is a common feature of most railway tasks where activities that appear simple become complex by needing to consider the interactions between *customer service, line control, signalling, engineering* and *safety*.

Appreciating this complexity requires that the appropriate range of expertise is engaged in the analytical process and making sure all responsible parties are able to express their concerns. It is often important first to carry out an analysis of the system in which tasks reside, before examining any particular task in detail. By considering the general function it becomes possible to anticipate the range of concerns to be accommodated. Figure 48.2 illustrates how functions associated with trackside maintenance may be related to each other.

'Maintaining trackside equipment' is clearly not a task for one person. To complete these activities successfully will entail line-control and signalling staff, maintenance staff, safety marshals, engineering depots and supervision. To conduct the analysis requires managers, engineers, supervisory and safety specialists to communicate with each other. Their concern will be to identify what needs to be done for safe and comprehensive maintenance in the context of the wider railway operation, then construct a system comprising people and equipment that will deliver these functions.

In addition, in a large organisation such as a railway business, there are often tasks that occur commonly, yet which vary in detail. For example, train operation in cabs of different design will follow a similar pattern and be subject to the same safety considerations, but will vary in the details of activities. Equally, station security will involve similar functions across stations, but local arrangements and concerns will mean that the finer detail of the analysis varies. It is often possible to capitalise on the similarity between instances whilst accounting for the differences. This is achieved by establishing the common functions of the task at the higher levels in the analysis, then providing variation at lower levels. It means that task experts representing different stations need to be recruited to the analyst team.

*Task Analysis Workshop*

In modelling a multifaceted task or system, such as represented in Figure 48.2, it is useful to convene a joint activity to enable interested parties to collaborate. This is an unusual way to conduct task analysis, but can be very effective. The analyst uses a white board, flip chart, or a laptop computer with projection facilities so that everyone can share the information and the emerging task description. The analyst operates as a *facilitator*, often focusing attention on one workshop member at a time, whilst taking account of comments made by others. In this way contributing parties can represent their discipline and appreciate why operating decisions are taken. It is noticeable in such workshops how different people, with their different perspectives, gain from this contact as they often start to understand aspects of the system they were unfamiliar with, and appreciate where compromise is necessary.

One such exercise was undertaken in the analysis of a system control task for London Underground. The computer control of the line offered novel opportunities to control staff, since many signalling and line-control functions were to be automated. Management wanted to take advantage of the flexibility offered by this

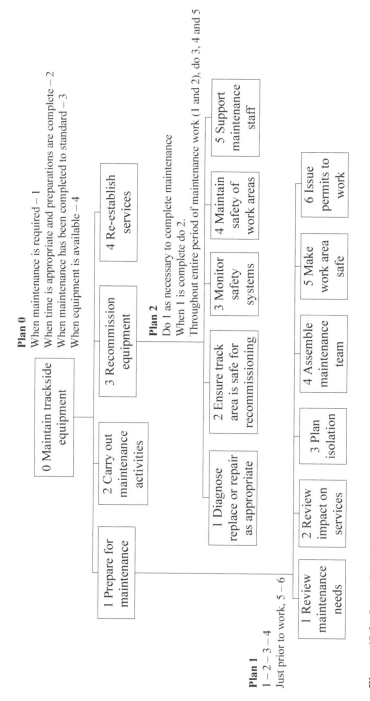

**Figure 48.2    Some functions associated with trackside maintenance in HTA format**

new technology and were confronted with the problem of how to construct jobs, job descriptions, training specifications and team organisation in ways which integrated signalling and line control duties to take account of the automation provided by the new system. The answer was to model the tasks that staff needed to carry out.

A number of task analysis sessions were held, attended by line controllers, signalling staff, managers, personnel staff and supervisors. It soon emerged that the line controllers and signalling staff were trying to 'shoehorn' the new task into a form with which they were familiar – assuming that some people on the proposed operating team would be line-controllers, while others would be signalers. However, management and personnel staff were at pains to resist this in order to take advantage of the investment involved in this innovation. The people attending these sessions contributed in different ways. The signalers and line controllers provided expertise at a *procedural* level, while management provided expertise at a broader *operational* level that reflected the requirements of the organisation in operating this new system. The sessions were used to shape and reshape drafts of the tasks. Eventually, different perspectives were understood and a composite activity emerged with individual roles for control-team members to act flexibly whilst observing proper arrangements for safe and effective signalling and line control. The task analysis also specified how the operating team should deal with system problems – diagnosing the problem, compensation while the problem was dealt with, and recovery. This task design allowed for the team size to be varied throughout the working day in accordance with changes in traffic as indicated by the timetable and the supervisor's judgment of workload. Indeed, the final task resembled that of *air traffic control*, where ATC officers are responsible for an area of airspace, to manage the expeditious passage of aircraft, and intervening only when safety is threatened. Just as ATC officers may be directed to supervise larger or smaller air sectors according to traffic requirements, so too could the staffing of the control centre be varied to cope with the changing demands of the timetable.

An important point to emphasise in both Figures 48.1 and 48.2 is that task representation focuses upon goals to be achieved by the railway system and not on human factors, despite the fact that human factors considerations are identified during the analysis. As these are *neutral* operational statements, it means that different solutions can be offered to deal with identified problems. This is another benefit of using a team of informants from different domains; not only do they bring together necessary considerations in defining the task, they can also trade ideas concerning overcoming problems. Because task analysis methods have emerged from the human factors discipline, it is often the case that human factors expertise is supplied by the analyst. Nonetheless, the method of HTA does not permit human factors to dominate proceedings but allows it to take its place on merit.

### Timeliness in Task Design

A major responsibility of project management in the design of complex systems is the timely integration of the different contributing disciplines. Many of these disciplines have evolved over several years and project management methods and experience have shown how these should work together effectively as part of a wider project. Unfortunately, it is rare for the human factor discipline to be accommodated in this way; it is often regarded as an unnecessary frill that can be considered only if time and resources permit. However, a design process, that does not take human factors into account, risks design decisions being made that do not properly support human performance – or will only do so at a cost that management might find unacceptable.

In the railway system control task discussed here, it emerged that team members had to communicate with each other during the course of their work. The task analysis had identified roles for different team members where they concentrated their efforts on a designated sector of line. However, a problem in any sector had implications for activity in other sectors; different team members needed to know how colleagues would be dealing with various issues before they could resolve their own problems in an optimal fashion. Communication between team members was essential. This work was done late on in system development, with many control room layout decisions already taken and implemented. The desk positions in the control room were already specified and the installation of electric cabling had already been completed.

Some simulation exercises were carried out to observe this mode of operation and it was clear this was a very noisy task. One member of this prototype team was seen to duck below his desk to avoid the gaze of people trying to attract his attention, in order to concentrate on what he had to do. Consequently, he avoided key information from colleagues. As an experiment, we changed the control room layout so that all controllers were sitting around a table in 'conference' style. The noise levels dropped almost immediately. Because all members of the team were in line of sight of each other, each team member could see how busy colleagues were. Because they were working closer together they better understood what colleagues were trying to do and appreciated how their information related to these decisions. They could refrain from interrupting unnecessarily but recognised that an interruption from a colleague was likely to be relevant and should be listened to. What emerged, therefore, was genuine teamwork.

This was not a comprehensive study and did not cause the company to review the engineering decisions that had been taken. But the point is still worth making that only by trying to understand how people will work, is it possible to provide facilities that will best suit requirements. Control room design had not been undertaken on the strength of any precise idea of tasks or how teams would be organised, but had been motivated more by how such things were usually done. Had tasks and team roles been considered earlier, then different room layouts could have been considered on their merits. Architectural issues and hard-wiring would not have been completed thus leaving greater freedom for innovative design to match the requirements of the team.

It is impossible to cover here the detailed way that human factors impinges on engineering decisions. Suffice to say, every engineering judgment has the potential to constrain the degrees of freedom available for human factors design, including: allocation of function; situating equipment; designing equipment, interfaces and the working environment. These are also inextricably connected with specifying teams and jobs, training and personnel support. To incorporate human factors properly, engineering decisions should be carefully reviewed to see how they constrain design for human performance and, therefore, where the human factors input is most timely.

## Conclusion – Organising for Human Factors

Resolving these problems of human factors integration is an organisational issue. The range of human factors inputs and their timeliness needs to be appreciated. Expertise to provide these inputs needs to obtained. There is an organisational development requirement to promote the integration of human factors and to prepare members of the design team to accommodate this sort of input.

The range of human factors methods include: (i) modelling or synthesising tasks, using task analysis, and providing guidance for making design decisions; (ii) using evaluation methods to evaluate stages of design in order to identify potential human factors problems sooner rather than later; and (iii) auditing how the overall system will work. Of these stages, the major human factors input, generally acknowledged, is the last one concerning audit. This is, undoubtedly, essential to 'prove' the system and because there may be emergent properties that cause operating problems that only become manifest when viewing the system as a whole. But the first two areas of input can aid substantially in the identification of problems that could be resolved to reduce the likelihood of serious difficulties emerging that will be costly to change later on.

It may be possible to provide guidance on the timeliness issue by indicating different stages of design and showing how these constrain human factors design. This would probably be difficult to achieve in a wholly convincing way. Instead, the principle of timeliness in human factors needs to be appreciated to the extent that project managers take this into account when planning project stages. Project input, concerning the timeliness of human factors should be a component of all project planning activities.

The issue of human factors expertise needs to be addressed. It will provide advice on when and what inputs are beneficial with regard to: priming designers to take account of human factors; providing standards and design advice; carrying out, teaching, supporting or evaluating task analysis; it can evaluate design stages and system performance, including contributing to safety audits and human error analysis. Ideally, organisations would employ and respect their own in-house expertise. In engineering development projects, it may be appropriate to buy in expertise for the duration of the project. Nevertheless, even here, it should be acknowledged that there is a need for all members of a development team to appreciate the role that human factors plays in contributing to system development.

Important in this respect is the specification of tasks by conducting task analysis from an operational perspective. It is in this way, that human factors can justify its input by complying with system goals. Thus, engineers and managers, responsible for system performance, can be satisfied that their real needs are being addressed.

## References

Annett, J., Duncan, K.D., Stammers, R.B. and Gray, M.J. (1971), *Task Analysis, Training Information Number 6*, London: HMSO.

Duncan, K.D. (1972), 'Strategies for the Analysis of the Task', in J. Hartley (ed.), *Programmed Instruction: An Education Technology*, London: Butterworth.

Duncan, K.D. (1974), 'Analytical Techniques in Training Design', in E. Edwards and F.P. Leeds (eds), *The Human Operator and Process Control*, London: Taylor and Francis.
Shepherd, A. (2001), *Hierarchical Task Analysis*, London: Taylor and Francis.

# Chapter 49

# Systems Engineering Tools for Task Modelling

Ged Morrisroe

## Introduction

In system development there is a need to understand the tasks that are carried out by the users. Traditionally for task definition, ergonomists have used structured analysis tools such as hierarchical task analysis and other forms of functional decomposition.

It is noted that task modelling – in its various forms of task allocation, task definition, workflow analysis – is a common and vitally important component of the systems engineer and ergonomist's job.

In the past, ergonomists and systems engineers have not used the same tools and techniques. Now, systems engineers have mature modelling tools that are commonly used for the definition and design of systems.

This chapter argues that there is much to recommend the use of systems engineering tools when they suit the purpose of ergonomists. This will facilitate the integration of ergonomics in the projects and improve the communications between the parties.

## Background

There are a number of challenges that face the deployment of ergonomics in development projects. The first risk is that it is not carried out at all. For the rail industry this risk is reduced by company standards requiring ergonomics integration plans for projects (cf., LUL and Network Rail). The second risk is that even when an ergonomics integration plan is agreed and resourced, the products of the work fail to make a real impact on the system. To be specific, the failure is not shaping and influencing the design decisions that have been made for the system.

Figure 49.1 shows a project team organisation and roles for a moderately large project.

There are two possible 'homes' for the members of the ergonomics team; either in the systems engineering group or in the safety and *ilities group (the *ilities are the quality and PRAMs – reliability, maintainability etc. – specialists).

The systems engineering function has the responsibility to achieve the system goals by the following activities and methods:

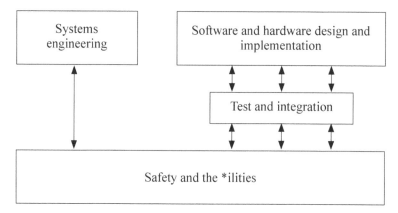

**Figure 49.1   Project organisation and roles**

- analyse – investigate, model;
- evaluate – perform trade studies, prototype and trial;
- specify – requirements, rules and specifications.

The safety and *ilities function assess the constraints and demands of their particular area and levy those as requirements on systems engineering. They are a separate function because they need to maintain a level of detachment and independence.

The role and activities of the systems engineering function and those of the ergonomists bear a striking resemblance. This, and the proposition that the ergonomics work is concerned with shaping and influencing design decisions, leads to a compelling need for the ergonomics work to be undertaken within systems engineering.

**What Language?**

If ergonomics is part of the systems engineering function, then it needs to be able to communicate successfully with the other members of the team. Ergonomists, in the main, use techniques and methods whose outputs are not mainstream and are not always easily assimilated by non-ergonomists.

For many years, systems engineering has been very vigorous in defining and developing tools and methods for modelling systems, and there is a good level of maturity and standardisation in the approaches. In particular, there is widespread adoption of UML (unified modelling language) developed by Booch, Jacobson and Rumbaugh (Booch, Jacobson and Rumbaugh, 1998; Fowler, 2003).

Models are created to be a simplification or shorthand representation of the existing (or future) real world. There is a danger in 'modelling for modelling's sake', that is, a 'define everything' and see what comes out speculative approach. It is wishful thinking to assume that someone's model of the system will address the specific questions of other readers with different needs. If we, as ergonomists, are to use UML models, we must choose and set up the models to investigate the properties of the system in which we are interested. That is, the information and evidence that we require determines the objective, scope and type of modelling activity that we undertake; this ensures that what is modelled will help us formulate and choose between design solutions.

There are broadly two types of models:

- static models;
- dynamic models.

From an ergonomics viewpoint:

- static models can be used to represent the conceptual model that users will need to learn, think and operate the system; and
- dynamic models can be used to describe the logical and temporal elements of the system such as work flow, user –system –user interaction, system state transitions, navigation models etc.

*Static Models*

Conceptual modelling is an important approach used by ergonomists. This involves the definition of the conceptual objects and the actions that can be performed on them. For instance, a simple railway conceptual model may contain the following conceptual objects:

- train;
- path;
- route;
- signal;
- point;
- track;
- territory;
- signaller.

Further, for each conceptual object we can describe the actions that can be performed, for instance, set point. In a similar way, we could use UML class diagrams to represent the same key entities in the system.

In Figure 49.2 is a class diagram that represents five entities. The diagram also described the relationships or associations between the signaller, path, point, signal and territory entities. The object (actor) signaller controls an area (territory), can request a path to be set and set signal to danger. The path object has been modelled to request the relevant signal and point to be set to achieve the path specified by the signaller. Finally, signal and point can either implement or reject the path setting request. There is feedback to the signaller regarding the success of the path setting request. (NB: there is a linguistic awkwardness here because an object is always referred to in the singular.)

Also of interest is the multiplicity (also known as cardinality) of the relationships. In the figure there is a rule stated here that one territory has one signaller and vice versa. If this relationship was to be changed so that a signaller could control more than one territory the cardinality will change to 'one or more' (denoted as 1..*).

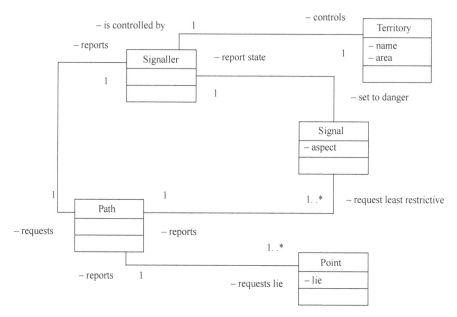

**Figure 49.2   Example class diagram**

If these class diagrams are produced, and if we ergonomists are unable to interpret and understand the implications of these condensed representations, the status and relevance of the ergonomist's conceptual model (objects and actions) prepared earlier must be in doubt. Therefore it is important the conceptual modelling (if one is needed) agrees with the static class diagrams.

Finally, the example model (Figure 49.2) is extremely conceptual. Systems and software engineers will want to produce much more computationally detailed models. If the top level models and definitions are agreed, there is a good chance that the more detailed, implementation level modelling will also be in alignment. This is why involvement in the early, top level modelling really pays off.

*Dynamic Models*

In the area of dynamic models, two techniques are particularly suited for task definition:

1   sequence diagrams;
2   state transition diagrams.

*Sequence diagrams*

The starting point for sequence diagrams is the definition of scenarios that are typical of, or critical to, the operation of the system. Table 49.1 lists a candidate scenarios for development.

**Table 49.1  Candidate scenarios for development**

| 0100 | Signaller makes point-to-point voice call to Train Driver |
|------|----------------------------------------------------------|
| 0110 | Driver makes call to signaller |
| 0120 | Driver and signaller communicate |
| 0130 | Signaller ends call to driver |
| 0140 | Driver ends call to signaller |
| 0150 | start up, registration, change mode, de-register and close down operations |
| 0200 | emergency calls |
| 0300 | priority and pre-emption calls |
| 0400 | radio functions (eg hold, transfer) |
| 0500 | formulation and recipt of text messages |
| 0600 | targetting signallers |
| 0700 | light duty working and combination and splitting territories |
| 0800 | SMS |
| 0900 | group, multi party and multi-driver calls |
| 1000 | failure modes and fallback behaviours |
| 1100 | shunting operations |

Once a set of operational scenarios that cover the range of operational conditions that may exist in the system have been determined and agreed (for example normal, disturbed/perturbed, emergency and migration), each can be described in the form of sequence diagrams.

Once the actors and system entities have been defined, a time-based activity line can be created as shown in Figure 49.3.

In the example in Figure 49.3 there are two human actors at the left and right hand extremes. In the central area there are three system objects: the signaller radio and the driver radio (including the user interface for both – there is a note to indicate the multiple methods for initiating a call). The object in the centre (called RT manager) provides the infrastructure for the call.

These sequence diagrams can be constructed with operational users and technical experts. Both are needed to make sure that the system is meeting the operational requirements and that the requirements are technically achievable.

Once the sequence diagram has been constructed, there are a number of insights to be made. Taking the example in Figure 49.3, we can ask/or observe:

1   what is heard and seen by the signaller as the call progresses through the various systems and whether these indications are consistent with user expectations;
2   the ringing indication for the signaller: the ringing tone does not indicate that the call is ringing at the driver's radio, just that the call has been requested by the RT manager. There is a likely failure case where the radio does not respond to the request (new scenario);
3   once the driver has answered the call there are three operations ('call established') prior to the channel for two way communication being opened. Does this appear

Scenario 0100: signaller makes point-to-point voice call to train driver

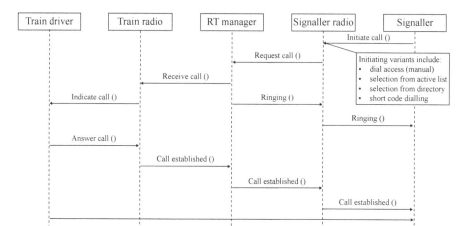

**Figure 49.3   Example sequence diagram**

as instantaneous to both parties? Are there failure modes that can occur and how would the user detect these?

*State Transition Diagrams*

In all systems there are system states that will affect the user. Depending on the reason for the state changes, it is likely that the user will be exposed to different methods of operation in the various states. Figure 49.4 shows an example of how cab radio modes are moved between by the user.

There is often a high degree of complexity encountered in innocuous user operations such as logging on and off, selecting user roles and user account management. In Figure 49.4, the transitions available to a user are modelled. The states are shown in the round cornered boxes and the operation (by the user in this case) is shown on the arrowed line.

The radio in its 'off' state is actually active even though the HMI is not powered up. When the driver turns the radio on, the radio HMI is powered up and the radio declares itself active to the network. On completion of this transition the radio has a subset of the full radio functionality available. Similarly when the driver selects the shunting or journey registered states, the appearance and functionality will change. The note reflects the debate about entering an active journey registered state from shunting.

**Conclusion**

It is argued that systems engineering goals and products are highly similar to those espoused by the project ergonomist. In addition, systems engineering has developed

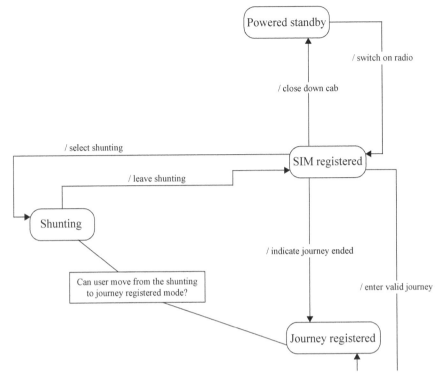

**Figure 49.4  Example state transition diagram**

and standardised on a core set of tools. There are good reasons for ergonomists to use these tools.

However, it should be remembered that analysis and modelling is not *design*. Any analysis method or modelling tool selected must provide evidence useful in making design decisions.

## References

Annett, J., Duncan, K.D., Stammers, R.B. and Gray, M.J. (1971), *Task Analysis, Training Information Number 6*, London: HMSO.

Booch, G., Jacobson, I. and Rumbaugh., J. (1998), *Unified Modeling Language User Guide*, Redwood City, CA: Addison-Wesley.

Fowler, M. (2003), *UML Distilled: A Brief Guide to the Standard Object Modeling Language*, 3rd edn, Redwood City, CA: Addison-Wesley.

# Index